International Conference on Advanced Engineering and Technology (ICAET 2020)

IOP Conference Series: Materials Science and Engineering
Volume 1117

Incheon, South Korea
18 - 20 December 2020

ISBN: 978-1-7138-3403-8
ISSN: 1757-8981

Printed from e-media with permission by:

Curran Associates, Inc.
57 Morehouse Lane
Red Hook, NY 12571

Some format issues inherent in the e-media version may also appear in this print version.

Institute of Physics
Dirac House, Temple Back
Bristol BS1 6BE UK

Phone: 44 1 17 929 7481
Fax: 44 1 17 920 0979

techtracking@iop.org

Additional copies of this publication are available from:

Curran Associates, Inc.
57 Morehouse Lane
Red Hook, NY 12571 USA
Phone: 845-758-0400
Fax: 845-758-2633
Email: curran@proceedings.com
Web: www.proceedings.com

TABLE OF CONTENTS

Preface

The ICAET 2020 is the 2020 7th International Conference on Advanced Engineering and Technology (7th ICAET) will be held in Incheon, Korea during December 18-20, 2020. But currently, the entire world is struggling against the virulent pandemic COVID-19, so that the ICAET 2020 conference changed the presentation method from Oral presentation to Remote Video Presentation to help with Global Quarantine Activities.

The Virtual of ICAET 2020 was held on Incheon National University at Video Conference Meeting Room on December 18-20, 2020. The ICAET Committee members attended conference at on-site and also authors attended the conference as Video Meeting. The conference program covered Invited Keynote Speech, Oral, and Poster Presentations, they presented their presentation 15 minutes including Q&A time.

The committee of ICAET 2020 expresses their sincere thanks to all authors for their high-quality research papers and careful presentations. All reviewers are also thanked for their careful comments and advices. Thanks are finally given to IOP Publication as well for producing this volume.

The Organizing Committee of ICAET 2020

Committee Chair

Prof. Jongwan Hu

Incheon National University

IOP Conf. Series: Materials Science and Engineering 1117 (2021) 011001 doi:10.1088/1757-899X/1117/1/011001

Conference Photograph

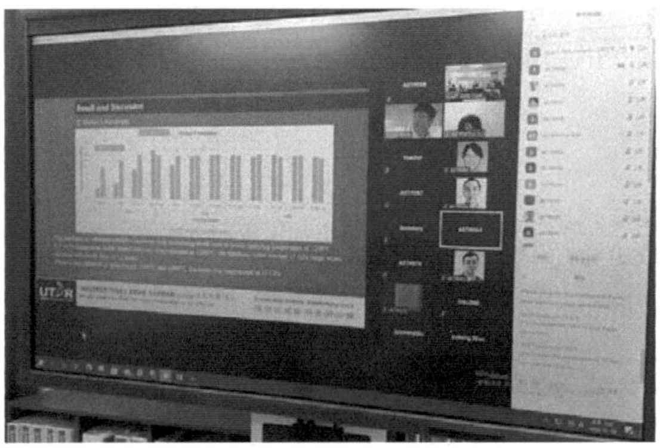

IOP Conf. Series: Materials Science and Engineering 1117 (2021) 011001 doi:10.1088/1757-899X/1117/1/011001

Sponsors

Incheon Disaster Prevention Research Center (IDPRC)	IDPRC 인천대학교 방재연구센터 Incheon Disaster Prevention Center

Recently, various efforts to prevent and prepare are vitally needed for prevention of Disasters and calamities. So we understand the necessity for technology of disaster and we built up the Incheon Disaster Prevention Research Center (IDPRC) in Incheon National University (INU) in 1997.

Accordingly, Incheon Disaster Prevention Research Center (IDPRC) in Incheon National University has progressed of research on the prevention of disasters and calamities through the various seminars, conference and lectures. These research could be conducted cooperation with Incheon National University (INU) in various fields' structure, soil, hydraulic and environment.

Incheon Disaster Prevention Research Center (IDPRC) will try to be a leader in the disaster of industry through the various research activities and global conference.

IOP Conf. Series: Materials Science and Engineering 1117 (2021) 011001 doi:10.1088/1757-899X/1117/1/011001

Lists of Committees

Committee name	Organizer name	Organization
Honorary Chair	Jongwan Hu	Incheon National University
Conference General Chairs	Mosbeh Kaloop	Incheon National University
Honorary Chair	Dongkeon Kim	Dong-A University
Technical Committees	J. W. Hu	Incheon National University

IOP Conf. Series: Materials Science and Engineering 1117 (2021) 011002 doi:10.1088/1757-899X/1117/1/011002

Peer review declaration

All papers published in this volume of *IOP Conference Series: Materials Science and Engineering* have been peer reviewed through processes administered by the Editors. Reviews were conducted by expert referees to the professional and scientific standards expected of a proceedings journal published by IOP Publishing.

- **Type of peer review:**
  All papers revise by three of committee members of the conference. The papers are divided based on the specialist of the committee members; and in the case of special topics, other reviewers invite for reviewing the selected manuscripts.

- **Conference submission management system:**
  Our conference has Secretary for management on conference papers. So, authors send their papers to Conference E-mail, and our secretary has managed it. The secretary sends the papers to the conference chairs to evaluate the suitability of submitted papers to the conference scopes.

- **Number of submissions received:**
  Especially in this time, only few of papers were submitted cause of COVID-19. The total number of submission papers were 146 papers.

- **Number of submissions sent for review:**
  The submission papers were revised from the management conference system and committee organizations and the papers that sent for review were 95 papers.

- **Number of submissions accepted:**
  The papers that accepted in the conference for publishing after review processing are 35 papers.

- **Acceptance Rate (Number of Submissions Accepted / Number of Submissions Received X 100):** 23 %

- **Average number of reviews per paper:**
  Different cycles review processing implemented for improving the papers quality. In average, three times review on per paper were processed.

- **Total number of reviewers involved:**
  Three of reviewers are took part in the review process. When we receive positive reports from two at least we can accept the paper after evaluating the reviewer report and authors responses.

- **Any additional info on review process:**
  The processing of review work of this conference was carried out as follows: (1) the secretary received the papers by emails and processed the initial checking of the papers, he revised the format, language and roughly check of papers contents; (2) the general chairs received the processed papers and they revised the papers scopes and

IOP Conf. Series: Materials Science and Engineering 1117 (2021) 011002 doi:10.1088/1757-899X/1117/1/011002

if these papers relate to the conference scopes it can be passed to the next processing, else the papers can be rejected. (3) all accepted papers divided into categories based on the papers contents and sent to three committee members who have a good experience in the paper contents. (4) after receiving the review reports from the committee members, the review reports sent to authors for sending their responses on the listed comments. (5) the reviewing cycling process divided into accept, minor and major revisions; and based on the review comments and suggestion, the papers can be processing up to three times to improve the manuscript contents. (6) finally, the authors should revise/argue the comments and send their response through email; and the processing cycling until organizer accept the paper according to the final review reports.

- **Contact person for queries:**
  : Jack Son (E-mail: icaet2020@icaet-conf.org)

Data analysis of the corrosion protection performance of aminobenzene concentrations on 1070 aluminum alloy in dilute HCl solution

Roland T. Loto[1], Cleophas Loto[1] and Tobi Daada[1]

[1]Department of Mechanical Engineering, Covenant University, Ogun state, Nigeria

Email: tolu.loto@gmail.com

Abstract. Gravimetric and statistical analysis was used to assess the protection performance of aminobenzene compound on 1070 aluminum alloy in 1 M HCl solution in volumetric concentrations of 2.5%, 5%, 7.5%, 10%, 12.5% and 15%. Data output shows aminobenzene performed poorly at lower concentrations (2.5% to 10% aminobenzene concentrations) with highest value of 61.3% at 10% concentration and 480 h of exposure time. However, at higher concentrations the corrosion and oxidation of the aluminium surface was effectively suppressed with peak value of 81.7% at 480 h of exposure. Aminobenzene inhibition performance was determined to be highly concentration dependent with limited variation with respect to observation time. Standard deviation data showed significant variation of protection performance data at lower aminobenzene concentrations due to variation of the protection performance data with respect to observation time and thermodynamic instability of the inhibition mechanism. At 12.5% and 15% aminobenzene concentration the standard deviation values visibly decreased due to improved inhibition performance. Data also showed that only 31.67% of the protection performance data is above 70% inhibition efficiency at margin of error of +8.32%. Data from analysis of variance shows inhibitor concentration is the only statistical relevant factor at 75.57% responsible inhibition output of aminobenzene on the aluminium alloy compared to observation time with statistical relevance value of 12.28%.

1. Introduction

Aluminum is an important structural engineering material whose extensive utilization ranks behind carbon steels. The production and application of aluminum has continued to expand geometrically for applications such as in aerospace, electrical conduction, energy generation, automotive parts, consumer products, shipping and marine etc. [1-3]. This is due to their relatively low density compared to ferrous alloys, light weight and corrosion resistance [4-7]. The mechanical, metallurgical and physical properties of aluminum can be altered and optimized for extreme structural applications [8]. The corrosion resistance of aluminum is due to the formation of an impenetrable, inert oxide layer which is self-healing and enables its extensive utilization in a variety of industrial environments and service conditions. This invariably extends their operational lifespan. However, aluminum being an amphoteric metal can be highly reactive in aqueous acid and alkaline solutions especially in the presence of Cl^- anions which accelerates the surface deterioration of the metal. The deterioration appears in the form of white discolorations and precipitates on the metal [9, 10]. The economic and industrial relevance of aluminum enables extensive scientific study of the metal in corrosive conditions to optimize their productivity and applications. The study is centered on the electrochemical and corrosion reaction mechanisms occurring the metal surface [11-15]. According to Solange *et al* [16], localized deterioration of aluminum alloy in chloride solution is strongly related to the particles intermetallic phase and composition. 1xxx series of aluminum are strain hardenable and utilized in applications where structural strength is not an important factor but corrosion resistance is of utmost importance. The corrosion resistance of aluminum can be augmented with the used of fluid derivatives known as corrosion inhibitors. These compounds act by forming a protective barrier on the metal, modification of the corrosive medium and formation of complexes with the corrosive anions [17-25]. Research on appropriate corrosion inhibitors for aluminum is ongoing. However, data analysis of the protection performance corrosion inhibitors is highly relevant to evaluate the advantage and possible limitation of the applied chemical compounds analogous to observation time and inhibitor concentration. Appropriate substantiation of their peak performance with to concentration, storage life, durability and observation time is extremely important. The manuscript focusses on the gravimetric analysis of the inhibition action of aminobenzene on 1070 aluminum in 1 M HCl solution.

2. Material and methods

2.1 Materials preparation

1070 aluminium alloy (A70) rod secured in Lagos, Nigeria was analysed in the Materials Characterization Laboratory at Covenant University, Ota, Ogun State, Nigeria. The elemental components of A70 are outlaid in Table 1. A70 was cut and machined into 7 independent exhibits. The exhibits were grinded and smoothened with emery papers. Aminobenzene (ANZ), a brownish solution was bought from Sigma Aldrich, USA. AZN was deposited into to 1 M HCl in cubic values of 2.5%, 5%, 7.5%, 10%, 12.5% and 15% per 400 mL of the acid.

Table 1. Atomic components of (wt. %) of A70

Element	Cu	V	Mn	Si	Fe	Zn	Ti	Mg	Al
Wt.%	0.04%	0.05%	0.03%	0.2%	0.25%	0.04%	0.03%	0.03%	99.7%

2.2 Gravimetric analysis

Gravimetric analysis of ANZ inhibition on A70 was executed at 24 h hiatus for a total of 480 h of exposure. Corrosion rate, C_R (mm/y) was enumerated from the equation below;

$$C_R = \left[\frac{87.6W}{DAt}\right] \tag{1}$$

W illustrates weight loss (g), D illustrates density (g/cm^3), A illustrates total surface area of A70 exhibit (cm^2), 87.6 illustrates corrosion rate constant and t illustrates time (h). Protection performance (P_P) of the ANZ was enumerated from the equation below;

$$P_P = \left[\frac{W_1 - W_2}{W_1}\right] \times 100 \tag{2}$$

W_1 and W_2 illustrates weight loss of A70 at specific ANZ concentrations.

2.3 Statistical analysis

Binary-factor single level anova analysis (F - test) was applied to calculated the statistical importance of ANZ concentrates and observation time on ANZ performance output. The test was done at confidence level of 95% (significance level of $\alpha = 0.05$) according to the equations below. The summation of column squares (observation time) was enumerated as follows;

$$SS_c = \frac{\sum T_c^2}{nr} - \frac{T^2}{N} \tag{3}$$

The summation of row squares (ANZ concentration) was enumerated from the equation below;

$$SS_r = \frac{\sum T_r^2}{nc} - \frac{T^2}{N} \tag{4}$$

Total summation of squares equals

$$SS_{Total} = \sum x^2 - \frac{T^2}{N} \tag{5}$$

3. Results and discussion

3.1 Gravimetric analysis

Table 2 and 3 shows A70 corrosion rate and ANZ protection performance data with respect to observation time. A70 corrosion rate data at 0% ANZ concentration varied from 1.104 mm/y at 24 h to 18.471 mm/y. Between 24 h and 120 h of exposure, progressive increase in corrosion rate occurred to peak value of 37.798 mm/y (120 h) due to electrochemical oxidation of A70 surface according to the equation below;

$$2Al(s)+6HCl(aq)\rightarrow 2AlCl_3(aq)+3H_2(g)$$

The Cl- anions weakens the protective Al2O3 oxide on A70 exposing the substrate alloy to accelerated corrosion. Beyond 120 h, decrease in carrion rate occurred due to weaken of the acid electrolyte. It must be noted that addition of ANZ compound at 2.5% and 5% concentration substantially increased A70 corrosion as shown in Table 2. At 24 h, corrosion rate of A70 has increased to 1.262 mm/y and 1.854 mm/y from 1.104 mm/y at 0% ANZ concentration. Instantaneous increase in corrosion rate results in values to 29.343 mm/y and 47.614 mm/y occurred at 48 h of exposure before gradually decrease with respect to observation time till 16.847 mm/y and 16.835 mm/y at 480 h. This shows ANZ is completely ineffective in suppressing the corrosion of A70. At 7.5% ANZ concentration and 24 h of exposure, the corrosion rate of A70 is negligibly below the value obtained at 0% ANZ. Peak value was attained at 120 h (20.919 mm/y) before gradual decrease to 9.253 mm/y at 480 h. This observation confirms ANZ compound at 7.5% concentration has strong electrochemical influence on A70 corrosion. The final corrosion rate value of 9.253 mm/y is lower than 16.847 mm/y and 16.835 mm/y at 2.5% and 5% ANZ concentration, and 18.471 mm/y. Further increase in ANZ concentration i.e. from 10% to 15% ANZ concentration decreases A70 corrosion. At 24 h, the corrosion rate values are 0.651 mm/y, 0.434 mm/y and 0.631 mm/y compared to 1.104 mm/y at 0% ANZ while at 480 h, A70 corrosion rate values from 10% to 12.5% ANZ concentration are 5.539 mm/y, 7.368 mm/y and 3.419 mm/y. The corrosion rate values at 480 h decreased proportionately with increase in ANZ concentration. The lowest value obtained is 3.419 mm/y at 15% ANZ concentration.

Further insight to the electrochemical behaviour of ANZ is shown in Table 3. The inhibition protection performance of ANZ varied significantly with its concentration as shown in the change in protection performance data with respect to ANZ concentration. At 24 h, poor protection performance data occurred from 2.5% to 7.5% ANZ concentration (-14.3%, -67.9% and 8.9%) due to inability of protonated ANZ molecules to stifle the redox reaction processes. As a result, the molecules are involved in accelerating the corrosion process. The corrosion rate values at this concentration varied with observation time to attain final values of 8.8%, 8.9% and 49.9% at 480 h. At 10%, 12.5% and 15% ANZ concentration, significant improvement in ANZ protection performance is observed with values of 60.7%, 41.1% and 42.9% at 24 h while at 480 h the values are 61.3%, 70.8% and 81.7%. The protection performance data shows ANZ compared effectively inhibited A70 corrosion at higher concentrations from 10% to 15% ANZ concentration while at lower concentrations ANZ performance is significantly poor signifying concentration dependent performance. Secondly, it was noted that ANZ protection performance data varied differentially with exposure time due to delayed inhibition effect on A70 and relative instability with respect to observation time. Comparative plots of ANZ concentration at 2.5% and 15% are shown in Fig. 1. The plot configurations shows ANZ attained relative corrosion inhibition stability at 120 h and 48 h of observation time (2.5% and 15% ANZ concentration) till 480 h. This shows variation in performance output is limited with respect to observation time as the inhibition effect of ANZ exhibits thermodynamic stability. However, the significant difference/variation between the plot configurations shows ANZ concentration strongly influences its performance output. Hence effective utilization of ANZ is possible at higher concentrations of 12.5% and 15% ANZ concentration.

Table 2. A70 corrosion rata data at specific ANZ concentration with respect to observation time

ANZ Concentration (%) / Observation Time (h)	0% ANZ	2.5% ANZ	5% ANZ	7.5% ANZ	10% ANZ	12.5% ANZ	15% ANZ
0	0	0	0	0	0	0	0
24	1.104	1.262	1.854	1.006	0.651	0.434	0.631
48	26.947	29.343	47.614	19.651	0.434	27.016	0.434
72	20.785	33.32	44.422	20.403	3.379	18.405	0.368
96	29.654	27.292	36.132	20.164	4.792	14.223	0.325
120	37.798	33.954	35.038	20.919	4.613	13.686	1.743
144	33.994	32.571	34.506	18.839	4.868	13.278	2.557
168	32.935	31.563	32.636	17.201	5.206	12.328	3.175
192	31.818	30.38	30.913	15.48	5.098	11.393	3.032
216	31.578	29.218	29.494	14.108	4.991	10.802	3.607
240	30.783	27.91	28.154	13.335	5.113	10.004	3.589
264	29.228	26.265	26.491	12.37	5.036	9.392	3.48
288	27.767	24.734	24.929	11.523	4.951	8.853	3.392
312	26.412	23.438	23.541	11.134	5.044	8.463	3.463
336	25.078	22.113	22.222	10.58	5.033	8.154	3.428
360	23.672	20.802	20.891	10.053	4.994	7.813	3.331
384	22.302	19.636	19.971	10.097	5.387	7.535	3.471
408	21.114	18.814	19.087	10.061	5.428	7.644	3.412
432	20.053	18.062	18.297	9.515	5.477	7.511	3.421
456	19.083	17.352	17.626	9.356	5.539	7.368	3.419
480	18.471	16.847	16.835	9.253	5.398	7.143	3.386

Table 3. A70 protection performance data at specific ANZ concentration with respect to observation time

ANZ Concentration (%) / Observation Time (h)	**2.5% ANZ**	**5% ANZ**	**7.5% ANZ**	**10% ANZ**	**12.5% ANZ**	**15% ANZ**
0	0	0	0	0	0	0
24	-14.3	-67.9	8.9	60.7	41.1	42.9
48	-8.9	-76.7	27.1	-0.3	98.4	98.4
72	-60.3	-113.7	1.8	11.5	83.7	98.2
96	-30.2	-21.9	32	52	83.8	98.9
120	8	-0.7	39.9	60.7	86.7	95
144	2.4	-1.5	42.1	60.9	85.7	92.5
168	4.2	0.9	44.6	62.6	84.2	90.4
192	4.5	3.7	47.8	62.2	84.2	90.5
216	7.4	6.6	51.3	65.8	83.4	88.6
240	9.3	8.5	55.3	86.2	82.8	88.3
264	10.1	9.4	56.7	67.5	82.2	88.1
288	10.9	10.2	57.7	68.1	81.3	87.8
312	11.3	10.9	58.5	68	80.9	86.9
336	11.8	11.4	57.9	67.5	79.9	86.3
360	12.1	11.8	57.5	67	79.8	85.9
384	12	10.5	54.7	66.2	75.8	84.3
408	10.9	9.6	52.4	63.8	74.3	83.8

432	9.9	8.8	52.6	62.5	72.7	82.9
456	9.1	7.6	51	61.4	71	82.1
480	8.8	8.9	49.9	61.3	70.8	81.7

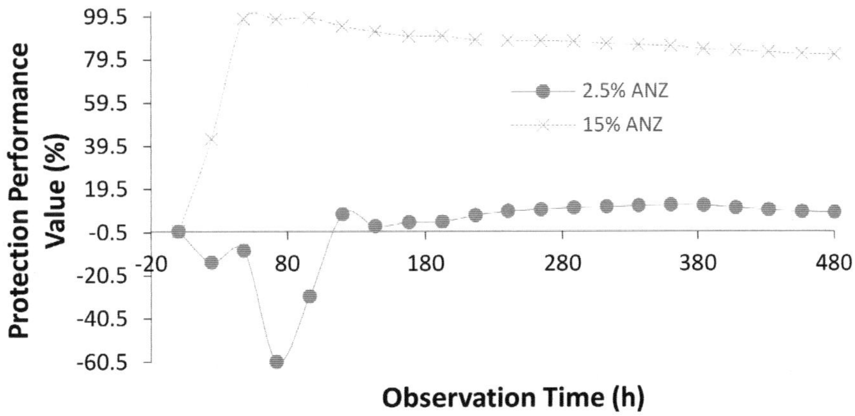

Figure. 1 Comparative plot of ANZ protection performance data at 2.5% and 15% ANZ concentration with respect to observation time

3.2 Statistical analysis

The mean, standard deviation (SD) and margin of error for ANZ inhibition performance on A70 are shown in Table 4. The mean data shows the average protection performance data with respect to observation time. The mean values shown in Table 3 vary with ANZ concentration. From 2.5% to 10% ANZ concentration, the mean values are significantly below the data for effective corrosion inhibition with respect to observation time. However, at 12.5% and 15% ANZ concentration the mean values in Table 3 depict effective corrosion inhibition with respect to exposure time. SD shows the extent of variation of ANZ protection performance data from mean value. The greater the variation of protection performance data; the greater the instability of ANZ performance with respect to observation time. This is disadvantageous to the overall inhibition performance of ANZ. The SD values from 2.5% ANZ to 10% ANZ are generally high signifying high variation. However, the values at 12.5% and 15% ANZ concentration are relatively lower with values of 14.88 and 16.48. The lower SD values signify ANZ inhibition performance is relatively stable with respect to observation time and presents a reliable concentration in industrial operating conditions. Table 3 also shows only 31.67% of ANZ protection performance data have values greater than 70% inhibition efficiency at margin of error of ±8.32%.

Table 4. Mean, standard deviation and margin of error for ANZ protection performance on A70 in 1 M HCl electrolyte

HCl						
Conc. (%)	2.5	5	7.5	10	12.5	15
SD	22.30	43.63	17.84	26.21	14.88	16.48
Mean	-7.79	-26.27	35.08	52.23	81.40	88.37
Margin of Error	±8.32%		**Data above 70% Protection Performance**		31.67%	

Statistical estimation by analysis of variance (ANOVA) was used to assess the statistical pertinence of ANZ concentration and observation time on the protection performance values of ANZ in HCl media. The statistical figures are laid out in Table 5 where the statistical relevance factor illustrates the valuated relevance of the origin of variation (ANZ Conc. and observation time) on the performance output of the ANZ compound. The mean square ratio illustrates the numerical value that must be greater than the theoretical significance factor in order to validate the dominance of the statistical relevance factor. The mean square ratio associated with ANZ concentration and observation time in HCl media exceeds the analogous theoretical significance factor. Hence ANZ concentration and observation time are both numerically relevant, significantly influencing the protection performance values of ANZ on A70. However, ANZ concentration dominates the performance output at 75.57% while the influence of observation time is minimal but relevant at 12.28%.

Table 5. ANOVA data for statistical importance of ANZ concentration and observation time on ANZ protection performance values

HCl			
Origin of Variation	**Mean Square Ratio (F)**	**Theoretical Significance Factor**	**Statistical Relevance Factor, F (%)**
ANZ Conc.	55.99	2.42	75.57
Observation Time	5.06	2.1	12.28

4. Conclusion

Aminobenzene effectively inhibited the corrosion of 1070 aluminum alloy at higher concentrations. Its inhibition performance was determined to be strongly dependent on its concentration. Data from statistical analysis shows significant variation from standard deviation at lower inhibitor concentration due to instability of the inhibition mechanism and subsequent poor inhibition performance. This contrasts the corresponding low variation at higher aminobenzene concentrations. The protection performance data exhibiting values above 70% inhibition efficiency is however less than 30% signifying concentration dependent inhibition performance. The effect of inhibitor concentration is above 70% from statistical analysis.

Acknowledgement

The author is grateful to Covenant University Ota, Ogun State, Nigeria for the sponsorship and provision of research facilities for this project.

References

[1] Birbilis, N., Muster, T., & Buchheit, R.G. (2011). Corrosion of aluminum alloys. corrosion mechanisms in theory and practice, 3rd. Ed., CRC Press, pp. 705-736.
[2] Davis, J.R. (1999). Corrosion of aluminum and aluminum alloys, ASM International, Materials Park, Ohio, USA.
[3] Makanjuola, O., Kayode, O., John, O., & Samuel, O. (2013). Inhibition of aluminium in HCl by amine modified epoxy resin. Journal of Materials, 479728. http://dx.doi.org/10.1155/2013/479728.
[4] Sivam, S.P., Kumar, A., Sathiya, K., & Rajendrakumar, S. (2016). Investigation exploration outcome of heat treatment on corrosion resistance of AA 5083 in marine application. International Journal of Chemical Sciences, 14(S2), 453-460.
[5] Bruhn, E.F. (1972). Analysis and design of flight vehicle structures, Jacobs Publishing Inc., Florida, USA.
[6] Wernick, S., Pinner, R., & Sheasby, P.G. (1987). The surface treatment of aluminum and its alloys, ASN International Finishing Publications Ltd., Teddington, UK.
[7] Bertila, A., Matilde, F., Oladis, T., Silagdy, A., Melani, C., Jomar, R., & Raúl, C. (2013). Corrosion costs in preventive and corrective maintenance in equipment and facilities in industry. Revista Tecnica de la Facultad de Ingenieria Universidad Del Zulia, 36(1), 23 – 33.
[8] Maria-Cristiana, E., Ileana, P.N., Raluca, Z., Alina, M., & Vasile, B. (2010). Influence of heat treatment on microstructure and corrosion behaviour of 7 series Al alloys, Proceedings of the 2nd

International Conference on Manufacturing Engineering, Quality and Production Systems (MEQAPS'10). Constantza, Romania.

[9] Hurlen, T., Lian, H., Odegard, O.S., & Valand, T.V. (1984). Corrosion and passive behaviour of aluminium in weakly acid solution. Electrochimica Acta, 29, 579-585.

[10] Mountarlier, V., Gigandet, M.P., Normand, B., & Pagetti, J. (2005). EIS characterisation of anodic films formed on 2024 aluminium alloy, in sulphuric acid containing molybdate or permanganate species. Corrosion Science, 47(1), 937-951.

[11] Lukovits, I., Kalman, E., & Zucchi, F. (2001). Corrosion inhibitors—Correlation between electronic structure and efficiency. Corrosion, 57, 3-9.

[12] Brett, C.M. (1992). On the electrochemical behaviour of aluminium in acidic chloride solution. Corrosion Science, 33(2), 203-210.

[13] Brett, C.M.A., Gomes, I.A.R., & Martins, J.P.S. (1994). The electrochemical behaviour and corrosion of aluminium in chloride media. The effect of inhibitor anions. Corrosion Science, 36, 915-923.

[14] El Abedin, S.Z. (2001). Role of chromate, molybdate and tungstate anions on the inhibition of aluminiumin chloride solutions. Journal of Applied Electrochemistry, 31, 711-718.

[15] Beck, T.R. (1998). Size distribution of etch pits in aluminium. Electrochimica Acta, 33(10), 1321-1327.

[16] Solange, P.D., Benjamín, H.P., Rosario, Á.G., & Mauro, B.V. (2011). Study of pitting corrosion of the commercial aluminium alloy AA3003 in saline environment. Revista Tecnica de la Facultad de Ingenieria Universidad Del Zulia, 34(1), 108 - 123.

[17] Hu, T., Shi, H., Wei, T., Liu, F., Fan, S., & Hana, E. (2015). Cerium tartrate as a corrosion inhibitor for AA 2024–T3. Corrosion Science, 95, 152-161.

[18] Sherif, E.-S.M. (2012). Effects of 3-amino-1,2,4-triazole-5-thiol on the inhibition of pure aluminum corrosion in aerated stagnant 3.5 wt.% NaCl solution as a corrosion inhibitor. International Journal of Electrochemical Science, 7 (6), 4847-4859.

[19] Catubig, R., Hughes, A.E., Cole, I.S., Hinton, B.R.W., & Forsyth, M. (2014). The use of cerium and praseodymium mercaptoacetate as thiol-containing inhibitors for AA2024-T3. Corrosion Science, 81, 45-53.

[20] Zor, S., & Sağdinç, S, (2014). Experimental and theoretical study of sulfathiazole as environmentally friendly inhibitor on aluminum corrosion in NaCl. Protection of Metals and the Physical Chemistry of Surfaces, 50(2), 244-253.

[21] Lakshmi, N.V., Arivazhagan, N., & Karthikeyan, S. (2013). The corrosion inhibition of aluminium in 3.5% NaCl by diisopropyl thiourea, International Journal of ChemTech Research, 5(4), 1959-1963.

[22] Snihirova, D., Lamaka, S.V., Taheri, P. Mol, J.M.C., & Montemor, M.F. (2016). Comparison of the synergistic effects of inhibitor mixtures tailored for enhanced corrosion protection of bare and coated AA2024-T3. Surface and Coatings Technology, 303(B), 342-351.

[23] Loto, R.T., and Loto, C.A. (2012). Effect of P-phenylediamine on the corrosion of austenitic stainless steel type 304 in hydrochloric acid. International Journal of Electrochemical Science, 7(10), 9423-9440.

[24] Loto, R.T. (2017). Anti-corrosion performance of the synergistic properties of benzenecarbonitrile and 5-bromovanillin on 1018 carbon steel in HCl environment. Scientific Reports, 7(1), 17555.

[25] Loto, C.A., Loto, R.T., & Popoola, A.P.I. (2011). Effect of neem leaf (azadirachita indica) extract on the corrosion inhibition of mild steel in dilute acids. International Journal of Physical Sciences, 6(9), 2249-2257.

IOP Conf. Series: Materials Science and Engineering 1117 (2021) 012002 doi:10.1088/1757-899X/1117/1/012002

Gravimetric and data analysis of the corrosion resistance behaviour and inhibition of C26000 brass in dilute HNO_3 solution

Roland T. Loto[1], Cleophas Loto[1] and Oshogbunu Ohwofasa[1]

[1]Department of Mechanical Engineering, Covenant University, Ogun state, Nigeria

Email: tolu.loto@gmail.com

Abstract. The corrosion resistance of as-received and heat treated C26000 brass (annealed and quenched) was evaluated by gravimetric analysis in 0.5 M, 1 M, 1.5 M, 2 M, 2.5 M, 3 M and 3.5 M HNO_3 solution. Generally, corrosion rate of C26000 increased with increase in HNO_3 concentration and varies significantly with respect to observation time. Results showed that heat treatment has no influence on the corrosion resistance of the brass heat treated brass compared to the as-received sample. Application of garlic extracts in HNO_3 solution at 0.5 M significantly decreased the corrosion rate of C26000 60% concentration with inhibition value of 79.46%. The garlic extract exhibited significant concentration dependent behaviour and its inhibition performance varies with observation time. Statistical evaluation showed the standard deviation values of as-received and heat treated C26000 varied with respect to mean inhibition data. However, variation increased with HNO_3 concentration due to thermodynamic instability with respect to observation time. In the presence of the garlic extracts, the high standard deviation values at the lowest and highest garlic concentration signify significant variation with observation time and thermodynamic instability. Evaluation through analysis of variance shows HNO_3 concentration and observation are statistically irrelevant to the corrosion rate output of as-received and heat treated C26000. In the presence of garlic extracts only the extract concentration is statistically relevant with value of 90.52%.

1. Introduction

Corrosion is the devolution of the electrochemical, physical, mechanical and metallurgical properties of metallic and non-metallic materials due to intermolecular reactions between the materials and the corrosive anions of their operating environments. Corrosion is more prevalent on metallic alloys due to the high tendency of the metallic surface to form alternating electrochemical cells were oxidation-reduction reactions predominates. The consequences of corrosion are enormous, diverse and in most cases complicated ranging from component damage in mechanical systems to industrial shutdowns and huge economic and financial losses. Significant reduction in metal thickness, evolution of destructive pits and collapse of structures have been the major cause of toxic seepage chemicals, high preservation cost and injury to industry operators. According to Electric Power Research Institute corrosion damage to the power generating establishment totals $17 billion with an estimated 22% of the being avoidable [1]. The foremost corrosion association in America (National Association of Corrosion Engineers) proved in a report that corrosion damage constitutes about 4% of the United States Gross National Product [2]. Globally, corrosion cost the world economy 4% of its Gross

Domestic Product ($2.5 trillion). Alloys of Cu are broadly utilized for the manufacture and construction condenser tubing, desalination units, parts of heat exchangers, hydraulic lines and piping material due to their unique mechanical and aesthetic attributes. The fabricability, thermal, electrical and antiseptic and corrosion resistant of Cu alloys is responsible for their ever increasing applications in most cases replacing ferrous alloys [3-6]. In mildly corrosive conditions the resistance of Cu alloys results from initiation and partial growth of an inert, resilience and transparent film on the alloy. However, in astringent temperature conditions coupled with the presence of corrosive anions the vulnerability of Cu alloys to degradation becomes apparent. Though this is significantly dependent on the concentration of corrosive anions, the relative constituents of Cu alloys, temperature considerations, flow rapidity etc. [7-11]. Cu alloy tends to be reactive in the presence of Cl^- anions resulting in localized deterioration of the alloy [12-17]. The alloys applied in petrochemical refineries and desalination plants tends to fail due to formation of corrosion pits [18-22]. C26000 (alpha brass) Cu alloys are utilized in applications substantially dangerous to ferrous alloys though there are conditions where corrosive anions selectively degrade Cu and Zn in C26000 Cu alloy [23-25]. Application of chemical compounds identified as corrosion inhibitors have been known to significantly reduce corrosion of metallic alloys [26-35]. This research focusses on the effect of metallurgical variation by heat treatment on C26000 corrosion in specific concentrations of HNO_3 solution, while further analysis of C26000 electrochemical behaviour was studied in HNO_3 at specific concentrations of garlic extracts.

2. Experimental methods

2.1 Materials preparation

C26000 brass alloy (C26) rod was acquired in Lagos, Nigeria and analysed in the Materials Characterization Laboratory at Covenant University, Ota, Ogun State, Nigeria. The elemental components of C26 are outlaid in Table 1. C26 was cut and machined into 3 sets of experimental exhibits with average dimension of 1 cm diameter and 2 cm length. The first main set consist of 24 exhibits of C26 which were further subdivided into 3 sets of 8 exhibits each. The first subset of C26 exhibits were subjected to heat-treatment (annealing) in a muffle furnace up to 425°C before gradual cooling to 37 °C ambient temperature. The second subset of C26 exhibits were subjected to heat-treatment (quenching) to 425 °C and instantaneously cooled to 37 °C ambient temperature. The second set main consists of 7 exhibits of C26 while the third set consists of 5 exhibits of C26. All the C26 exhibits were grinded and smoothened with emery papers for gravimetric analysis. Potassium gluconate (PG), a crystalline white powder was secured from Sigma Aldrich, USA. 1Kg of garlic (GLC) was cut into numerous smaller slices and immersed with 2.5 L of C_2H_5OH for 576 h after which the GLC slices were sieved to retain the solution of C_2H_5OH and GLC. The solution was segregated with a rotary evaporator which removed C_2H_5OH from the solution and retaining the garlic extract.

Table 1. Elemental constituents (wt. %) of C26000 Cu alloy

C26000						
Element Symbol	Pb	Zn	Fe	Ni	Mn	Cu
% Composition	0.07	28.38	0.05	-	-	71.5

2.2 Gravimetric analysis

The first main set of C26 exhibits consisting of 3 subsets separately underwent gravimetric analysis in 0.5 M, 1 M, 1.5 M, 2 M, 2.5 M, 3 M and 3.5 M HNO_3 solution for 504 h while measurements were taking every 24 h. The second main set of C26 exhibits underwent gravimetric analysis in 0%, 0.5%, 1%, 1.5%, 2%, 2.5% and 3% PG cubic concentration per 200 ml of 0.5 M HNO_3 solution for 312 h. The third set of C26 exhibits underwent gravimetric analysis in 20%, 40%, 60% and 80% of GLC extract per 200 ml of 0.5 M HNO_3 solution for 384 h. Corrosion rate, C_R (mm/y) for C26 was enumerated from the equation below;

$$C_R = \left[\frac{87.6W}{DAt}\right] \quad (1)$$

W illustrates weight loss (g), D illustrates density (g/cm^3), A illustrates total surface area of C26 exhibit (cm^2), 87.6 illustrates corrosion rate constant and t illustrates time (h). The protection performance (P_P) of the PG and GLC were enumerated from the equation below;

$$P_P = \left[\frac{W_1 - W_2}{W_1}\right] \times 100 \quad (2)$$

W_1 and W_2 illustrates weight loss of C26 at specific PG and GLC concentrations.

2.3 Statistical analysis

Binary-factor single level Anova analysis (F - test) was applied to calculated the statistical importance of HNO_3, PG and GLC concentrates and observation time on PG and GLC performance output, and corrosion rate data in the presence of HNO_3 concentrations (0.5 M to 3.5 M HNO_3). The test was done at confidence level of 95% (significance level of α = 0.05) according to the equations below. The summation of column squares (observation time) was enumerated as follows;

$$SS_c = \frac{\sum T_c^2}{nr} - \frac{T^2}{N} \quad (3)$$

The summation of row squares (HNO_3, PG and GLC concentrations) was enumerated from the equation below;

$$SS_r = \frac{\sum T_r^2}{nc} - \frac{T^2}{N} \quad (4)$$

Total summation of squares equals

$$SS_{Total} = \sum x^2 - \frac{T^2}{N} \quad (5)$$

3. Results and discussion

3.1 Gravimetric analysis

Tables 2 and 3 shows the corrosion rate data for untreated, annealed and quenched C26 Cu alloy in 0.5 M – 3.5 M HNO_3 solution with respect to 504 h of observation time. Observation of the corrosion rate data for untreated C26 shows that at 504 h increase in HNO3 concentration results in increase in corrosion of untreated C26 due to oxidation of C26 by NO_3^- ions according to the equations below;

$3Cu + 8HNO_3 \rightarrow 3Cu(NO_3)_2 + 2NO + 4H_2O$

$3Zn + 8HNO_3 \rightarrow 3Zn(NO_3)_2 + 2NO + 4H_2O$

Increase in concentration of HNO_3 accelerated the corrosion of untreated C26 resulting in higher surface oxidation and degradation of the alloy. However, it must be noted that untreated C26 exhibits differential reaction to variation in HNO_3 corrosion with respect to observation time. At 0.5 M HNO_3, corrosion rate of untreated C26 at 24 h is 4.37 mm/y. This value decreased to 0.64 mm/y at 504 h. However, from 1 M to 2 M HNO_3 concentration the corrosion rates of untreated C26 initiated at 2.03 mm/y, 2.78 mm/y and 2.78 mm/y (24 h) and progressively increased to 3.99 mm/y, 4.8 mm/y and 5.92 mm/y at 504 h whereas from 2.5 M to 3.5 M HNO_3 concentration the corrosion rate of untreated C26 initiated at 12.31 mm/y, 21.34 mm/y and 31.23, and decreased to 7.79 mm/y, 8.21 mm/y and 7.57 mm/y at 504h. Annealing C26 altered its metallurgical structure, hence it corrosion reaction behaviour as shown in the corrosion rate data in Table 2. The corrosion rate values of annealed C26 alloy at 24 h of observation time significantly differs from the values obtained for untreated C26 whereas at 504 h variation of annealed corrosion rate values with HNO_3 concentration and observation time are quite similar to the observation of untreated C26 showing that annealing has limited or negligible influence on the corrosion behaviour of annealed C26 compared to untreated C26. The corrosion rate values of quenched C26 in Table 3 exhibits similar observation showing as earlier confirmed that heat treatment does not influence the corrosion behaviour of C26 Cu alloy.

Further analysis of the corrosion behaviour of C26 Cu alloy was performed by assessing the corrosion protection behaviour of GLC extracts on C26 Cu alloy in 0.5 M HNO_3 solution for 312 h as shown in Table 4. The performance of GLC on C26 at 312 h is marginal with optimum corrosion protection value of 79.46% at 60% GLC concentration. Observing the inhibition performance of GLC with respect to observation time shows that at 24 h of exposure, GLC extract effectively inhibited the corrosion of C26 with values of 99.7%, 106.6%, 103.6% and 102.85% at 20%, 40%, 60% and 80% GLC concentration. The values decreased progressively to 60.41%, 67.20%, 79.46% and 58.63% signifying significant variation with respect to observation time whereas inhibition performance did not vary proportionally with GLC concentration. GLC extract exhibited peak performance from 40% to 60% concentration. Secondly, the progressive decrease in protection performance value shows GLC has weak adsorption properties and its corrosion inhibition behaviour is thermodynamically unstable.

Table 2. Corrosion rate data of untreated and annealed C26 Cu alloy in specific concentrations of HNO_3 solution with respect to observation time

	Untreated C26							**Annealed C26**						
HNO_3 Conc. (M) / Observation Time (h)	**0.5 M**	**1 M**	**1.5 M**	**2 M**	**2.5 M**	**3 M**	**3.5 M**	**0.5 M**	**1 M**	**1.5 M**	**2 M**	**2.5 M**	**3 M**	**3.5 M**
24	4.37	2.03	2.78	2.78	12.31	21.34	31.23	0.04	1.89	1.43	8.35	11.39	6.19	29.28
48	2.45	2.3	1.81	3.43	11.05	20.13	34.39	0.04	1.29	1.48	8.56	12.06	9.57	30.42
72	1.66	2.28	1.99	4.61	12.55	21.34	32.19	0.25	1.32	1.57	9.69	12.13	11.74	26.91
96	1.26	2.43	2.29	5.45	12.7	21.55	29.64	0.26	1.06	1.52	10.37	12.66	12.83	25.15
120	1.03	2.31	2.35	7.28	14.15	19.92	26.85	0.35	0.95	1.77	10.32	12.1	12.86	23.17
144	0.87	2.44	2.47	6.84	13.47	20.1	24.35	0.4	0.81	1.76	10.29	11	13.2	21.12
168	0.75	3.09	2.91	7.96	15.37	19.28	22.35	0.42	1.01	2.19	10.37	10.36	13.75	19.53
192	0.66	2.92	3.29	7.53	14.3	17.9	19.88	0.38	1.15	2.39	9.86	10.04	13.33	18.11
216	0.59	3.13	3.78	7.19	13.56	16.85	17.67	0.46	1.21	2.74	9.67	10.76	13.15	17.23
240	0.54	3.17	3.42	7.01	12.74	16.4	15.9	0.44	1.26	3.56	9.57	11.01	12.83	16.5
264	0.52	3.38	3.41	6.8	12.2	15.67	14.46	0.52	1.44	3.78	9.45	10.96	12.52	15
288	0.49	3.64	3.69	6.59	11.56	14.36	13.25	0.52	1.52	4.07	9.06	10.57	12.12	13.75
312	0.46	3.52	3.57	6.21	10.92	13.26	12.23	0.52	1.66	4.37	8.77	10.54	11.7	12.69
336	0.46	3.56	3.77	6.14	10.41	12.31	11.36	0.54	1.75	4.36	8.62	11.43	11.45	11.79
360	0.43	3.75	3.71	6.04	9.95	11.49	10.6	0.54	1.88	4.36	8.51	10.67	10.69	11
384	0.42	3.74	3.8	5.87	9.48	10.77	9.94	0.59	1.91	4.39	8.13	10.01	10.02	10.31
408	0.43	3.69	5.61	5.67	9.08	10.14	9.34	0.59	2.11	4.33	7.82	9.42	9.43	9.71
432	0.41	3.57	4.69	5.78	8.76	9.58	8.83	0.61	2.27	4.29	7.53	8.89	8.91	9.17
456	0.51	3.58	4.61	5.84	8.48	9.07	8.37	0.89	2.55	4.21	7.39	8.43	8.44	8.69
480	0.57	3.8	4.76	5.89	8.18	8.06	7.95	1.18	2.81	4.24	7.19	8	8.02	8.25
504	0.64	3.99	4.8	5.92	7.79	8.21	7.57	1.42	3.08	4.26	7.01	7.62	7.64	7.86

Table 3. Corrosion rate data of quenched C26 Cu alloy in specific concentrations of HNO_3 solution with respect to observation time

Quenched C26							
HNO_3 Concentration (M) / Observation Time (h)	**0.5 M**	**1 M**	**1.5 M**	**2 M**	**2.5 M**	**3 M**	**3.5 M**
24	0.29	0.16	0.89	3.6	14.41	20.91	21.27

48	0.21	0.15	0.9	3.86	10.85	19.9	21.24
72	0.36	0.44	0.81	5.43	11.68	19.84	21.74
96	0.48	0.5	0.78	5.7	11.8	19.29	20.42
120	0.5	0.4	1.07	6.95	11.69	18.38	20.47
144	0.52	0.34	1.26	7.53	11.52	17.78	19.41
168	0.55	0.37	1.57	7.31	11.97	16.38	17.84
192	0.53	0.45	1.72	7.51	11.98	15.81	16.65
216	0.48	0.44	1.84	7.9	12.02	15.6	16.26
240	0.45	0.43	2.1	7.72	11.39	14.58	15.38
264	0.43	0.49	2.31	7.57	11.22	13.96	14.65
288	0.41	0.57	2.8	7.71	10.72	13.57	13.43
312	0.39	0.68	3.15	7.72	10.22	12.52	12.4
336	0.37	0.74	3.65	7.55	9.98	11.63	11.51
360	0.38	0.82	3.62	7.63	9.82	10.85	10.74
384	0.38	0.88	3.89	7.47	9.69	10.18	10.07
408	0.57	0.93	4.04	7.23	9.12	9.58	9.48
432	0.7	1.07	3.97	7.15	8.61	9.04	8.95
456	0.77	1.14	3.91	6.95	8.16	8.57	8.48
480	0.85	1.2	3.9	6.71	7.75	8.14	8.06
504	0.96	1.29	4.97	6.6	7.38	7.75	7.68

Table 4. C26 corrosion rate and GLC protection performance data at specific GLC concentrations in 0.5 M HNO_3 solution with respect to observation time

	Inhibition Efficiency (%)			
GLC Concentration (%) / Observation Time (h)	**20%**	**40%**	**60%**	**80%**
24	99.70	106.60	103.60	102.85
48	98.33	102.3	101.71	103.86
72	91.20	98.30	97.30	100.00
96	87.25	96.71	94.00	93.82
120	81.91	95.09	91.48	88.76
144	75.54	92.35	90.15	82.52
168	72.29	88.48	87.72	78.70
192	70.16	86.57	85.83	74.65
216	68.65	83.84	84.56	71.66
240	64.66	75.47	82.11	67.90
264	63.42	73.36	81.20	61.91
288	61.95	70.69	79.88	59.67
312	60.41	67.20	79.46	58.63

3.2 Statistical analysis

Statistical data showing the mean and standard deviation (SD) for untreated, annealed and quenched C26 corrosion rate data with respect to exposure time and HNO_3 concentration is shown in Table 5 and 6 while Table 7 shows the mean and SD data for GLC protection performance data with respect to exposure time and GLC concentration. The mean corrosion rate data for untreated C26 varied proportionally with increase in HNO_3 concentration. At 0.5 M HNO_3, the SD value is 1.19. Decrease occurred at 1 M HNO_3 before gradual increase with respect to HNO_3 concentration till 3.5 M HNO_3. The higher the SD value, the greater the variation of untreated C26 corrosion rate from mean value signifying significant thermodynamic instability occurring on C26 surface in the presence of NO3- ions. The highest SD value for untreated C26 was obtained at 3.5 M HNO_3 (6.43) while the lowest value was obtained at 1 M HNO_3 (0.42) signifying stable electrochemical behaviour at this concentration. The mean values for annealed C26 exhibited similar trend compared to untreated C26 Cu alloy though the values for annealed C26 where generally higher. However, the SD values obtained are generally lower. Observation of the corrosion rate data in Tables 2 and 3 shows heat treatment has limited influence on the corrosion reaction behaviour of C26 Cu alloy. However, SD values obtained are relatively lower, as a result annealed C26 alloy are thermodynamically more stable than untreated C26 in dilute HNO_3 solution but similar to quenched C26 alloy (Table 6).

The mean values in Table 7 show effective inhibition performance generally with respect to observation time. The SD values in Table 7 vary with GLC concentration but not linearly. The SD values at 20% and 80% GLC concentration are significantly higher than the values obtained at 40% and 60% GLC. The higher SD values obtained specific to GLC concentration is due to significant variation of GLC protection performance data from mean value signifying thermodynamic instability and relative unstable inhibition behaviour. At 40% and 60% GLC concentration lower SD value is due to stable inhibition behaviour of GLC with respect to observation time. As shown in Table 7, 62.5% of GLC protection performance data are greater than 85% inhibition efficiency at margin of error of $\pm$15%.

Table 5. Mean and standard deviation data for untreated and annealed C26 Cu alloy with respect to HNO_3 concentration and observation time

C26	**Untreated**							**Annealed**						
HNO_3 Conc. (M)	**0.5**	**1**	**1.5**	**2**	**2.5**	**3**	**3.5**	**0.5**	**1**	**1.5**	**2**	**2.5**	**3**	**3.5**
SD	1.19	0.42	0.64	1.83	1.22	1.86	6.43	0.16	0.29	0.79	0.73	0.86	2.34	5.04
Mean	1.42	2.61	2.71	6.01	13.22	19.48	25.45	0.30	1.20	2.04	9.71	11.35	11.95	22.74

Table 6. Mean and standard deviation data for quenched C26 Cu alloy with respect to HNO_3 concentration and observation time

C26	**Quenched**						
HNO_3 Conc. (M)	**0.5**	**1**	**1.5**	**2**	**2.5**	**3**	**3.5**
SD	1.04	2.37	1.57	1.65	1.65	2.76	7.12
Mean	0.44	0.37	1.29	6.35	11.93	17.85	19.07

Table 7. Mean and standard deviation data for GLC protection performance output on C26 Cu alloy with respect to GLC concentration and observation time

GLC Conc. (%)	**20**	**40**	**60**	**80**
SD	12.63	9.25	7.26	13.30
Mean	80.97	92.57	91.85	86.47
Margin of Error	$\pm$15%			
Data above 85% Inhibition	62.5%			

Analysis of variance (ANOVA) was used to evaluate the statistical importance of HNO_3 concentration and observation time on the corrosion rate data of untreated, annealed and quenched C26 alloy. The data are shown in Tables 8 and 9. Table 10 shows the data from analysis of variance for GLC protection performance data with respect to GLC concentration and observation time. The statistical relevance factor shows the percentage value of the statistical importance of the sources of variation on the corrosion rate output of untreated, annealed and quenched C26, and GLC protection performance output. The theoretical significance factor is the threshold value wherewith the mean square ratio is measured against and must be greater than for the statistical relevance factor to be established. Tables 8 shows observation time overwhelmingly influenced the corrosion rate output of untreated and annealed C26 Cu alloy compared to HNO_3 which is also relatively high. However, the mean square ratio is greater than the theoretical significance factor thus they are statistically irrelevant despite their influence on the corrosion rate output of untreated and annealed C26 Cu alloy. Similar value trend was observed in Table 9 for the statistical data for quenched C26 Cu alloy. Table 10 shows that GLC concentration is the only statistical relevant factor influencing the protection performance output of GLC extract. This is further established from the mean square ration value which is greater than the theoretical significance factor. However, observation time as shown the table is statistical irrelevant for reasons earlier explained.

Table 8. ANOVA data for the corrosion rate output of untreated and annealed C26 Cu alloy with respect to HNO_3 concentration and observation time

	Untreated C26			**Annealed C26**		
Source of Variation	**Mean Square Ratio (F)**	**Theoretical Significance Factor**	**Statistical Relevance Factor, F (%)**	**Mean Square Ratio (F)**	**Theoretical Significance Factor**	**Statistical Relevance Factor, F (%)**
HNO_3 Conc.	-6.70	2.42	97.11	-5.31	2.42	96.47
Observation Time	-6.13	2.1	133.36	-6.13	2.1	167.12

Table 9. ANOVA data for the corrosion rate output of quenched C26 Cu alloy with respect to HNO_3 concentration and observation time

Source of Variation	**Mean Square Ratio (F)**	**Theoretical Significance Factor**	**Statistical Relevance Factor, F (%)**
GLC Conc.	-10.44	2.42	97.5
Observation Time	-6.18	2.1	86.54

Table 10. ANOVA data for the GLC protection performance output with respect to GLC concentration and observation time

Source of Variation	**Mean Square Ratio (F)**	**Theoretical Significance Factor**	**Statistical Relevance Factor, F (%)**
GLC Conc.	22.95	2.42	90.52
Observation Time	-2.20	2.1	-26.02

4. Conclusion

Gravimetric analysis of the corrosion resistance of as-received and heat treated C26000 Cu alloy specific concentrations of HNO_3 solution showed that heat treatment no influence on the electrochemical characteristics of C26000 alloy. Data showed increase in HNO_3 concentration strongly influenced the corrosion resistance of as-received and heat treated C26000 alloy while the corrosion rate was observed to vary with observation time. Further analysis of as-received C26000 in the presence of garlic extracts at the lowest HNO_3 concentration showed the extract performed marginally at most concentrations with optimal protection performance at 60% extract concentration. Corrosion resistance of C26000 vary minimally from mean values with respect to observation time. However, in the presence of the extracts variation increased significantly. Statistical data showed the extract concentration was the major determinant factor influencing the corrosion behaviour of C26000 whereas in specific HNO_3 concentration without the extracts corrosion resistance exhibited no statistical relationship with observation time and HNO_3 concentration.

Acknowledgement
The author is grateful to Covenant University Ota, Ogun State, Nigeria for the sponsorship and provision of research facilities for this project.

References

[1] Syrett, B.S. (2002). Cost of Corrosion in the Electric Power Industry, EPRI Report 1004662.
[2] Koch, G.H., Brongers, M.P.H., Thompson, N.G., Paul Virmani, Y., & Payer, J.H. (2002) Corrosion Costs and Preventive Strategies in the United States, Supplement to Materials Performance, pp. 4-9
[3] Ateya, B.G., Ashour, E.A., & Sayed, M.S. (1994). Corrosion of α-Al bronze in saline water. Journal of Electrochemical Society, 141, 71-78,
[4] Powell, C. Corrosion and Biofouling Resistance Evaluation of 90-10 Copper-Nickel. https://docplayer.net/21504306-Corrosion-andbiofouling- resistance-evaluation-of-90-10-coppernickel.html (27. 5. 2019.)
[5] Davis, J.R. (2001). Alloying: Understanding the basics, ASM International, USA
[6] Aljinović, LJ., Gudić, S., & Šmith, M. (2000). Inhibition of CuNi10Fe corrosion in seawater by sodiumdiethyl- dithiocarbamate:an electrochemical and analytical study. Journal of Applied Electrochemistry, 30, 973-979,
[7] Badawy, W.A., Ismail, K.M., & Fathi, A.M. (2005). Effect of Ni content on the corrosion behavior of Cu–Ni alloys in neutral chloride solutions. Electrochimica Acta, 50, 3603-3608.
[8] Kievits, F.J., & Ijsseling, F.P. (1972). Research into the corrosion behaviour of CuNi10Fe alloys in seawater. Materials and Corrosion, 12, 1084-1096.
[9] Ijsseling, F.P. (1974). The application of the polarization resistance method to the study of the corrosion behaviour of CuNi10Fe in sea-water. Corrosion Science, 14, 97-110.
[10] Schleich, W. Application of copper –nickel alloy UNS C70600 for seawater service. https://www.copper.org/applications/marine/cuni/alloys/u ns_c70600.html (10. 7. 2019.)
[11] Hack HP and Pickering HW (1991) AC impedance study of Cu and Cu-Ni alloys in aerated saltwater J. Electrochem. Soc. 138 690-695.
[12] Babouri, L., Belmokre, K., Kabir, A., El Mendili, Y., & Abdelouas, A. (2015). Study of the stability of the passive film formed on copper-based binary alloys in 3% NaCl solution, Proceedings of the 17th Topical Meeting of the International Society of Electrochemistry, Saint-Malo, France.
[13] Lu, H., Gao, K., & Chu, W. (1998). Determination of tensile stress induced by dezincification layer during corrosion for brass. Corrosion Science, 40(10) 1663-1670.
[14] Gouda, V.K., Selim, I.Z., Khedr, A.A., & Fathi, A.M. (1999). Pitting Corrosion Behaviour of Monel-400 Alloy in Chloride Solutions. Journal of Materials Science and Technology, 15, 208.
[15] Morales, J., Fernandez, G.T., Esparza, P., Gonzalez, S., Salvarezza, R.C., & Arvia, A.J. (1995). Corrosion Science, 37, 211-225.
[16] Beccaria, A.M., Mor ED., Poggi, G., & Mazza, F. (1987). A study of the corrosion products of aluminum brass formed in sodium sulfate solution in the presence of chlorides. Corrosion Science, 27, 363-372.
[17] Fontana, M.G., & Greene, N.D. (1967) Corrosion Forms, Corrosion Engineering, McGraw-Hill, New York, pp. 41 – 44.
[18] Suleiman, M.I., Al-Masri, A.S., Padmalayan, S., & Al Musharfy, M. Corrosivity of UAE seawater and brine under statistic and dynamic conditions, Takreer Research Centre, Abu Dhabi Oil Refining Company. http://eurocorr.efcweb.org/2017/abstracts/8/88556%20.pdf. (27. 5. 2020.)

[19] Al-Subaie, K.Z., & Hodgkiess, T. (2003). Corrosion of copper-nickel alloys in simulated vapourside environments. Desalination, 158, 43–50.
[20] Asrar, N., Malik, A.U., Ahmed, S., Al-Sheikh, A., Al- Ghamadi, F., & Al-Thobiety, M. Desalinatio,n 116, 135–144.
[21] Al-Sum, E.A., Aziz, S., Al-Radif, A., Samir, M., & Heikel, O. (1993). Proc. International Desalination Association (IDA) World Conference, Yokohama, Japan, pp. 501–511.
[22] Malik, A., Kutty, M., Ismail, N., & Al-Fozan, S. (1995). Materials performance and failure evaluation in SWCC MSF plants. Desalination, 97, 171–187.
[23] Morales, J., Fernandez, G.T., Esparza, P., Gonzalez, S., Salvarezza, R.C., & Arvia, A.J. (1998). A comparative study of the passivation and localized corrosion of α-brass and β-brass in borate buffer solutions containing sodium chloride: III. The effect of temperature. Corrosion Science 40, 177-190.
[24] Burzynska, L., Maraszewska, A., & Zembura, Z. (1996). The corrosion of Cu–47.3 at % Zn brass in aerated 1.0 M HCl Zn brass in aerated 1.0 M HCl. Corrosion Science, 38, 337-347.
[25] Zou, J.Y., Wang, D.H., & Qiu, W.C. (1997). Solid-state diffusion during the selective dissolution of brass: chronoamperometry and positron annihilation study. Electrochimica Acta. 43, 1733-1737.
[26] Mazumder, M.A.J., Al-Muallem, H.A., Faiz, M., & Ali, S.A. (2014). Design and synthesis of a novel class of inhibitors for mild steel corrosion in acidic and carbon dioxide- saturated saline media. Corrosion Science, 87, 187 – 198.
[27] Brycki, B.E., Kowalczyk, I.H., Szulc, A., Kaczerewska, O., & Pakiet, M. (2017). Organic Corrosion Inhibitors, IntechOpen. http://doi.org/10.5772/intechopen.72943.
[28] Li, Y., Zhao, P., Liang, Q., & Hou, B. (2005). Berberine as a natural source inhibitor for mild steel in 1 M H_2SO_4. Applied Surface Science, 252, 1245-1253.
[29] Quartarone, G., Ronchin, L., Vavasori, A., Tortato, C., & Bonaldo, L. (2012). Inhibitive action of gramine towards corrosion of mild steel in deaerated 1.0 M hydrochloric acid solutions. Corrosion Science, 64, 82-89.
[30] Ashassi-Sorkhabi, H., Majidi, M. R., & Seyyedi. (2004). Investigation of inhibition effect of some amino acids against steel corrosion in HCl solution. Applied Surface Science, 225, 176-185.
[31] ÖZcan, M. (2008). AC impedance measurements of cysteine adsorption at mild steel/sulphuric acid interface as corrosion inhibitor. Journal of Solid State Electrochemistry, 12, 1653-1661.
[32] Loto, R.T. (2017). Anti-corrosion performance of the synergistic properties of benzenecarbonitrile and 5-bromovanillin on 1018 carbon steel in HCl environment. Science Reports, 7(1), 17555
[33] Loto, R.T., & Tobilola, O. (2018). Corrosion inhibition properties of the synergistic effect of 4-hydroxy-3-methoxybenzaldehyde and hexadecyltrimethylammoniumbromide on mild steel in dilute acid solutions. Journal of King Saud University Engineering Science, 30(4), 384-390
[34] Loto, R.T., & Loto, C.A. (2012). Effect of P-phenylediamine on the corrosion of austenitic stainless steel type 304 in hydrochloric acid. International Journal of Electrochemical Science, 7(10), 9423-9440.
[35] Loto, R.T., & Oghenerukewe, E. (2016). Inhibition studies of rosmarinus officinalis on the pitting corrosion resistance 439LL ferritic stainless steel in dilute sulphuric acid. Oriental Journal of Chemistry, 32(5), 2813-2832.

IOP Conf. Series: Materials Science and Engineering 1117 (2021) 012003 doi:10.1088/1757-899X/1117/1/012003

Assessment breaking strength the chassis of Multi-purpose forest fire fighting vehicle

Nguyen Thanh Tung, Luong Van Van*, Nguyen Thai Van

Vinh Long University of Technology Education, Vietnam

* Corresponding Author: vanlv@vlute.edu.vn, Telephone +84902933271

Abstract. Chassis is a main component of the automobile, during the operation of the chassis, it is fail to fatigue damage or break. Therefore, in the process of designing the chassis, it is important to study the durability to destroy the chassis. The paper presents the results of analysing breaking strength chassis of multi-purpose forest fire fighting vehicle by Ansys Workbench software to determine stress, strain in some cases of road movement. The results of the study underlie a scientific for completing the structure of multi-purpose forest fire fighting vehicle.

Keywords: Stress, Strain, Chassis, Multi-purpose forest fire fighting vehicle.

1. Introduction

Multi-purpose forest fire fighting vehicle manufactured by Vietnam include a combination of fire fighting equipment: cut the tree, clean the litter, high-pressure water pump, create corridor fire insulation cutting machine, vacuum and high wind speed bowling machine, extinguish the fire sand blast apparatus that is mounted basic on Ural 4320, active three axles vehicle.

Figure. 1 Multi-purpose forest fire fighting vehicle

The chassis is a beam system supporting installation details cargo loads, receive and transmit reaction during vehicle operation with different road conditions. Furthermore, chassis is also influenced by vibrations from the engine, the drivetrain. Due to the installation of specialized fire fighting equipment, the load of the vehicle increases, which affects the durability of the chassis. The application of finite element method (FEM) to determine the locations of stress and deformation to predict life and optimize the structure of the chassis. [1,2,3]. From those requests, the author analyzed the durability of multi-

IOP Conf. Series: Materials Science and Engineering 1117 (2021) 012003 doi:10.1088/1757-899X/1117/1/012003

purpose forest fire fighting vehicle chassis when the chassis was subjected to specific loads using Ansys Workbench.

Mohd Azizi Muhammad Nora and the authors [4] stress analysis of 35 ton truck frame, I-beam structure. Ebrahim Ebrahimi, Alimohamad Borghei and Morteza Almasi [5] built a truck chassis model and analyzed stress and deformation components. Analysis has shown the maximum stress and strain of the chassis. Akash Singh Patel, Jaideep Chitransh [6] built the 3D model of TATA 2518 TC truck chassis with Catia software and simulated by Ansys software. The analysis results show that stress and deformation values appear on the frame when changing the main bar section of the frame. Goolla Murali, Subramanyam.B, Dulam Naveen [7] studied the determination of torsional and bending stiffness of truck chassis by finite element method, improve the rigidity by changing geometry size and structural characteristics. Chintada.Vinnod babu, Chiranjeeva Rao.Seela, Vykunta Rao.Matta [8] analyzed the stress and strain of the chassis when changing different materials. Kiran Ghodvinde, S. R.Wankhade [9] analyzed the chassis stress under static conditions. In the above studies, the author realized that this work has not yet been studied on the breaking strength the chassis of multi-purpose forest fire fighting vehicle. Therefore, it is necessary to study the durability of the chassis by finite element method to improve the design of multi-purpose forest fire fighting vehicle.

2. Chassis model building of multi-purpose forest fire fighting vehicle

2.1. Multi-purpose forest fire fighting vehicle structual

Multi-purpose forest fire fighting vehicle chassis has length of 7370mm, width of 832mm, including 2 main bars and 9 horizontal bars.

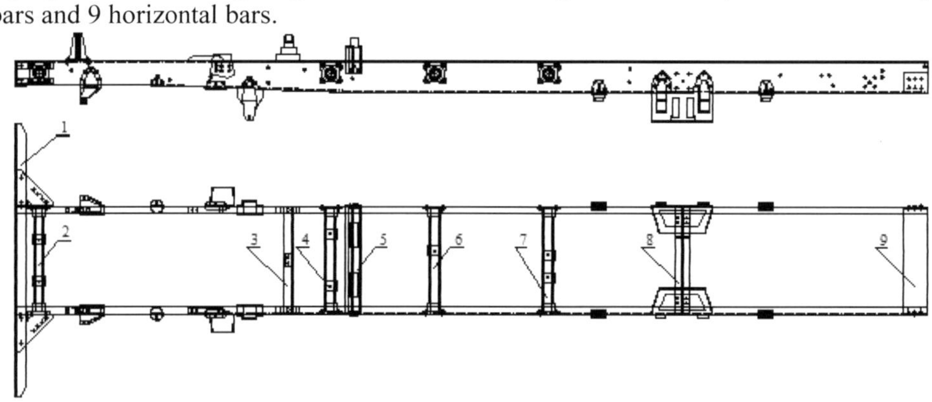

Figure. 2 Chassis design 2-D of multi-purpose forest fire fighting vehicle

The main bar structure on each side includes two bars with a thickness of 8mm that are nested, the outer main bar has a length of 7370mm, the inner main bar has a length of 5760mm from back the chassis.

IOP Conf. Series: Materials Science and Engineering 1117 (2021) 012003 doi:10.1088/1757-899X/1117/1/012003

2.2. Model building

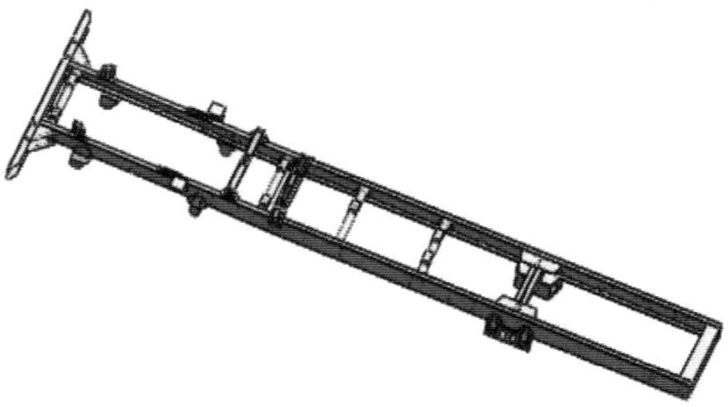

Figure. 3 3D chassis model

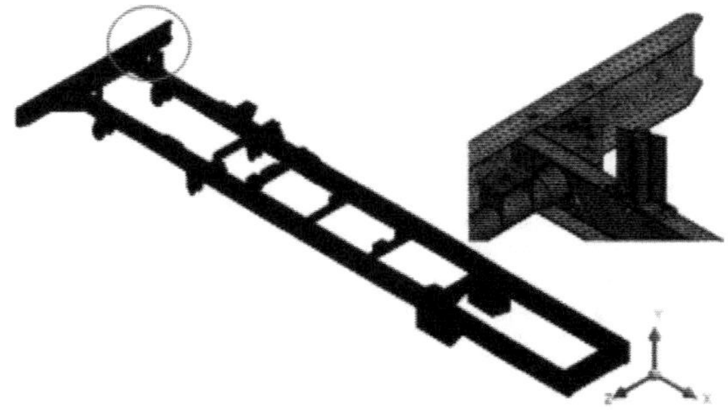

Figure. 4 Finite element model

Model of chassis multi-purpose forest fire fighting vehicle was built by Solidworks software According to the actual size of the chassis. Chassis was created by 40CR steel material with specific gravity is 7850 kg/m^3, the elastic modulus is 205 Gpa, the yield stress is 785 Mpa, the breaking stress limit is 980 MPa and the Possion coefficient is 0,29.

The building process finite element models has an important role in design analysis problems. The meshing of the chassis model is based on the structural characteristics of the chassis and the model control criteria of Ansys, this allows to determine the accuracy of the finite element model that we use. In this study, the hexahedral (Hex 20) and tetrahedral elements (Tet 10) were combined to reduce the size of the problem and increase the accuracy of the model. The finite element model of the chassis has 174223 elements, 376982 nodes and the average element quality reaches 0,718, shown in Figure 3b.

3. Assessing the breaking strength chassis

3.1. Chassis supporting to maximum static load

In this case, the loads applyed on the chassis include: Engine load (12258N), cabin load (8340N), container load (118500N), load of tree cutting mechanism (9180N) and lawn cutting mechanism (7848N).

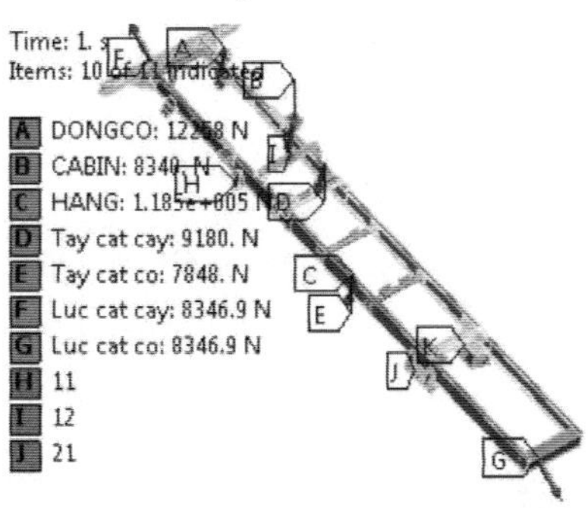

Figure. 5 Position of force on the chassis when subjected to the maximum static load

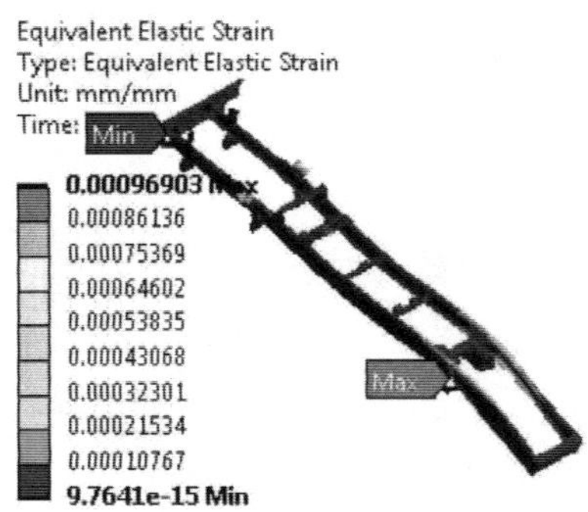

Figure. 6 Equivalent Elastic Strain

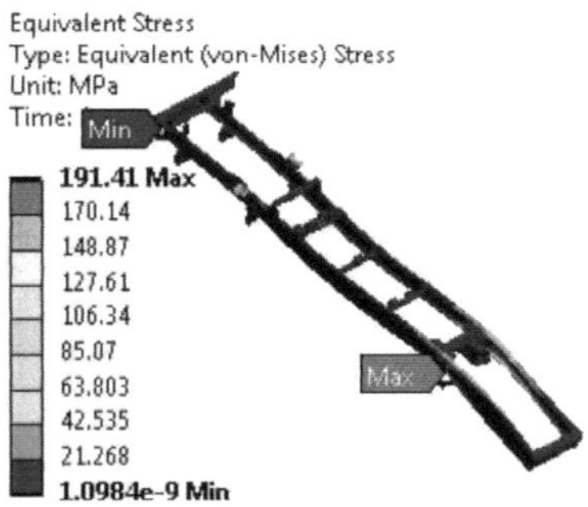

Figure. 7 Equivalent stress (Von Mises)

Table 1. Maximum strain and stress

	Strain (mm/mm)	Stress (MPa)
Maximum	0,00096903	191.41

In this case, the analysis results show that the maximum stress is 191.41 MPa, lower than the yield stress and ultimate stress of materials is 785 MPa and 980 MPa, so the chassis ensures conditions the breaking strength.

3.2. Chassis supporting load when the right front wheel pass surface roughness

In this case, the right front wheels through the bad road and road surface malfunction (bumping height (h) varies from 0.1m to 0.4m). In addition to static load, the chassis is also subject to the load from the road surface acting up through the suspension.

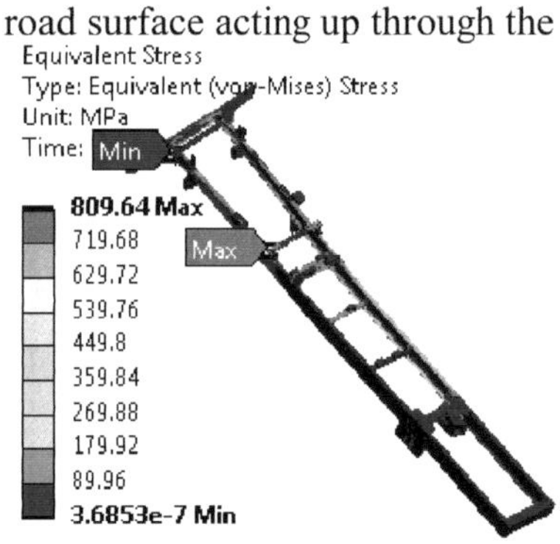

Figure. 8 Equivalent stress (h = 0.1m)

Figure. 9 Equivalent stress (h = 0.4m)

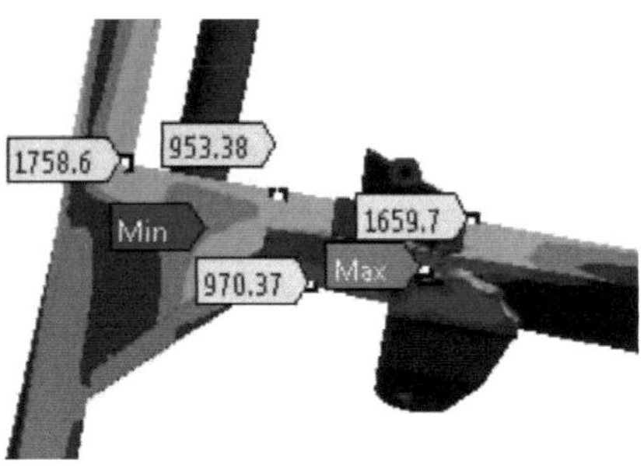

Figure. 10 Position where maximum equivalent stress appears (h = 0.4m)

Table 2. Maximum strain and stress

		Strain (mm/mm)	Stress (MPa)
Bumpy height	0.1m	0.0080162	809.64
	0.2m	0.009104	1190.5
	0.3m	0.010359	1735.9
	0.4m	0.01361	2280.9

The results of the analysis when the right front wheel has to be bumped (the height of the pavement changes) shows that the maximum stresses are greater than the yield limit of the material is 785 MPa, when the height of the pavement road is 0.1 m, the equivalent stress is 809.64 MPa, which is almost equal to the destruction limit and 980 MPa, so the chassis does not guarantee condition breaking strength.

Because there is only one rod in the front main bar where there is a large stress, it is necessary to insert a cylinder like the rear main bar or increase the thickness of the main bar to ensure durability.

3.3. Chassis supporting load when the the rear wheel has to move

In this case, the right rear wheel through the bad road and road surface malfunction (bumping height (h) varies from 0.1m to 0.4m).

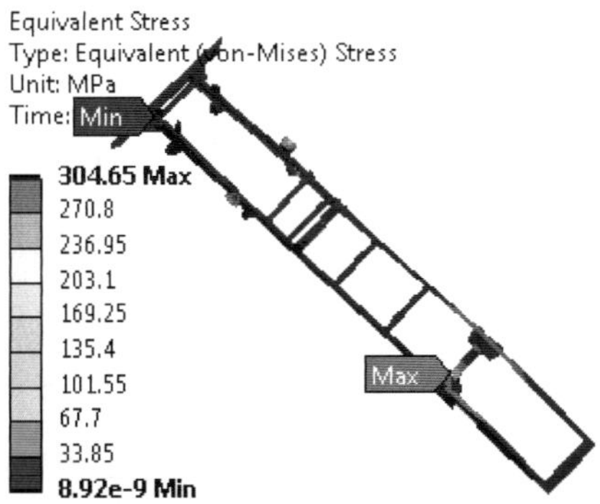

Figure. 11 Equivalent stress (h = 0.1m)

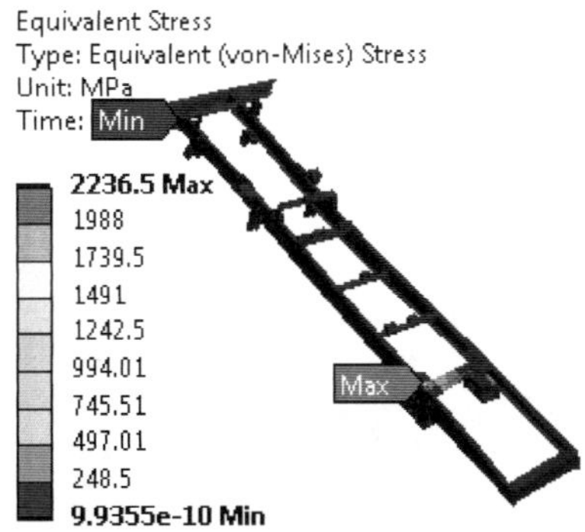

Figure. 12 Equivalent stress (h = 0.4m)

Figure. 13 Position where maximum equivalent stress appears (h = 0.4m)

Table 3. Maximum strain and stress

		Strain (mm/mm)	**Stress (MPa)**
Bumpy height	0.1m	0.0015794	304.65
	0.2m	0.0042238	830.62
	0.3m	0.0077836	1533.6
	0.4m	0.011352	2236.5

The results of the analysis show that most of the largest stresses greater than the yield limit values of the material are 785 MPa (except the height of the pavement pavement 0.1 m). When the height of the road is 0.1m and 0.2m, the equivalent stresses are respectively 304.65 MPa and 830.62 MPa. When the rear wheel is bumped on the road surface with a height of 0.1m, the chassis still ensures durable conditions, in other cases, the chassis is not stable. The position where the maximum equivalent stress value appears when the wheel is subjected to a height road 0.4 m is at the horizontal bar described in figure 13. To reduce the stress in this case, it is necessary to increase the cross-sectional area of the horizontal bar.

3.4. Chassis supporting load when the the two cross wheels are pulled out

In this case, the chassis load bearing from the road surface acting on the rear wheels F_{z32}, front wheels F_{z11} (bumping height (h) varies from 0.1m to 0.4m).

Figure. 14 Equivalent stress (h = 0,1m)

Figure. 15 Equivalent stress (h = 0,4m)

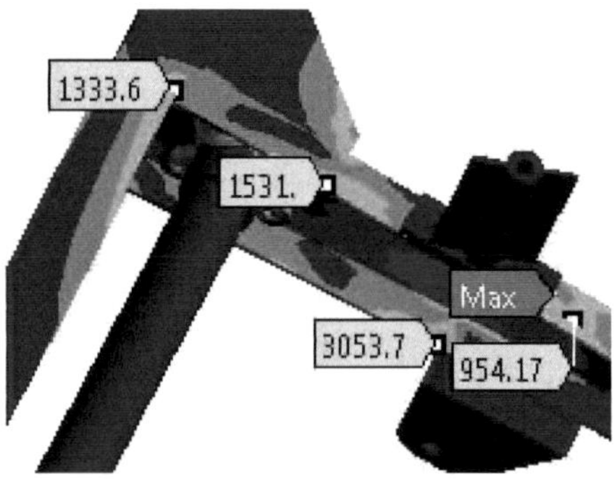

Figure. 16 Position where maximum equivalent stress appears (h = 0,4m)

Table 4. Maximum strain and stress

		Strain (mm/mm)	**Stress (MPa)**
Bumpy height	0.1m	0.0055687	612.04
	0.2m	0.0098846	1927
	0.3m	0.,014449	2817
	0.4m	0.019014	3707

When the height of the surface roughness is 0.1 m, the maximum stressesis 612.04 MPa. The stresses are mainly concentrated at the corners of the chassis. When the height of the surface roughness is greater than 0.1m, the chassis is not guaranteed to be durable.

4. Conclusion

The paper presents the results of analyzing the durability of the chassis multi-purpose forest fire fighting vehicle using Ansys Workbench. The analyzed results determine the stress and strain values of the chassis in proportion the maximum under load modes.

The result is the first step in the durability study to complete the structure of the chassis multi-purpose forest fire fighting vehicle. In order to make a more actual assessment, validation experiments are needed. In addition, more research is needed under the influence of dynamic load when the vehicle is moving on the forest road to assess the vibration and durability of the chassis more completely.

References

[1]. Patil H B Kachave S D and Deore E R 2013 *IOSR Journal of Mechanical and Civil Engineering* March – April 2013(1) 44–49.
[2]. Rahman R A Tamin M N and Kurdi O 2008 *Jurnal Mekanikal* December 2008(26) 76–85.
[3]. Karaoglu C and Kuralay N S 2005 *Elsevier Science B.V. Finite Elements in Analysis and Design* 38 1115–1130.
[4]. Mohd Azizi Muhammad Nora, Helmi Rashida, Wan Mohd Faizul Wan Mahyuddin, *Stress Analysis of a Low Loader Chassis*, Elsevier Ltd. Sci Verse Science Direct Procedia Engineering 41 (2012) 995 – 1001.

[5]. Ebrahimi A Borghei and Morteza Almasi 2010, *Research Journal of Applied Sciences, Engineering and Technology* 2(3) 2040-7467.
[6]. A S Patel, J Chitransh 2016 *International Journal of Engineering Sciences & Research Technology*, 5(9): September, 692-713.
[7]. G Murali, Subramanyam.B, D Naveen 2013, *Altair Technology Conference*. 1-6
[8]. C Vinnod , C R Seela, V R Matta 2015, *International Journal of Engineering Trends and Technology*, 22 (7)- 315 – 318.
[9]. K Ghodvinde, S. R.Wankhade 2014, *International Journal on Mechanical Engineering and Robotics*, 2 (6) 2321-5747.
[10]. AUTOMOBILES (1999), *Ural-4320-10, Ural-4320-31*, and their MODIFICATIONS.

Seismic Analysis of Concrete Arch Dam Considering Material Failure Criterion

Mohammad Hajmohammadian Baghban*[1], Iman Faridmehr [2], Reza Goldaran[3], Roozbeh Safaeian Amoly [3]

[1]Department of Manufacturing and Civil Engineering, Norwegian University of Science and Technology (NTNU), 2815 Gjøvik, Norway
[2] South Ural State University, 454080 Chelyabinsk, Lenin Prospect 76, Russian Federation, Russia
[3]Civil Engineering Department, Faculty of Engineering, Girne American University, Girne, N. Cyprus

*Correspondence: mohammad.baghban@ntnu.no; Tel.: +47-48-351-726

Abstract. Since large dams collapse can lead to catastrophic consequences on human's lives and properties, the reliable performance of such structures during a strong earthquake is extremely important. Therefore, applying reliable analysis methods for designing large dams and seismic evaluation of the existing dam is crucial. The present study investigates the behavior of a concrete arch dam during the maximum credible earthquake "MCL" through non-linear time history analysis. A 30-meter height concrete arch dam was simulated in SAP2000-academic version- using 3D solid elements. Rayleigh damping and plasticity-based five-parameter Willam-Warnke model, as a means of failure criterion, were taken into account during non-linear time history analysis. The results indicated that the seismic load combination had a significant impact on final stress distribution in which high tensile cracks penetrated over the upstream heel.

1. Introduction

Hydraulic structures especially large dams, have been amongst the first structures for which seismic analysis was taken into account. Being earthquake-resistant is of high importance for large dams, where collapses causing irreparable human causalities and financial losses. The majority of large dams throughout the world are located within high seismicity areas, which were affected by strong motions in the past decades. Seismic analysis of large dams is an increasing demand in engineering societies and has a great contribution to the inclusive events for seismic risk reduction. The foremost objective in seismic analysis of large dams is to facilitate understanding of the dynamic behavior and failure mechanism under seismic loading.

The seismic analysis method of large dams proposed by Westergaard in the 1930s [1] has found international acceptance amongst international earthquake engineering communities. This pseudo-static analysis approach accounts for both hydrodynamic pressure and the inertial effects of mass concrete. This conventional "static" design approach has been widely employed to analysis concrete dams however in some cases, this method is based on unrealistic assumptions, and those dams designed according to this theory have experienced unacceptable damage during strong earthquakes. Meanwhile, in the current design approach, the linear elastic concept is taken into account in which the criteria of maximum principal stresses are implemented to vulnerability assessment of dams, where the non-linear behavior of mass concrete (damage-cracking) are ignored. Concrete, as a quasi-brittle material, behaves as linear elastic only when it is subjected to limited normal loads, while damage-cracking will occur as a result of tensile stresses exceed the concrete strength during strong earthquakes. Therefore, in the seismic vulnerability of large dams during the earthquake, it is necessary to evaluate the damage-cracking development in the dam body.

Seismic analysis of large dams has been the center of attention for researchers and dam's experts [2-7]. However, few, if any, researches have focused on the material nonlinearity of mass concrete, i.e. damage- cracking phenomenon [8]. Initial studies of material nonlinearity mainly consider fracture mechanics [9, 10], and the discrete crack model [9] to detect crack development in the mass concrete. As an alternative approach, the continuum method using the crack band model was also developed [11] to study the cracking of the concrete gravity dams, in which it takes into account the strain-softening characteristic of concrete. The crack band method labels the crack as a band comprising concentrated parallel fissures, whereas the material beyond the crack band remaining linear elastic[12]. This approach substitutes the orthotropic properties with isotropic constitutive relation within the failure zone. As soon as the concrete reaches its maximum tensile strength, the material located at the crack zone shows strain-softening performance and follows the equivalent softening constitutive relation.

In this paper, the seismic performance of an existing concrete arch dam was investigated considering damage-cracking development as a result of material nonlinearity under a strong earthquake. The plasticity-based five-parameter Willam-Warnke model [13] used for mass concrete and both cracking and crushing failure modes considered due to a multi-axial stress state. Besides, DCR (Demand Capacity Ratio) - CID (Cumulative Inelastic Duration) method was employed to evaluate the extent of the damage on the dam's body.

2. Research Methodology

2.1. Load Type and Combination

In the design of arch dams, it is essential to determine the loads required in the stability and stress analysis. In this study, the following loads were considered: Dead load, Water load (hydrostatic pressure), and hydrodynamic pressure due to Maximum Credible Level (MCL) ground motion. The dead load is calculated based on the weight of the concrete used in the construction of the dam body act on the center of gravity. The unit weight of concrete is generally assumed to be 24 KN/m^3.

The forces acting on arch dams due to water can be divided into two categories as hydrostatic and hydrodynamic Hydrostatic forces are static forces by the upstream water applied to arch dams. The magnitude of hydrostatic loads acting on arch dams can vary due to the rise or fall of the upstream during a flood or low rainfall months. Hydrodynamic forces are generally formed as a result of strong waves hitting to arch dams during extreme loading conditions i.e. an earthquake. The earthquake load consists of two types: inertia force due to the horizontal acceleration of the dam and hydrodynamic forces resulting from the interaction of the reservoir water and dam body.

During an earthquake, if the dam reservoir is filled with water, the distribution of the hydrodynamic forces can be calculated by Westergaard's theory. Then the hydrodynamic force is added to the hydrostatic force. In Westergaard's theory, it is possible to estimate the hydrodynamic forces at any water level of the dam reservoir.

$$F_{hd} = \gamma_\omega \, (\alpha) \; C_e\sqrt{H_u \, y} \tag{1}$$

F_{hd}, additional total water load down to depth y (Westergaard's theory), KN/m^2;
y, Depth from the reservoir water surface to the point of action of hydrodynamic pressure, (m);
C_e, factor depending principally on the depth of water and the earthquake vibration period T_e in second;

Westergaard's approximate equation for Ce, which is sufficiently accurate for all usual conditions calculated from the following equation:

$$C_e = \frac{0.9}{\sqrt{1 - 7(\frac{H_u}{1000\,T_e})^2}} \tag{2}$$

in which T_e is the period of the first mode.

According to the International Committee On Large Dams ICOLD [14], there are three seismic levels with different Peak Ground Acceleration (PGA) (Table 1). In the Design Basic Level (DBL), there is a possibility of a strong earthquake during the useful lifetime of the structure. At this level of shaking, the structure is assumed to resist lateral loading adequately without any structural damage. The percentage of risk-taking at this level is considered to be between 20 and 64%, for the useful lifetime of 100 years, and it is in proportion with the return period of 100 to 500 years. In the Maximum Design Level (MDL), the possibility of occurrence strong earthquake is low during the useful lifetime of the structure and it may experience damage to some extent; however, it should be able to continue operating. The percentage of risk at this level is between 10 and 20 %, which is in proportion with the return period of 500 to 1000 years. At the Maximum Credible Level (MCL), the structure is allowed to perform non-linearly. Damage to the structure might be severe, but these damages should be considered in the design process that guarantees reservoir safety without any human casualties in downstream. For many existing dams, the Construction Level (CL) with a return period of 50 years was considered at the time of project construction. In this research, MCL was taken into account to evaluate the hydrodynamic performance of the arch dam under consideration.

Table 1. Peak ground acceleration (PGA) for different seismic levels

Seismic design level	Return period (in years)	Maximum peak ground acceleration (g)	
		Horizontal (PGA H)	Vertical (PGA V)
CL	50	0.13g	
DBL	500	0.29 g	0.18 g
MDL	1000	0.39 g	0.27 g
MCL	Deterministic	0.54 g	0.44 g

In the MCL loading condition, the structure can have serious/un-repairable structural damages without the sudden release of the reservoir's water. Based on the previous reports provided [15], the most critical earthquake that the existing dams in MCL are exposed to, is the 1990 Manjil earthquake. Longitudinal components of Majil ground acceleration at MCL which was used in this analysis are shown in figure 1.

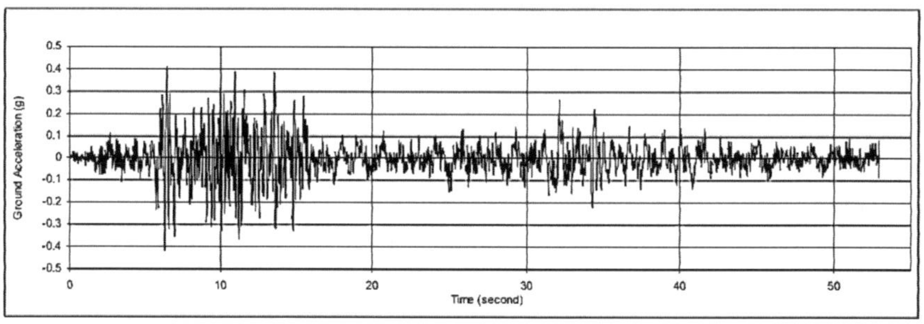

Figure. 1 Scaled acceleration for Manjil earthquake at Abbar station (Component L, MCL)

Table 2 shows the load combinations were considered in this study based on the ICOLD recommendation. To avoid analysis complexity, silt and thermal forces were not taken into account. Additionally, since the water was not directly modeled, the Westergaard theory was employed to simulate hydrodynamic forces. The earthquake force applied to the dam is proportional to the concrete mass.

Table 2. Load combinations

Load combination	Description
SU1	DL+Hs
SU2	DL+Hs+Hd
DU1	DL+Hs+Hd+E

DL is dead load, Hs is hydrostatic force, Hd is hydrodynamic force, and E is earthquake force

2.2. Material Properties

The modulus of elasticity, Poisson's ratio, and the unit weight are 40 GPa, 0.2, 24 KN/m^3 respectively for mass concrete. The actual compressive strength of concrete is assumed as 35 MPa. To calculate the tensile strength of concrete, the Raphael equation is used [16], equation (3), which states the true tensile strength of concrete which is close to the Brazilian test results [17].

$$f'_t = 0.34 f'_c{}^{2/3} = 3.4MPa \tag{3}$$

Table 3. summarizes the compressive and tensile stress, safety factors, and their relevant allowable stresses.

Table 3. compressive and tensile stress, safety factors, and relevant allowable stresses.

	Compressive stresses safety factor	Tensile stresses safety factor	Allowable compressive stresses (MPa)	Allowable tensile stresses (MPa)
Static condition	2	1	17.5	3.4
Dynamic condition	1.5	2	23.3	1.7

2.3. Failure Criterion

The Willam-Warnke model (plasticity-based five-parameter) was used for mass concrete failure criterion. Both crushing and cracking failure modes were taken into account. The criterion for mass concrete failure resulting from multi-axial stress state can be determined as follows:

$$\frac{\Omega}{f_c} - \gamma \geq 0 \tag{4}$$

where Ω is a function of the principal stress state, γ is the failure surface expressed in terms of principal stresses and the five input parameters which are described in Table 4.

Table 4. input parameters for failure criterion

Parameter	Description
f_t	Ultimate uniaxial tensile strength
f_c	Ultimate uniaxial compressive strength
f_{cb}	Ultimate biaxial compressive strength
f_1	Ultimate compressive strength for a state of biaxial compression superimposed on the hydrostatic stress state
f_2	Ultimate compressive strength for a state of uniaxial compression superimposed on the hydrostatic stress state

If equation (4) is satisfied, cracking or crushing of mass concrete occurs. In that case $_if$ one of the principal concrete stresses is tensile, the crack occurs. On the other hand, if all principal stresses are compressive, crushing occurs. The failure surface can be specified with two parameters f_c and f_t. The other parameters in the Willam-Warnke model can be calculated by default as follows:

$$f_{cb} = 1.2\, f_c \tag{5}$$

$$f_1 = 1.45\, f_c \tag{6}$$

$$f_2 = 1.725\, f_c \tag{7}$$

However, these default values are valid only for stress states where the following condition is satisfied:

$$|\sigma_n| \leq \sqrt{3}\, f_c \tag{8}$$
$$\sigma_n = \frac{1}{3}(\sigma_{xp} + \sigma_{yp} + \sigma_{zp})$$

Both function Ω and failure surface γ are expressed in terms of principal stresses denoted as σ_1, σ_2, and σ_3 where;

$\sigma_1 = \max(\sigma_{xp}\ ,\ \sigma_{xp}, \sigma_{xp})$ and $\sigma_3 = \min(\sigma_{xp}\, , \sigma_{xp}, \sigma_{xp})$ where $\sigma_1 \geq \sigma_2 \geq \sigma_3$

The failure of concrete is categorized into four domains:

$0 \geq \sigma_1 \geq \sigma_2 \geq \sigma_3$ (compression - compression - compression)
$\sigma_1 \geq 0 \geq \sigma_2 \geq \sigma_3$ (tensile - compression - compression)
$\sigma_1 \geq \sigma_2 \geq 0 \geq \sigma_3$ (tensile - tensile - compression)
$\sigma_1 \geq \sigma_2 \geq \sigma_3 \geq 0$ (tensile - tensile - tensile)

The general function Ω and the failure surface γ can be determined using independent relationships using concrete element consisting of the following assumptions and limitations:

i. Cracking at each Gaussian point in three orthogonal directions is allowed;
ii. If cracking occurs at a Gaussian point, the crack is modeled directly by modifying material properties;
iii. It is assumed that concrete is isotropic initially.

Cracking occurs when principal tensile stress in any direction lies outside the failure surface. In this model, cracking is permitted in three orthogonal directions at each Gaussian point. The presence of a crack at a Gaussian point and in a special direction represents through modification of stiffness matrix by exerting shear transfer coefficient in the cracked plane.

The reduction coefficient of tensile stress is assumed to be 0.6 in both static and dynamic conditions. In the utilized procedure, if the crack is closed, all compressive stresses orthogonal to crack plane can be transmitted and only a shear transfer coefficient βc ,in order to reduction of shear strength relative to the un-cracked case, is applied to the un-cracked matrix. The value of this parameter in the present study is equal to 0.9 for closed crack, and the strain-stress element matrix is as equation (9) that the superscript ck states that the strain-stress relationship is in the coordinate system parallel to the direction of principal stresses.

$$D_c^{ck} = \frac{E}{(1+\nu)(1-2\nu)}\begin{bmatrix} (1-\nu) & \nu & \nu & 0 & 0 & 0 \\ \nu & (1-\nu) & \nu & 0 & 0 & 0 \\ \nu & \nu & (1-\nu) & 0 & 0 & 0 \\ 0 & 0 & 0 & \frac{\beta_t(1-\nu)}{2} & 0 & 0 \\ 0 & 0 & 0 & 0 & \frac{1-2\nu}{2} & 0 \\ 0 & 0 & 0 & 0 & 0 & \frac{\beta_c(1-\nu)}{2} \end{bmatrix} \tag{9}$$

It is worth noting that the relationship between β_t and $\beta_c$is always as follows:

$$0 < \beta_t < \beta_c < 1 \tag{10}$$

In the present study, the value of the parameter β_t is taken as 0.2. Finally, the cracked element matrix is transferred to the element coordinate system by transfer matrix, $[T^{ck}]$, so that:

$$[D_c] = [T^{ck}][D_c^{ck}][T^{ck}] \tag{11}$$

The above transfer matrix is a function of the crack strain of concrete. It is noticed that if at a point the concrete in the uni-axial, biaxial or tri-axial case is fractured, concrete is considered as crushed. At this stage, the crushed element is eliminated from the stiffness matrix and its force is allocated to adjacent elements. Compressive stress-strain curvatures of concrete in dynamic conditions is shown in Figure 2.

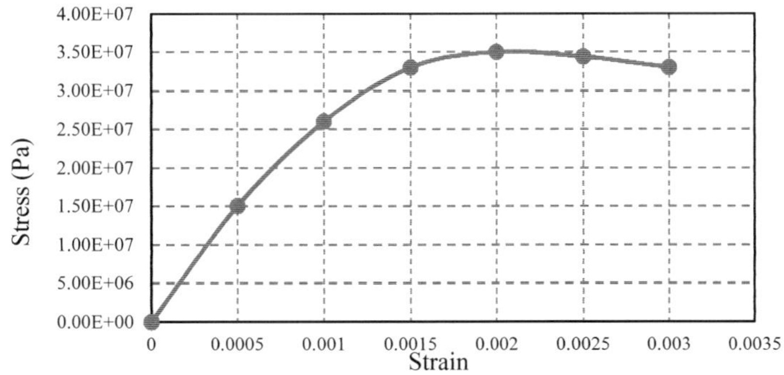

Figure. 2 Compressive stress-strain curve of concrete in dynamic conditions

2.4. Structural Damping

Structural damping is defined as a percentage of critical damping and is considered in the analysis as 5%. For the direct time integration, it is necessary to specify viscous damping in terms of Rayleigh damping with the mass and stiffness proportional to damping coefficients α_M and β_K, respectively. These parameters are chosen for two characteristic frequencies. The following relation between the damping ratio, the frequency and, the Rayleigh damping coefficients are used:

$$d = \frac{\alpha_M}{(2 \cdot \omega)} + \frac{\beta_K \cdot \omega}{2} \qquad (12)$$

In the case of Rayleigh damping, it is only possible to obtain the desired damping at two frequencies. In this study, two frequencies were taken into account as 2 and 6 Hz. Accordingly, damping parameters based on equation 12 are: $\alpha_M = 0.94248 \sec$ *and* $\beta_K = 0.00199\ 1/sec$

3. Model Description

In the present study, a 30-meter height concrete arch dam having a cross-sectional thickness of 4 meters at the crest and 12 meters at foundation along with crest arch length of 62.83 meters was numerically investigated. The finite element model used for the structural analyses has been developed with the software SAP 2000.

4. Results and Discussion

4.1. Shape modes and displacement

The first four mode shapes of the dam are shown in figure 3 for full reservoir condition. The mode shape represents the characteristic dynamic oscillation of the structure for the respective eigenfrequency.

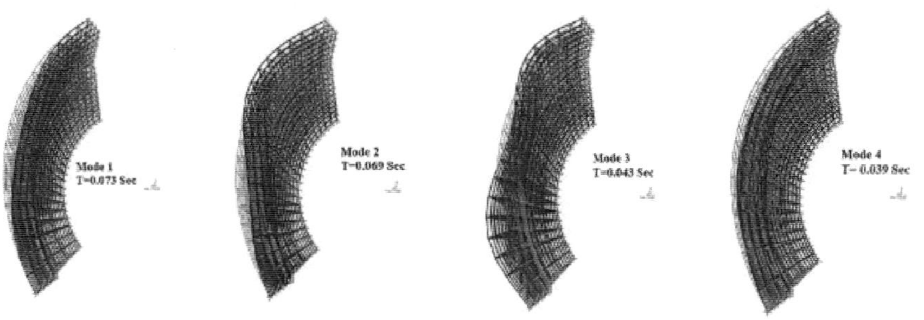

Figure. 3 The first four mode shapes of the case study

The result of Maximum crest displacement during Manjil MCL ground motion is shown in figure4. From figure 4, it can be observed that in the case of a dam with a rigid base, the maximum crest displacement is 1.8 cm.

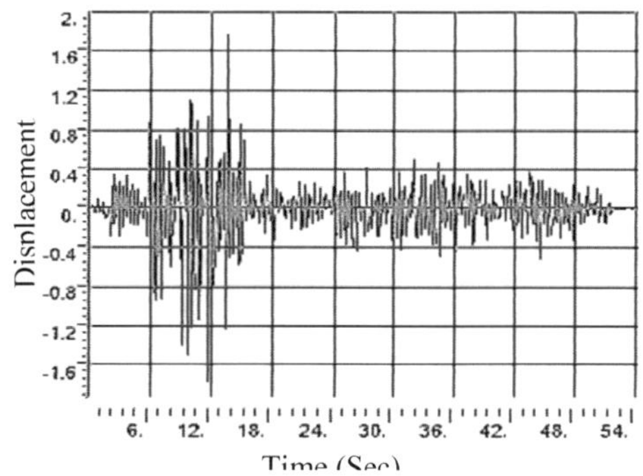

Figure. 4 Relative dynamic displacement of the crest (cm)

4.2. Principal stresses considering material nonlinearity at MCL level

In the following, the stresses are considered over the extent of the dam body. Tensile (+) and compressive (-) stresses are only reported in terms of principal stresses (Smax and Smin), while shear stresses represent the element shear in its local coordinates. For seismic load cases, the maximum and minimum principal stresses are taken from a critical time step during the analysis. Table 5 shows a summary of the absolute maximum or minimum value occurring in any node of upstream or downstream elements. These results are therefore indicative of extreme limits of the stress distributions.

Table. 5 Summary of stresses over dam body (in MPa)

	Upstream			Downstream		
Load Combination	SU1	SU2	DU1	SU1	SU2	DU1
Principal Compression	8	12.2	14	5.5	14.3	18.7
Principal Tension	2.3	3.5	4.4	1.8	2.4	3.8
Maximum Element Shear	3	0.9	1.05	2.7	0.4	1.3

The distribution of maximum principal stresses on the upstream and downstream faces of the dam at MCL level (time 10 sec) is shown in figure 5. A zone of high tensile stress, less than 4MPa, can be seen in the upstream face (upstream heel). Detailed consideration of the distribution of the maximum (tensile) principle stresses shows that the tensile stress developed in areas of the upstream face gradually reduces through the dam in the direction of the downstream face.

Overall, based on given results in Table 5, maximum compressive stress is about 19 MPa and remains well below the allowable compressive strength of the concrete for the extreme load scenario (allowable compressive stresses 32 MPa). However, high tensile stress exceeds the allowable tensile strength of the concrete, 1.7 MPa. The average tensile stress on the upstream face of the dam is around 4.0MPa with the maximum values occurring close to the (upstream heel) interface.

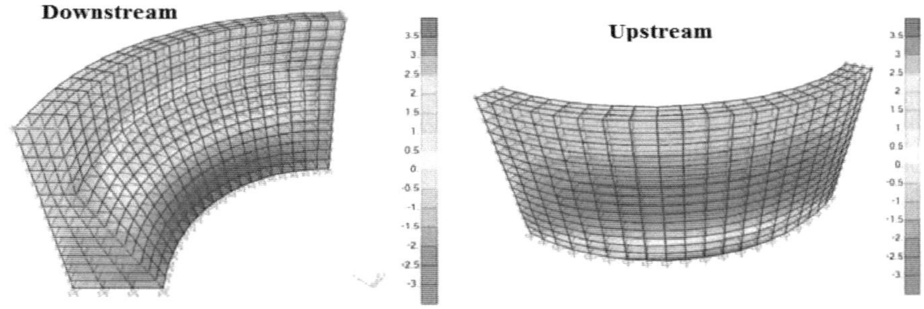

Figure. 5 Envelop values of Maximum Stresses at MCL level (time 10 sec)

Figure 6 shows the cracked Gaussian points on upstream faces at MCL considering material nonlinearity. Figure 6 shows that the most severely damaged region is in the bottom portion of the dam, where the cracks penetrate the small portion of the upstream heel with a maximum damage variable.

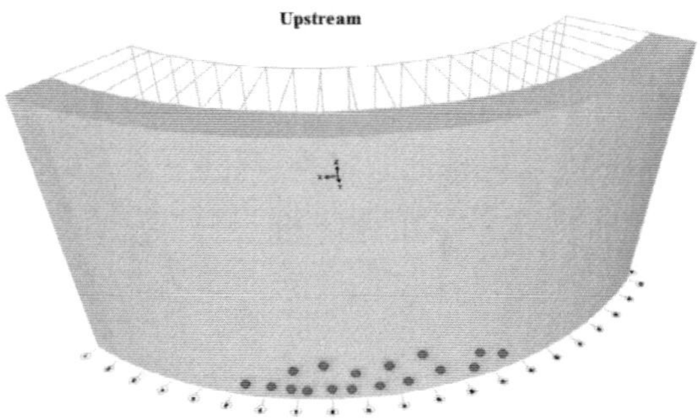

Figure. 6 Cracked Gaussian points

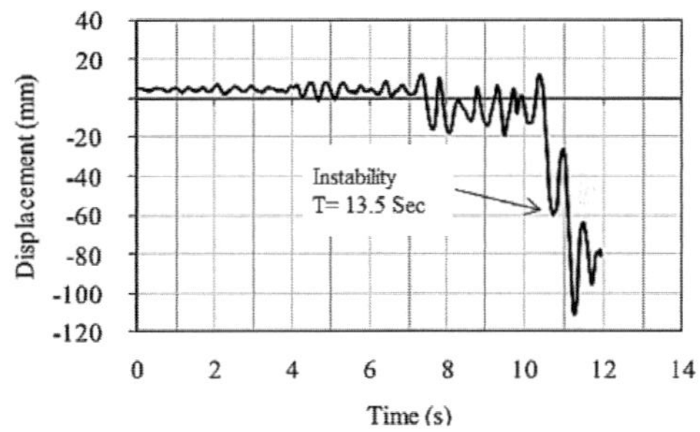

Figure. 7 Displacement time-history

In the present model, cracking starts about at the second 9th in the upstream face. In the second about 11, blocks perform almost separately and in 13.5s severe displacement is shown leads to the failure mechanism of the block as shown in figure 7. Considering the un-convergence of the block after 7 seconds of the beginning of earthquake loading, a significant displacement in the block leads to decide that dam is unsafe in the MCL earthquake.

4.3. Damage Evaluation using Demand Capacity Ratio- Cumulative Inelastic Duration

To assess dam stability under seismic events, the non-linear time-history method is used and is considered as one of the most reliable and prevalent methods of assessment. This method includes the DCR (Demand Capacity Ratio) - CID (Cumulative Inelastic Duration) method described in USACE Code [18]. According to this method, simple stress checks acquired from the non-linear analysis is used to assess the seismic performance of a dam. In general terms, in this method tensile stress should be lower than the allowable tensile strength of mass concrete. Nevertheless, a limited stress excursion beyond the allowable tensile strength is accepted for dynamic loadings. The performance evaluation and the assessment of damage level are expressed by relying on magnitudes of DCR, and cumulative duration of stress excursions beyond the tensile strength of the concrete and spatial extent of overstressed regions. Cumulative inelastic duration is roughly calculated by multiplying the number of stress excursions exceeding a certain DCR value by analytical time step. A performance curve presents the tolerable level of damage based on linear elastic analysis (Figure 8). The maximum allowable DCR for non-linear seismic analysis in large dams is 2. This value is equivalent to a stress demand twice the tensile strength of the concrete. in the proposed technique minor or no probability of damage can be expected if the calculated DCR is lower than or equal to 1.0. The dam may experience nonlinear behavior in the form of cracking or opening of joints if the calculated DCR exceeds 1.0. The level of nonlinearity or concrete cracking assume low to moderate damage if the DCR is less than 2.0.

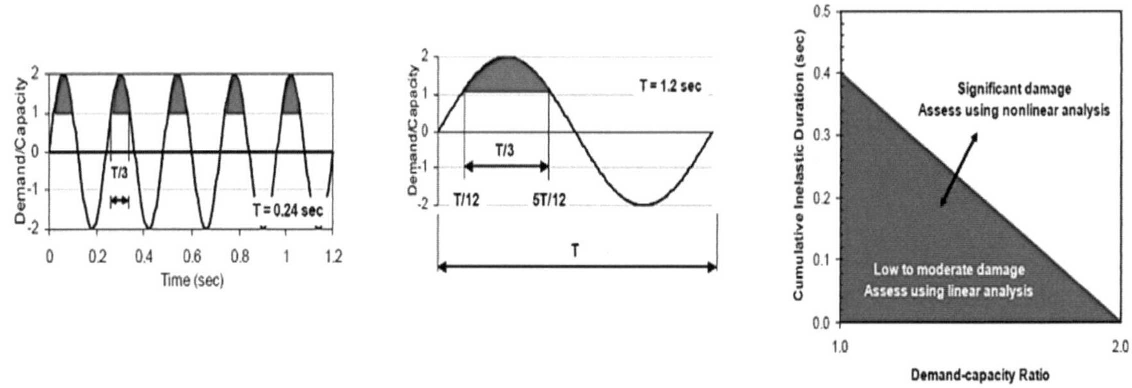

Figure. 8 Concept of Demand Capacity Ratio-Cumulative Inelastic Duration method [19]

Figure 9 shows that the performance curve due to the Manjil earthquake record where it is above the target line (in a limited area); demonstrates that the severe damages are occurred due to tensile stress.

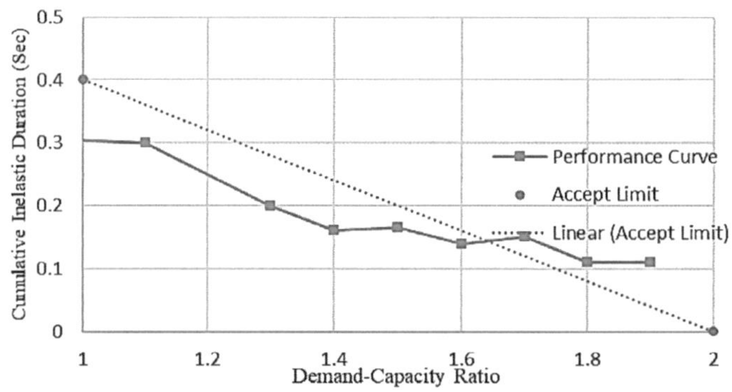

Figure. 9 Performance curve for case study subjected to Manjil earthquake

5. Conclusion

This study presents a seismic analysis of the arch concrete dam using non-linear time history analysis. The conclusions drawn from the above study are as follows:

i. The results show that seismic loading has a significant role in determining the final stress distribution within the dam body. In particular, the seismic load combinations appear to be more critical than those loads associated with the hydrostatic combinations.

ii. The maximum crest displacement was 18 mm when subjected to the MCL earthquake. Meanwhile, a band of high tensile stress is developed on the upstream face (upstream heel), with maximum values varying from about 3.5 MPa to about 4.4 MPa. As a result, cracks in the regions with high tensile stress could potentially be anticipated.

iii. The proposed material non-linearity failure criteria method predicted the crack patter adequately in conformity with principal stress distribution.

iv. The damage evaluation of the dam body was assessed utilizing a simplified demand capacity ratio- cumulative inelastic duration approach. The results indicated that the dam failed to address demand capacity ratio requirement and severe damages were expected as a result of high tensile stress.

References

[1] Han, R.P., *A simple and accurate added mass model for hydrodynamic fluid—Structure interaction analysis.* Journal of the Franklin Institute, 1996. **333**(6): p. 929-945.

[2] Hall, J.F. and A.K. Chopra, *Two-dimensional dynamic analysis of concrete gravity and embankment dams including hydrodynamic effects.* Earthquake Engineering & Structural Dynamics, 1982. **10**(2): p. 305-332.

[3] Calayir, Y. and M. Karaton, *Seismic fracture analysis of concrete gravity dams including dam–reservoir interaction.* Computers & structures, 2005. **83**(19-20): p. 1595-1606.

[4] Zhang, S. and G. Wang, *Effects of near-fault and far-fault ground motions on nonlinear dynamic response and seismic damage of concrete gravity dams.* Soil Dynamics and Earthquake Engineering, 2013. **53**: p. 217-229.

[5] Mirzabozorg, H., R. Kianoush, and M. Varmazyari, *Nonlinear behavior of concrete gravity dams and effect of input spatially variation.* Structural Engineering and Mechanics, 2010. **35**(3): p. 365-377.

[6] Sevim, B., A.C. Altunisik, and A. Bayraktar, *Structural identification of concrete arch dams by ambient vibration tests.* Advances in concrete construction, 2013. **1**(3): p. 227-237.

[7] Aminfar, M.H., et al., *Shape Optimization of Concrete Gravity Dams by ESO Method.* World Applied Sciences Journal, 2008. **5**(6): p. 675-679.

[8] Azmi, M. and P. Paultre, *Three-dimensional analysis of concrete dams including contraction joint non-linearity.* Engineering Structures, 2002. **24**(6): p. 757-771.

[9] Ayari, M.L. and V.E. Saouma, *A fracture mechanics based seismic analysis of concrete gravity dams using discrete cracks.* Engineering Fracture Mechanics, 1990. **35**(1-3): p. 587-598.

[10] Lee, J. and G.L. Fenves, *A plastic-damage concrete model for earthquake analysis of dams.* Earthquake engineering & structural dynamics, 1998. **27**(9): p. 937-956.

[11] Calayir, Y. and M. Karaton, *A continuum damage concrete model for earthquake analysis of concrete gravity dam–reservoir systems.* Soil Dynamics and Earthquake Engineering, 2005. **25**(11): p. 857-869.

[12] Bažant, Z.P. and B.H. Oh, *Crack band theory for fracture of concrete.* Matériaux et construction, 1983. **16**(3): p. 155-177.

[13] Hariri-Ardebili, M., S. Seyed-Kolbadi, and M. Kianoush, *FEM-based parametric analysis of a typical gravity dam considering input excitation mechanism.* Soil Dynamics and Earthquake Engineering, 2016. **84**: p. 22-43.

[14] Dams, T.I.C.o.L., *International Commission on Large Dams.* 2018.

[15] Hariri-Ardebili, M. and S. Seyed-Kolbadi, *Seismic cracking and instability of concrete dams: Smeared crack approach.* Engineering Failure Analysis, 2015. **52**: p. 45-60.

[16] Ahmed, M., J. Mallick, and M.A. Hasan, *A study of factors affecting the flexural tensile strength of concrete.* Journal of King Saud University-Engineering Sciences, 2016. **28**(2): p. 147-156.

[17] *López*, C.M., I. Carol, and A. Aguado, *Meso-structural study of concrete fracture using interface elements. II: compression, biaxial and Brazilian test.* Materials and structures, 2008. **41**(3): p. 601-620.

[18] *Hariri*-Ardebili, M. and V. Saouma, *Collapse fragility curves for concrete dams: comprehensive study.* Journal of Structural Engineering, 2016. **142**(10): p. 04016075.

[19] *Wang*, G., et al., *Integrated duration effects on seismic performance of concrete gravity dams using linear and nonlinear evaluation methods.* Soil dynamics and earthquake engineering, 2015. **79**: p. 223-236.

IOP Conf. Series: Materials Science and Engineering 1117 (2021) 012005 doi:10.1088/1757-899X/1117/1/012005

Erbium Related Photoluminescence of Silicon: Influence of Co-doping with Oxygen and Hydrogenation

Aynul Islam [1] and Anika Tasnim Aynul [2]

[1]Electronic Engineering, Bangor College, Central South University Forestry and Technology, China, Bangor University, United Kingdom.
[2]Department of Physics and Astronomy, University College of London (UCL), London, United Kingdom.

E-mail: a.islam@bangor.ac.uk; aynul@live.at

Abstract. This experiment deals with the influence of co-doping with oxygen and hydrogen and also annealing parameters, temperature and excitation power on the Er related photoluminescence (PL) of silicon. The ultimate goal is to optimise the PL intensity of the Er^{3+} internal transition to make at room-temperature luminescence possible. Silicon is a very inefficient light emitter, because of the low radiative recombination rate due to the indirect band gap. However, by adding optically active impurities such as erbium, Si can be made luminescent. Silicon is an ideal material for the fabrication of optical waveguides that are compatible with optical telecommunication technology at 1.54μm, because of its high transparency and high refractive index at this wavelength. Co-doping with oxygen and hydrogen can enhance the initial Er luminescence. To investigate the influence of oxygen and hydrogen on the diffusion process and the luminescence intensity, samples were doped with erbium, oxygen and hydrogen at different concentrations that were implanted both on Float Zone (FZ) and Czochralski (CZ) silicon wafers. In FZ silicon, the samples implanted with both oxygen and hydrogen showed the highest luminescence yield with a six times higher peak intensity as compared to samples implanted with only erbium and a two times higher peak intensity as compared to samples with erbium and oxygen, in accordance with previous results. However, identified by the line position and contrary to previous results on CZ-Si, the luminescence stems from an erbium-oxygen impurity complex and not from the so-called cubic centre. In CZ samples doped with erbium, oxygen and hydrogen show a 3 times higher intensity as compared to samples doped only with erbium and oxygen. Although the PL lines of the cubic centre are visible, they do not exhibit the strongest luminescence. In samples doped only with erbium and hydrogen and annealed at 900°C, we observe luminescence from only the cubic centre. Although the appearance of the cubic centre in these samples is most likely due to out-diffusion of erbium in the absence of oxygen, hydrogen again enhances the luminescence intensity by a factor 2.

1. Introduction:

1.1 Optical Technology

Optical technologies are becoming more and more important in areas that have traditionally been the domain of electronics. Optical telecommunication technology is one of the key research areas in Photonics. Research in design and manufacturing of optical fibres was started in the early 1980s [1]. In the late 1980s, this effort led to the development of manufacturing process for rare-earth-doped fibres, with the laboratory's first Erbium Doped Fibre Amplifier (EDFA) coming out in 1991. The most important rare-earth for application in photonics is erbium, which has a strong luminescence band around 1535 nm, corresponding to the low-loss window in silica optical fibres. At present, the Erbium Doped Fibre Amplifier (EDFA) dominates the field of optical amplification, and is now a mature technology [1,2]. The possible wavelength ranges that can be used for transmission are determined by dispersion and absorption in the fibre. There are three windows with low absorption, as can be seen in Figure 1. The first window between 800 and 900 nm is mainly due to technological reasons, since powerful semiconductor emitters based on the AlGaAs/GaAs material system can be produced at low cost for this wavelength range. For short wavelengths, attenuation is mainly determined by Raleigh-scattering ($\alpha \sim \lambda^{-4}$). The dispersion in silica vanishes around 1.3 μm, which is the second transmission window. The third window is defined by the minimum of attenuation, which is at 1.5 μm. If the profile of the refractive index of the fibre is shaped in a special way[2] (quadruple cladding), however, the range of maximal bandwidth at 1.3 μm (dispersion minimum) can be shifted to 1.5 μm [4,5].

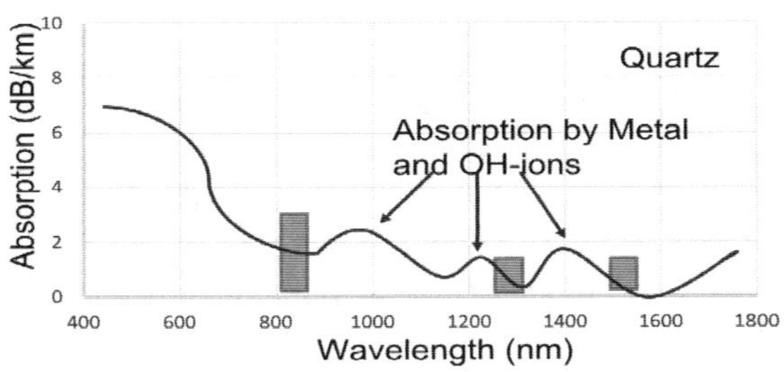

Figure 1. Schematic drawing for silica in which damping depends on wavelength. The arrows terms indicate the absorption due to metal and OH-ions.

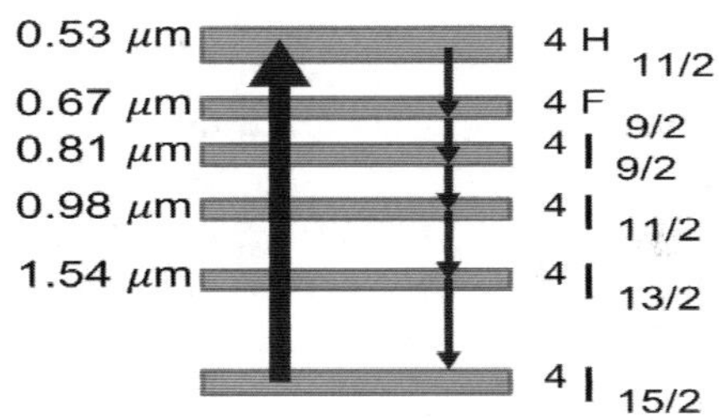

Figure 2. Stark-split Er3+ energy levels, showing emission at 1.54 μm, symbols are given in Russel-Saunders [3, 4].

1.2 Silicon Technology

Silicon is the dominant semiconductor material in the microelectronic industry, and has been called the "engine" behind the information revolution. In 1983 Ennen [6] et al. pointed out the potential of rare earth ions in semiconductor materials for the development of light-emitting diodes and lasers. In particular, erbium is a promising candidate because it shows a strong luminescence at 1.54μm. This luminescence was found to be fairly independent of the host materials [7]. Figure 2 shows a schematic level diagram of the Stark-split Er^{3+} energy levels due to the presence of the host material, labelled using Russel-Saunders notation. The transition from the first excited state ($^4I_{13/2}$) to the ground state ($^4I_{15/2}$) occurs at a wavelength of approximately 1.54 μm. The emission wavelength is relatively insensitive to the host material, because the 4f shell is shielded from its surroundings by the filled 5s and 5p shells. The optical transition occurs between the two lowest spin-orbit levels, $^4I_{13/2} \rightarrow {}^4I_{15/2}$, of Er^{3+} ($4f^{11}$). Since the erbium luminescence is due to an intra-4f shell transition, the influence of the crystal field of the host lattice is weak. Therefore, erbium as an impurity in silicon can be expected to show luminescence at room temperature.

2. Experimental Methods:

2.1 Photoluminescence

Photoluminescence (PL) is simple, versatile, and non-destructive. PL is the spontaneous emission of light from a material under optical excitation. Eventually, these excitations relax and the electrons return to the ground state. If radiative relaxation occurs, the emitted light is called PL [8,13]. This can be investigated to characterize a variety of material parameters. A typical PL set-up is shown in Figure 3.

IOP Conf. Series: Materials Science and Engineering 1117 (2021) 012005 doi:10.1088/1757-899X/1117/1/012005

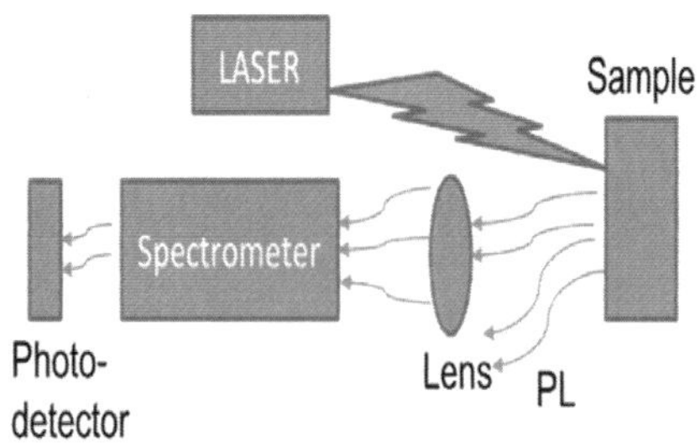

Figure 3. Typical experimental set-up for PL measurements.

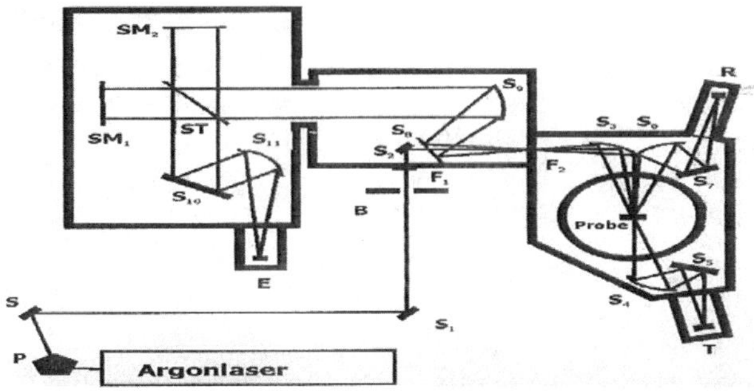

Figure 4. The arrangement of measurement with Fourier Spectrometer and sample chamber, there is a possibility of emission reflection and transmission geometry.

2.2 Fourier Spectroscopy

Photoluminescence measurements were performed with a Bomem DA8.23 Fourier-transform spectrometer. A Fourier spectrometer consists essentially of an optical part to process the light signal and a computer for the Fourier transformation. The optical part is based on a Michaelson interferometer; a brief description of the function is given in the Figure 4. There are two mirrors (SM1, SM2) – one moveable, another fixed and a semi-transparent beam splitter (ST), which are the core of the Fourier Spectrometer [9,10]. Photoluminescence (PL) measurements were performed by pumping with the 514-nm line of an Ar^+ ion laser, which has a penetration depth of about 1μm in silicon. The pump power was varied between 10 and 200 mW. The luminescence signal was analysed with a Fourier spectrometer and detected with a liquid-nitrogen-cooled Ge detector.

3. Experimental Results and Discussions (FZ-Si, CZ-Si):

3.1 Photoluminescence in Float Zone – Silicon (FZ-Si)

To investigate the influence of the respective impurities on both the enhancement of the luminescence and positioning of the Er-ions, we prepared four different sets of samples in FZ-Si material. The respective preparation parameters can be found in Table 1 below.

FZ-Samples	Implanted Impurities	Impurities Dosis (cm-2)	Annealing Temp.캜	Duration of Annealing (min.)	Energy at(keV)
FZ1	Er	1e14 cm^{-2}(Er)	600	15	300
FZ2	Er+O	1e14 cm^{-2}(Er) 1e15 cm^{-2}(O)	600/900	15/20	300/40
FZ3	Er+H	1e14 cm^{-2}(Er) 1e14 cm^{-2}(H)	600/900	15/20	300/30
FZ4	a)Er+O+H b)Er+O+H c)Er+O+H d)Er+O+H e)Er+O+H	1e14 cm^{-2}(Er) 1e15 cm^{-2}(O) 1e14 cm^{-2}(H) 3e14 cm^{-2}(H) 1e15 cm^{-2}(H) 3e15 cm^{-2}(H) 1e16 cm^{-2}(H)	600/900	15/20	300/40/30

Table 1. Data of respective preparation parameters for FZ-Si substrate.

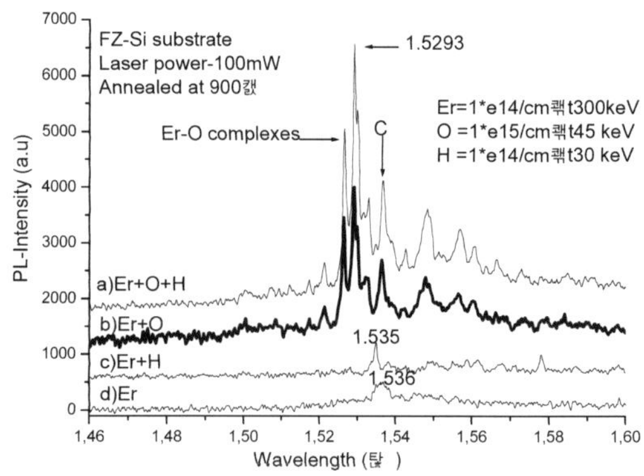

Figure 5. Photoluminescence spectra measured at 7K of FZ - Si sample implanted with a) Er+O+H, b) Er+O, c) Er + H, and d) Er, annealed at 900°C for 20 Min.

Figure 5 shows the Er^{3+}-related PL spectra for the FZ-Si samples in the spectral range of 1.5-1.6 μm. Spectra were taken at low temperatures, typically 7K, and with a laser pump power of 100 mW. In all cases, the spectrum shows a clear peak around 1.54 μm, characteristic of transitions between the $^4I_{13/2}$

IOP Conf. Series: Materials Science and Engineering 1117 (2021) 012005 doi:10.1088/1757-899X/1117/1/012005

and $^4I_{15/2}$ manifolds in Er^{3+}. The PL spectra show the typical features of Er-O centres formed, with the most intense peak at 1.53 μm, and related peaks at higher wavelengths [14]. The oxygen co-doped samples exhibits a three times higher peak intensity (curve 'b') than the samples implanted with only erbium (curve 'd'). The samples implanted with both O- and H- show the highest luminescence yield with a six times higher peak intensity (curve 'a') than the PL intensity of samples implanted with only erbium (curve 'd'). From curve 'c', we see that implanting hydrogen alone only slightly enhances the luminescence of erbium. Co-doping with oxygen (curve 'b') enhances the luminescence significantly, which is to be expected, since oxygen plays an important role in the formation of the optically active erbium centres.

3.2 Temperature Dependence in Float Zone – Silicon (FZ-Si) Substrate

The Er-O doped samples were implanted with the three different hydrogen contents such as a) 1x16cm^{-2}, b) 3x15cm^{-2} c) 1x10^{15} cm^{-2}, d) 3x10^{14} cm^{-2} and e) 1x10^{14} cm^{-2}. After implantation, the samples were annealed at 900°C in N^2 ambient for 20 minutes. Photoluminescence (PL) spectra were measured using the 515 nm line of an Ar^+ ion laser as the excitation source at a power of 100mW.

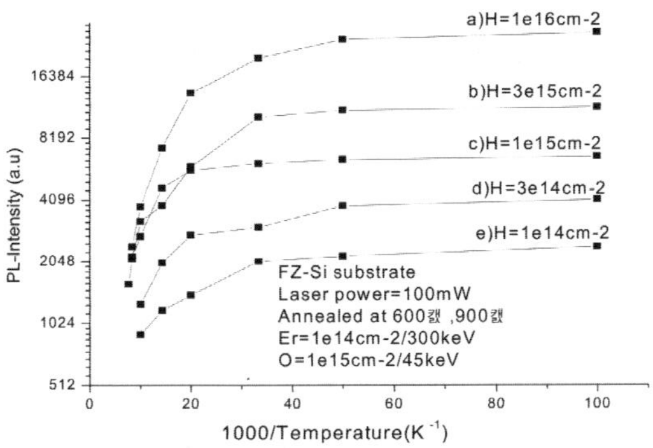

Figure 6. Arrhenius plot of the temperature-dependent PL intensity of Er-O doped Si samples with different hydrogen contents in FZ-Silicon.

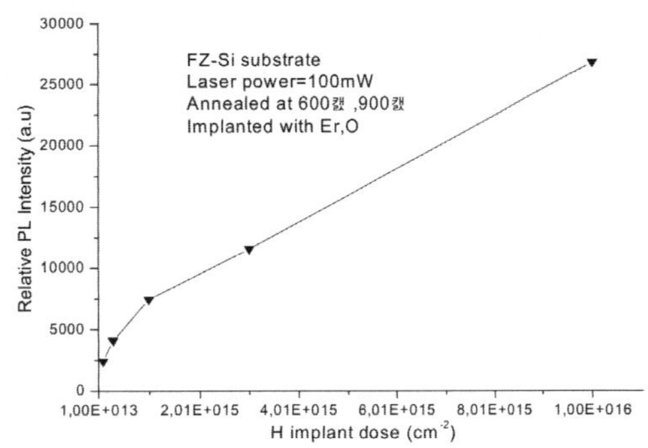

Figure 7. Peak value of the PL intensity (FZ-Si) as a function of the implanted H dose.

Figure 6 shows the temperature dependence of the integrated Er^{3+} luminescence intensity [curve (a), (b), (c), (d), (e)] of the dominant peak at 1529nm. The Er^{3+} emission at 1.53 μm exhibits a thermal quenching and becomes very weak at high temperatures. We measured the PL intensity at different temperatures between 10K and 150K. We see [Fig. 6] a weak quenching of the intensity to 20K, where a much stronger quenching sets in. The luminescence intensity rises for lower temperatures up to 15K.

In general, the quenching of the luminescence intensity in Er-doped silicon, as the temperature is increased, can be a result of two distinct processes: (1) a reduction in the excitation rate of Er^{3+} due to detrapping of electrical carriers which trapped at the Er^{3+}-related level in the Si, or (2) an increase in the de-excitation rate of excited Er^{3+} due to coupling to defect levels in the Si band gap, e.g., to the Er^{3+}-related defect.

In our cases, it can be assumed that the erbium luminescence intensity decrease is mainly caused by a reduction of the Er^{3+} luminescence efficiency due to a non-radiative energy back-transfer process. The Er^{3+} luminescence efficiency can decrease as a result of energy back-transfer from excited erbium to a trapped carrier state. Such back-transfer, followed by either exciton dissociation, or exciton recombination at the trap, decreases the Er^{3+} excited state life time, and with it the Er luminescence efficiency. Figure 7 shows the concentration dependence. We find a linear increase in the PL intensity with different concentration of hydrogen. Above this concentration, it can be assumed that the luminescence intensity will either saturate or slightly decrease. Even with more hydrogen, the amount of erbium that can be implanted without decreasing the luminescence intensity is limited to about 10^{19}

cm^{-3} [15]. This limitation on the erbium and oxygen content also limits the beneficial effect of hydrogen. Still, the saturation limit was obviously not reached with the dose applied in this work, and it might be of interest to find the concentration limit for hydrogen co-doping.

3.3 Photoluminescence in Czochralski – Silicon (CZ-Si)

In the case of CZ growth we used Si as substrate with high oxygen content (10^{18} cm^{-3}). The implanted ions are erbium, oxygen and hydrogen in different concentrations. We prepared five different samples for the CZ-Si material. The respective preparation parameters can be found in Table 2:

CZ-Samples	Implanted Impurities	Dosis of Impurities (cm-2)	Annealing Temp.캜	Duration of Annealing (min.)	Energy at (keV)
CZ1	Er+O	1e14 cm^{-2} (Er) 1e15 cm^{-2} (O)	600/900	15/20	300/45
CZ2	Er+H	1e14 cm^{-2} (Er) 1e15 cm^{-2} (H)	600/900	15/20	300/30
CZ3	a)Er+O+H b)Er+O+H c)Er+O+H	1e14 cm^{-2} (Er) 1e15 cm^{-2} (O) 1e14 cm^{-2} (H) 3e14 cm^{-2} (H) 1e15 cm^{-2} (H)	600/900	15/20	300/45/30

Table 2. Data of respective preparation parameters for CZ-Si substrate.

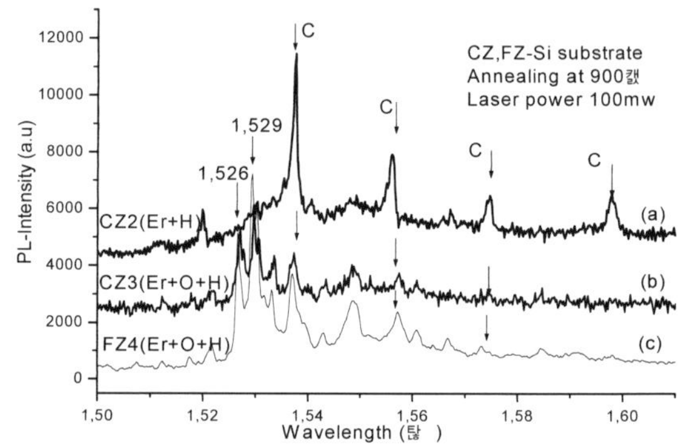

Figure 8. Photoluminescence spectra measured at 7K of CZ-Si sample with a) Er+H, b) Er+O+H, c) FZ-Si substrate with Er+O+H. The solid (labelled C) arrows indicate Er centres with cubic symmetries.

Figure 8 (a) shows the PL spectrum at 7K of CZ-Si sample implanted at 300 keV with the highest ion dose applied (1x14 cm^{-2}) and a hydrogen ion dose of(1x15 cm^{-2}) at 30 keV. The sample was annealed at 900°C for 20 min. The solid (labelled C) arrows indicate the positions of Er^{3+} centres with cubic site symmetry. We observed the maximum PL intensities in CZ-Si substrates in these samples (CZ3).
Figure 8 (b), (c) depicts the PL spectrum of a CZ-Si sample and FZ-Si sample. In FZ-Si we see more intense PL lines visible at smaller wavelengths as compared to the main cubic transition (6504.8 cm^{-1}(1.54 μm), denoted with C), which are related to Er^{3+} centres with lower site symmetry. The same emission lines appear also in CZ-Si. The difference of intensities in FZ and CZ material can be explained by the high oxygen content of approximately 10^{18}cm^{-3} in CZ material and the low oxygen content of approximately 10^{16}cm^{-3} in FZ material. It is assumed that the high optical activity of Er implanted into CZ-Si is related to erbium forming complexes with oxygen.

Figure 8 (a) shows the PL spectrum with high intensities of a CZ-Si substrate. The arrows labelled C indicate the positions of four transitions related to the Er^{3+} centre with cubic site symmetry. There are two possible explanation, such as 1) either erbium is diffused during annealing, lower erbium concentration in deeper regions of sample are responsible for luminescence and can explain appearance of cubic centre or 2) presence of interstitial oxygen can explain appearance of cubic centre. This presence of interstitial oxygen – although at a lower concentration than in CZ-Si - can explain these results: hydrogen mobilises the interstitial oxygen and when oxygen moves out of the interstitial lattice site, the lattice is locally widened and the big Er^{3+} ion can move into the interstitial sites.

4. Conclusion:

Samples were prepared on FZ and CZ silicon wafers with a resistivity of 10 – 24 Ωcm. The samples were doped with Er, O and H with different concentrations to investigate the influence of hydrogen on

IOP Conf. Series: Materials Science and Engineering 1117 (2021) 012005 doi:10.1088/1757-899X/1117/1/012005

the luminescence and diffusion process. In FZ silicon, the samples implanted with both oxygen and hydrogen showed the highest luminescence yield with a six times higher peak intensity as compared to samples implanted with only erbium and a two times higher peak intensity as compared to samples with erbium and oxygen, in accordance with previous results. However, identified by the line position and contrary to previous results on CZ-Si [15,16], the luminescence stems from an erbium-oxygen impurity complex with lower symmetry and not from the cubic symmetry site. Measurements of the luminescence intensity on excitation power revealed an increased number of optically active erbium centres. Investigations on the temperature dependence of the luminescence showed a rise in intensity up to 15 K, a weak quenching up to 20 K, at which point a stronger quenching sets in. These results are attributed to the much lower impurity concentration in FZ silicon, especially to a much lower concentration of interstitial oxygen as compared to CZ Si. The additionally implanted oxygen in these samples has a statistical distribution of positions in the lattice and might not be mobilised during the out-diffusion of hydrogen. The observed enhancement of the Er^{3+}-related PL can be due to another effect of hydrogen, the formation of complexes with oxygen at impurities. Future work would involve finding the concentration limit for hydrogen co-doping of the FZ silicon sample to attain the position of Er^{3+} centres with cubic site symmetry as was found in the CZ silicon sample.

References:

[1] P. Kiiveri , S. Tammela , " Spectral gain and noise measurement system for fibre amplifiers", Optical Engineering , vol. 34 (1995) no: 9, 2592-2595.

[2] Rare Earth Doped Semiconductors , edited by G. S. Pomrenke, P. B. Klein, And D. W. Langer. MRS Symp. Proc. No. 301 (Materials Research Society, Pittsburgh, 1993).

[3] Sven Lanzerstorfer, Er- dotierte Materiellen für optische and optoelektronische Anwendungen im Infrarot-Bereich bei einer Wellenlänge von 1.54µm , PhD thesis , Linz, Univ. of Linz, 1999.

[4] P. J. Mears , L. Reekie , I. M. Iauncey , and D. N. Payne , Electron. Lett. 23, 1026 (1987).

[5] Rare Earth Doped Semiconductors ll , edited by S. Coffa, A. Polman, and R.N. Schwartz, MRS Symp. Proc. No. 422 (Materials Research Society, Pittsburgh, 1996).

[6] E. Desurvire , R. J. Simpson , and P.C. Becker , Opt. Lett. 12,888 (1987).

[7] H. Ennen, J. Schneider, G. Pomrenke, and A. Axmann, Appl. Phys. Lett. 43, 943 (1983).

[8] Timothy H . Gfroerer " Photoluminescence in Analysis of Surfaces and Interfaces" Davidson College, Davidson , USA.

[9] Manual Book of Spectrometer / Spectrograph , SPEX, Industries, INC. 3880 PARK AVENUE, METUCHEN, N. J. 08840.

[10] G. Kocher – Oberlehner, " Herstellung von und PL-Untersuchung an mit Erbium dotiertem porösem Silizium", Dilomarbeit, Univ. Linz, 1997.

[11] W. Jantsch, G. Kocher, L. Palmetshofer, H. Prazbylinska, M. Stepikhova, H. Preier, Mat. Sci. Eng. B B81(2001) 86.

[12] H. Prazbylinska, W. Jantsch, Yu. Suprun – Belevitch, M. Stepikhova, L. Palmetshofer, G. Hendorfer, A. Kozanecki, R. J. Wilson, B. J. Sealy, Phys. Rev. B 54 (1996) 2532.

[13] S. M. Sze, Semiconductor devices, Physics and Technology, John Wiley and Sons, New York, 1985.

[14] H. Przybylinska, G. Hendorfer, M. Bruckner, L. Palmetshofer, W. Jantsch, Appl. Phys. Lett. 66 (1995) 490.

[15] Gudrun Kocher-Oberlehner „ Investigations on erbium doped silicon for possible optical and optoelectronic applications at a wavelength of 1.54 µm" , PhD thesis, Linz, Univ. of Linz, 2003.

[16] G. Steinbacher, Strukturierung und Charakterisierung von Si/SiGe und Si/SiC Heterostrukturen, Diploma thesis, Institute of Semiconductor Physics, University of Linz, 1997.

IOP Conf. Series: Materials Science and Engineering 1117 (2021) 012006 doi:10.1088/1757-899X/1117/1/012006

Finite element simulations of stresses in bone implants made by three-dimensional printing

V M Mikushev[1], A I Samarkin[2] and A S Khomutova[1]

[1] Department of Solid State Physics, St Petersburg State University, 7/9 Universitetskaya Emb., St Petersburg, 199034, Russia
[2] Department of Medical Informatics and Cybernetics, Pskov State University, 2 Lenin Square, Pskov, 180000, Russia
E-mail: v.mikushev@spbu.ru

Abstract. The article is devoted to the analysis of the strength properties of a compact bone tissue in order to produce subsequently the equivalent in strength individual implant. An example of a fragment of a skeleton bone shows the fundamental possibility of solving the problem using a 3d modelling finite element analysis software package.

1. Introduction

In 2013, 1518 patients with bone tumors were identified in Russia [1]. At the same time, in almost sixty percent of cases, they are localized in the extremities, a greater number of such patients are shown crippling treatment as an operation. In the USA, 8680 only new cases of CMT are diagnosed annually, in 60% of cases these tumors are localized in the extremities [2]. Every year in the world, about 3100 people die from malignant tumors of soft tissues, the latency during the first year after diagnosis is 32.0-34.5% [3]. Due to the low efficiency of chemoradiation treatment, the main method of treatment is surgical [4]. It consists of wide excision of the tumor in a single block with all the vessels, which allows removing the tumor in full (see figure 1).

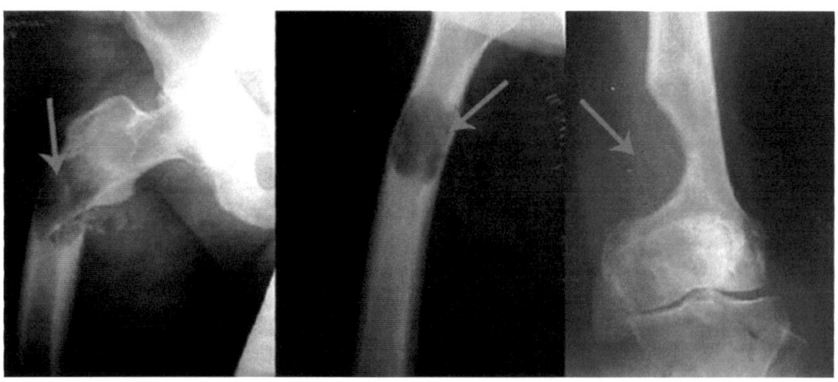

Figure. 1 Examples of bone tumors. Arrows indicate the localization of the distal focus of the metastatic bone.

To avoid amputation requires a complex reconstructive stage using synthetic prostheses [5], preferably made for each specific patient. At the same time, the structure of bone tissue is very complex both structurally and in terms of its genesis, including the effect of neoplasms on its strength.

Thus, a number of problems arise, mainly of a technical nature, namely: modelling the bone tissue and its strength properties, assessing the stresses between the bone and the installed implant, achieving a favorable character of these stresses for the patient.

2. Methods

2.1. About bone tissue modeling by the finite element method

The solution of complex problems associated with the strength calculations of individual elements and their systems is currently carried out almost exclusively by finite element methods (FEA). At the same time, modelling of materials with a complex structure (composites, lattice structures by FEA methods is difficult, due to the high computational costs of calculating the interaction between structural elements. In this regard, we are forced to consider the issue of the structure of bone tissue and methods of its modelling. 80% from the mass of all the bones of the body forms a compact bone tissue [6], in the structure of which osteons are allocated [7]. This structural element consists of cylinders consisting of bone plates inserted into each other. The number of plates varies from 3 to 25 [8]. The thickness of the bone plate is 4-15 microns. Each bone plate contains a large number of osteocytes, which are located in the bone lacunae. The osteocyte has a stellate shape with branching processes. The diameter of the osteon is 0.3-0.4 mm. The Havers canal is located in the centre of the osteon, in which two small blood vessels pass. The diameter of one capillary is 6-20 microns, and in channel n there are two capillaries, so the diameter of the Havers channel can be taken as 40 μm. [8, 9]

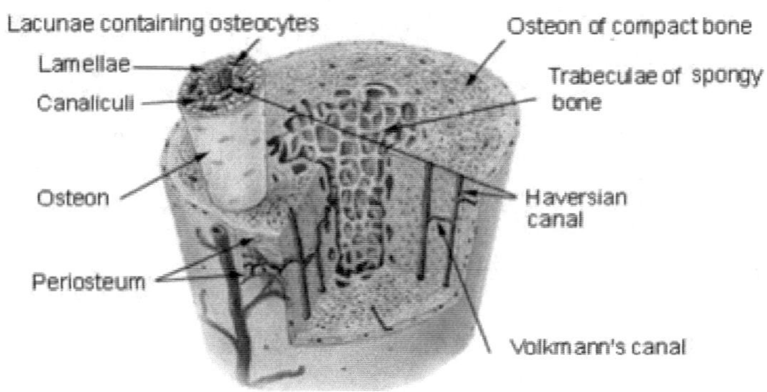

Figure. 2 Compact bone tissue structure (from [7]).

For modelling bone tissue, it is proposed to use the same approach that the authors used in [10] to analyze lattice structures in three-dimensional printing, namely, to simulate a single osteon (with or without osteocytes modelling). Based on the data obtained, calculate the averaged characteristics and then use them in the calculations. Despite some simplifications, the results of such modelling are in good agreement with the more complex techniques described, for example, in [11]

2.2. Bone and implant modeling methods

The peculiarities of the patient's body involve the use of personalized bone models and the individual construction of the affected area to be replaced by the implant. These tasks can be solved by importing the results of CT and MRI scans into the modelling environment. The presented models are created in Stereo Lithography (STL) format. Despite the known shortcomings of the format, it is quite common, exported and readable by most simulators, and relatively easy to fix.

Certain difficulties are caused by the modelling of neoplasms and their effect on the strength of the surrounding bone tissue. Further, it is assumed that the removed part of the neoplasm completely covers the weakened zone.

2.3. Increased compliance of the implant

Implant design is also an important issue. Despite the presence of examples of the implementation of self-dissolved constructions [12], at present, bone replacement elements made of titanium are more often used. In this case, difficulties arise with the manufacture of implants with a geometry unique for each patient and a problem with contact stresses in the area of the conjugation of the bone tissue with the titanium plate, since the strength of titanium alloys is significantly higher than the strength of bone tissue.

Customized implants can be produced using 3D printing methods. To reduce contact deformations, lattice implants should be made. The degree of attenuation can be calculated by analyzing a regular lattice structure or by topological optimization of the implant design.

3. Modeling

3.1. Simulation of single osteon

To simulate a single osteon, it is proposed to use the parametric model developed by the authors and presented below. The model uses the idea of bone plates as shells. Thus, the parameters of the model are the number of shells and the distance between them. The improved model includes lacunae that simulate osteocytes. Osteocytes are modelled as a parameterized 3D spiral array, as shown in figure 3, where F – axial load.

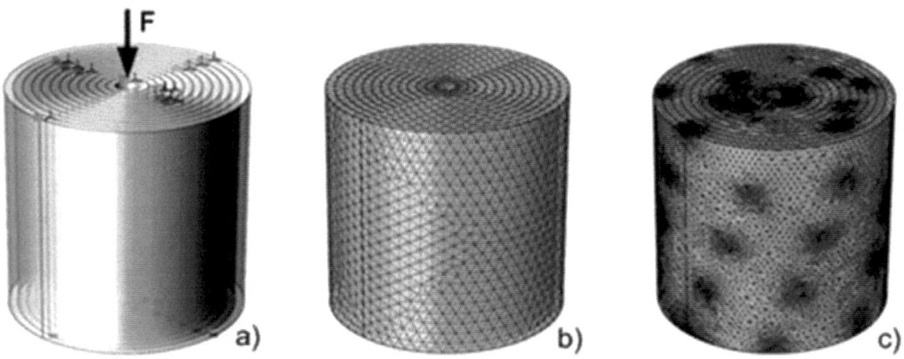

Figure. 3 Shell-based parametric osteon model (a – load scheme, b – simple shell model, c – shell model with 3d array of osteocytes).

The computational model provides for the creation of a mesh of finite elements, for which an automatic refinement of the size around the osteocytes is provided (refinement factor 2).

Let us set the following values of the initial parameters for carrying out model strength calculations: 1. Length of the model bone fragment (cylinder) $L = 0.4$ mm; 2. Fragment diameter $D = 0.4$ mm; 3. Axial load $F = 0.2$ H. Note that in both models of bone tissue, the deformation values are practically the same - 6.49 μm and 6.33 μm, and the equivalent stresses for the model with osteocytes are $4.2 / 2.8 = 1.33$ times higher, with the localization area stress is the zone of fixation of the base in the area of its intersection with the osteoblast cavity, as shown in figure 4.

IOP Conf. Series: Materials Science and Engineering 1117 (2021) 012006 doi:10.1088/1757-899X/1117/1/012006

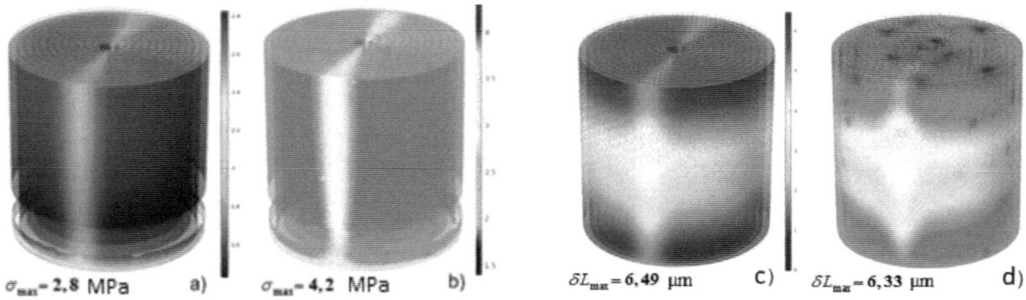

Figure. 4 Von Mises equivalent stresses (a, c) and deformation (b, d), where a and b is solid-shell model, c and d – with osteoblasts.

Thus, under the accepted assumptions, we can assume that the compact bone substance has the following characteristics (table 1):

Table 1. Mechanical parameters of compact bones tissue.

N	Parameter	Value
1.	Young's module	98 MPa
2.	Density	2100 kG/m^3.
3.	Poisson's module	0,33

3.2. Simulation of the behavior of healthy and tumor-affected bone under stress

The presented results of three-dimensional scanning of skeleton bones by computed tomography methods serve as a bone model. Note that the directly available scan results need serious revision by averaging coordinates, excluding small polygons and stitching cavities with patch polygons to achieve a closed bone geometry. Modern CAE analysis environments have automatic and semi-automatic tools for correcting raw geometry. Below in figure 5 the finite element mesh of the corrected healthy bone model (a) and visually simplified version with applied boundary conditions (b) is presented.

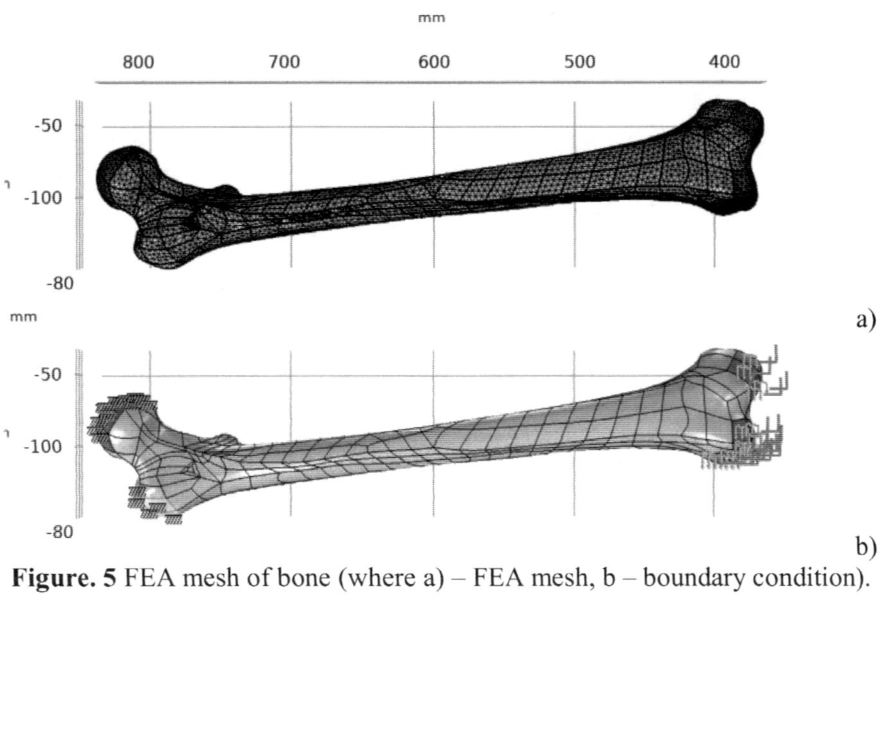

Figure. 5 FEA mesh of bone (where a) – FEA mesh, b – boundary condition).

As the maximum load, consider the force acting on the bone when the patient jumps on one leg. Let the patient's mass be $m = 120\,Kg$, the dynamic coefficient $kd = 1.7$ (taking into account the damping properties of tissues and cartilage of the skeleton), $g = 9.81\,m/s^2$, then the load can be estimated in $F = kd\,mg = 2000\,N$ and is directed along with the axis bones (horizontally in the picture above).

Neglecting the effect on the strength of the cancellous part of the bone, one can set the parameters of compact bone tissue discussed above (see Table 1).

We use a sequence of calculations typical for CAE systems: import of geometry and its correction, setting material parameters, creating a calculation scheme and determining boundary conditions, building a finite element mesh, numerically solving the resulting model and post-processing the results.

Figure 6 shows the calculation results: von Mises stress diagram, original mesh, deformed bone with a zone of increased stress.

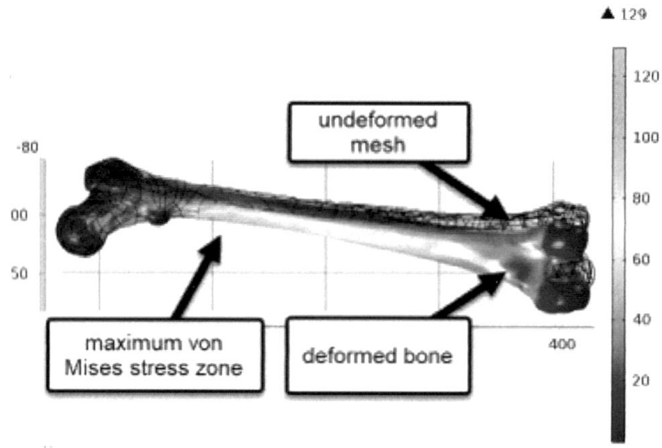

Figure. 6 Calculated von Mises stress in whole bone

With the previously adopted bone strength limit of 160 MPa, we obtain a safety factor of about 25%, which corresponds, for example, to the information given in [7].
Consider the influence of tumors on bone strength. Suppose the affected area is ball-shaped and completely removes bone. In this case, the position of the sphere and its size can be specified parametrically (see in figure 7a sample of tumor sphere and in figure 7 b sample of results).

Modelling was performed with bone strength three positions tumors (in the upper, middle and lower parts) while varying the radius of the sphere of the tumor, as shown below in table 2.

Table 2. von Mises equivalent stress (MPa) according by size and place of tumor zone.

Bone part	Radius of tumor sphere, mm		
	5	10	15
Near the knee	129	129	132
Middle part of the bone	355	267	400
Near the neck of the thigh	347	542	880

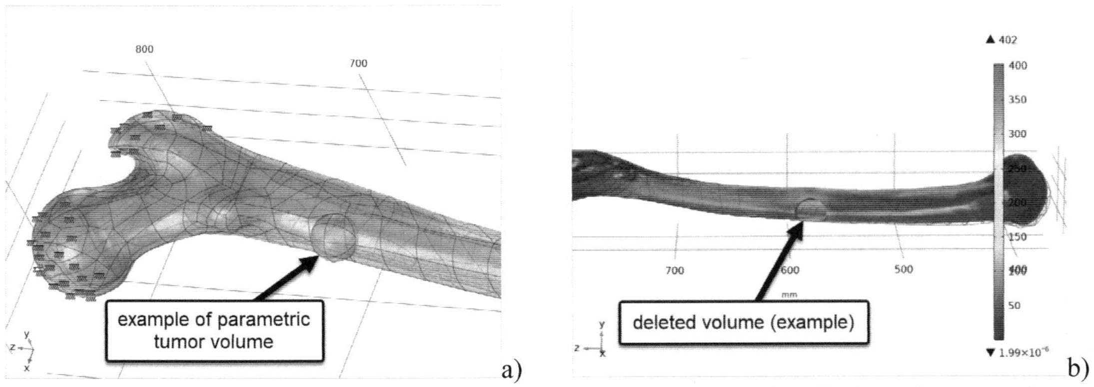

Figure. 7 Sample of modelling tumor zone as sphere (a) and results of calculation von Mises equivalent stress on bone with excluded volume

Calculations show that with damage to the bone tissue near the knee joint (at a distance of about 100 mm along the axis of the bone), the maximum stresses increase insignificantly up to 129 MPa, however, in the middle part of the bone and near the femoral neck, even an insignificant neoplasm causes terminal consequences for the bone. Let us pay attention to a decrease in stresses in the middle part of the bone with an increase in the radius of the sphere to 10 mm. This is due to the concentration of stresses on cavities with a small radius.

Consider a "sparing" implant design that replaces the removed bone tissue. In view of the limited volume of the article, we will limit ourselves here only to the localization of the implant in the middle part of the bone (see in figure 8).

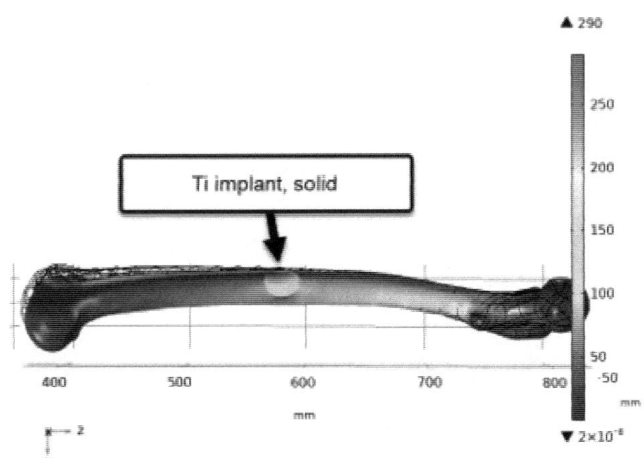

Figure. 8 Modelling of equivalent stress bone with implant

The maximum stresses of 290 MPa are observed in the contact zone of the plate with the bone along the loading axis and are of a local nature (already at a distance of 3-5 mm they do not exceed 75 MPa). However, this phenomenon can potentially lead to bone damage. These undesirable phenomena can be eliminated by increasing the compliance of the implant to values close to the bone material. In this work, an example of an implant is given, the increased stiffness of which is achieved due to the holes. Nevertheless, there is a more promising way - the manufacture of a lattice structure implant using the 3D printing method.

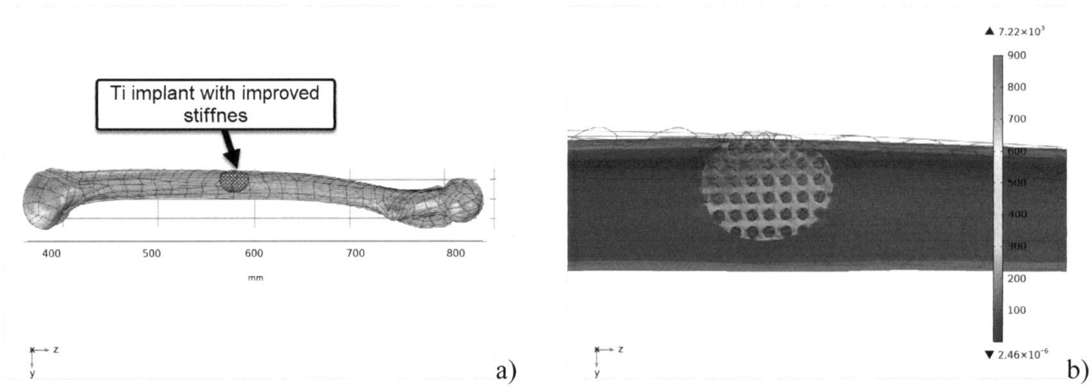

Figure. 9 Modelling of "sparing" implant (a – geometric model, b – equivalent stress)

When using an implant, the flexibility of the material of which is increased due to the presence of holes, it is possible to redistribute stresses evenly over the entire material of the implant plate. In this regard, the stresses in the implant increase, but the contact stresses on the transition surface from the implant to the bone are reduced to 64 MPa on average, and local stresses are reduced to 103 MPa.

4. Conclusions

Thus, the use of three-dimensional modeling technologies, individual manufacturing of prostheses of affected areas of bone tissue using three-dimensional printing methods reduce bone stress and can allow patients to lead a better lifestyle.

Measures to increase the compliance of the implant allow a more even distribution of stresses in the bone-implant pair, which reduces the risk of bone destruction in the area of contact with the prosthesis.

5. References

[1] Kononovich N A, Popkov A V, Popkov D A and Shastov A L 2018 *Scientific notes Kazan Bauman state academy of veterinary medicine* **236** 105–11

[2] Van der Graaf W T A *et al* 2012 *The Lancet* **379** 1879–86

[3] Ek E W, Rozen W M, Ek E T and Rudiger H A 2011 *Archives of orthopaedic and trauma surgery* **131** 487–95

[4] Bielack S *et al* 1999 *Klinische Padiatrie* **211** 260–70

[5] Wei F-Q, Sun W, Wong T-S, Gao W, Wen Y-H, Wei J-W, Wei Y and Wen W-P 2016 *Journal of experimental & clinical cancer research : CR* **35** 18

[6] Burt L A, Greene D A and Naughton G A 2017 *Pediatric exercise science* **29** 456–64

[7] Bilezikian J P, Martin T J, Clemens T L and Rosen C J (eds) 2020 *Principles of bone biology* (London, United Kingdom: Academic Press, an imprint of Elsevier)

[8] Netter F H 2014 *Atlas of Human Anatomy, Professional Edition: Including NetterReference.com Access with Full Downloadable Image Bank (Netter Basic Science)* 6th edn (Saint Louis: Elsevier Health Sciences)

[9] Drake R L, Vogl W, Mitchell A W M and Horn A 2020 *Gray's anatomy for students* (Philadelphia, PA: Elsevier)

[10] Samarkin A I, Samarkina E I, Mikushev V M and Plohov I V 2019 *AIP Conference Proceedings* **2188** 40004

[11] Narayan R (ed) 2020 *Rapid prototyping of biomaterials: Techniques in additive manufacturing (Woodhead Publishing series in biomaterials)* (Duxford: Woodhead Publishing, an imprint of Elsevier)

Biosynthesis of silver nanoparticles by cell-free extract from *Trichoderma reesei* - study on the influence of growth media

O Ts Gemishev[1,4], M I Panayotova[2,3], V T Panayotov[2]

[1]Faculty of Biology, Sofia University "St. Kliment Ohridski", Bulgaria
[2] University of Mining and Geology, Sofia, Bulgaria
[3]Author to whom any correspondence should be addressed.
[4]The National Science Fund of Bulgaria is appreciated for the financial support of the project DN-17/20

Abstract. *Trichoderma reesei* is a non-pathogenic microorganism able to reduce the toxic for the microorganism silver ions (Ag^+) to non-toxic silver nanoparticles (AgNPs). The effect of components of the nutrient medium on which the biomass from *T. reesei* was obtained on the biosynthesis of AgNPs from Ag^+ was monitored. Five media were studied. The main medium (1) contained glucose-2%, NH_4Cl-0.1%, $CO(NH_2)_2$-0.3%, $(NH_4)_2SO_4$-0.1%, KH_2PO_4-0.2%, $(NH_4)_2SO_4$-0.14%, $MgSO_4.7H_2O$-0.03%, $CaCl_2.2H_2O$-0.04%. The other were medium 1+0.1% yeast extract (2), medium 1+0.1% corn steep liquor (3), medium 1+0.1% peptone (4), medium 1+0.1% casamino acids (5). After culturing the strain, the resulting biomass was subjected to aqueous extraction, and the cell-free extract - CFE (containing biologically active metabolites) was used for extracellular biotransformation (performed with 10% extracted biomass and 10 mM $AgNO_3$) of Ag^+ to AgNPs. The AgNPs formation was monitored by UV-Vis. The isolated nanoparticles were characterized by TEM, ICP-AES and FTIR-ATR. The slowest reduction rate of Ag^+ was obtained with the CFE from biomass grown in medium 1 and the highest - with CFE from biomass grown in medium 3. Nearly mono disperse spherical AgNPs (2-6 nm) were synthesized with the aid of CFE of fungi cultivated in media 1 and 3. The AgNPs obtained with CFE from the other media were 4-11 nm.

1. Introduction

The use of silver nanoparticles (AgNPs) is considered a promising approach for solving different medical [1, 2] environmental [3] and agricultural [4, 5] problems.

AgNPs can be synthesized by chemical route, employing reagents to reduce the silver (Ag^+) ions and stabilize the nanoparticles formed. Production and use of these reagents can cause health and environmental risks. AgNPs can be produced by physical methods but usually this is linked with high energy consumption and related environmental problems. The mentioned issues have led to enhanced interest in biogenic synthesis (biosynthesis) methods. Biosynthesis of AgNPs can be carried out by using plants extracts, microorganisms (bacteria, fungi) or their metabolites, which operate as reducing agents and AgNPs stabilizing reagents. These methods are relatively simple and clean and can be sustainable and economical if optimal conditions are found for microorganisms' growth. Fungi, with their high tolerance to metals, good biomass production, easy handling and suitability for large-scale syntheses represent very attractive agents for biosynthesis of AgNPs [6]. Moreover, according to some studies, by optimizing conditions for fungi cultivation it is possible to influence the metabolism of fungi thus to vary the properties of the fungi themselves or of their metabolites used as reducing and capping reagents – with the final result - production of AgNPs with desired characteristics [7]. Biosynthesis of AgNPs by fungi can be realized by intracellular or extracellular mechanism. In the former the metal precursor is added to the mycelia culture and the nanoparticles are formed inside the fungal biomass. As a result, extraction of the nanoparticles is obligatory after the synthesis, applying centrifugation, filtration and chemical treatment to disrupt the biomass cells and release the nanoparticles. In the extracellular synthesis, the precursor Ag^+ ions are added to the aqueous cell-free extract (CFE) containing only the bio-molecules released from the carefully washed fungal biomass. This method is widely used. Some authors apply the term extracellular synthesis for nanoparticles production by using only filtered media where the fungi were cultivated [7, 8]. However, in this case strict difference has to be made between the action of the bio-molecules released by the fungal biomass and the action of the cultivation medium components.

In order to attain mono-disperse and stable AgNPs the parameters of the synthesis process have to be optimized. The following parameters have been widely studied in the literature: (i) during the fungi cultivation - fungus type, temperature, agitation, and time, and (ii) during the nanoparticles synthesis process - temperature, agitation, time, light availability, biomass to liquid mass ratio, concentration of the metal precursor and pH [2, 6, 9,10]. Influence of the cultivation media composition on the AgNPs production was less studied and mainly – in the case when in the extracellular synthesis the filtered media where the fungi were cultivated was used [7 - 10]. Such studies have been carried out for different microorganisms, such as *Fusarium oxysporum* [9], *Sclerotinia sclerotiorum* MTCC 8785 [2], *Klebsiella pneumoniae, E. coli* top 10 strain [8], *Escherichia coli and Pseudomonas jessinii* [7, 10]. The findings of different authors imply that the media for microorganisms' growth play an important role in the synthesis of metallic nanoparticles in terms of yield and nanoparticles size and shape. It is expected that the impact of microorganisms' cultivation medium will be more pronounced in the case of extracellular synthesis when the precursor Ag^+ ions are added to the aqueous CFE containing only the bio-molecules released by the carefully washed fungal biomass. It is also known that the biological synthesis of AgNPs is a reaction catalyzed by the biomolecules [9]. It is recognized that in different culture media compositions and under different circumstances microbial cells secrete diverse metabolites and specific kinds of proteins. That is why it can be anticipated that the chemical nature and concentration of those bio-molecules will depend on the cultivation medium composition.

Seventy percent of fungi which have been investigated for their use in AgNPs biosynthesis are pathogenic for plant, human or animals [11]. Where these fungi are used, measures and the corresponding costs are needed to deal with their eventual negative effects on the environment and humans. In order to produce nanoparticles on large scale, at preserving the environment, non-pathogenic fungi are needed. The fungus *Trichoderma reesei (T. reesei)* is a non-pathogenic, environmentally friendly microorganism, able to produce a variety of biologically active substances, including extracellular metabolites at high scale [12]. *T. reesei* for its own survival secrets extracellular reductase enzymes which are able to reduce toxic Ag^+ ions to nontoxic AgNPs [11]. The fungus is well-known for its high growth rate, easy management in large-scale production and inexpensive production procedures. Different authors used different media to cultivate *Trichoderma* species with the aim to use the bio-molecules released by the fungal biomass in water as reducing and capping reagents. Some examples are given in table 1. Studies on *Trichoderma*-based synthesis of AgNPs are still incomplete [13]. Especially this holds true for the studies on the impact of fungi growth medium on the biosynthesis of AgNPs. On the other hand, the feasibility of AgNPs biosynthesis depends on the successful production of the fungal metabolites in large scale [5].

As it can be seen in the table 1, glucose is most often used carbon source, yeast extract, peptone and corn steep liquor are often used as additives to the "classical" medium based on inorganic salts. In the pioneering work [17] on the application of *T. reesei* for synthesis of AgNPs by using the fungus mycelium the fungus was grown in medium composed of 0.5 % glucose and 0.4 % casein hydrolysate. However, no data are presented in that paper on the AgNPs synthesis by use of CFE.

In the study presented here, an extracellular biosynthesis of AgNPs from $AgNO_3$ solution by using the CFE of *T. reesei* biomass has been performed. The generalized effect of components of the nutrient medium on the *T. reesei* metabolite profile, that influences the biosynthesis of AgNPs from Ag^+, was monitored. According to our knowledge, until now such study is not available.

Table 1. Media used to cultivate *Trichoderma* species.

Fungus	Medium	Ref
T. inhamatum	Liquid broth containing 0.3% w/v malt extract, 1.0% w/v glucose, 0.3% w/v, yeast extract and 0.5% w/v peptone	[14]
T. harzianum	Glucose nutrient broth medium	[4]
T. viride	Liquid medium containing malt extract (0.3%), yeast extract (0.3%), glucose (1.5%) and peptone (0.5%).	[1]
T. viride	Liquid broth containing (g/L) KH_2PO_4 – 7; K_2HPO_4 – 2; $MgSO_4 \cdot 7H_2O$ – 0.1; $(NH_2)SO_4$ – 1; yeast extract – 0.6; glucose – 10.	[15]
T. atroviride	Potato dextrose broth (PDB)	[13]
T. harzianum	Liquid broth containing, (g/L) KH_2PO_4 (7), K_2HPO_4 (2), $MgSO_4 \cdot 7H_2O$ (0.1), $(NH_2)SO_4$ (1); yeast extract (0.6), glucose (10).	[16]
T. harzianum SYA.F4	Medium[a] containing, in g/L: 9.0 Oat, 5.0 yeast extract, 1.0 $NaNO_3$, 1.0 KH_2PO_4, 1.0 peptone, 0.3 $MgSO_4 \cdot 7H_2O$, pH 5.5, 4.2% (w/v) $(NH_2)SO_4$, 2% (w/v) glucose	[5]

[a]According to the authors this medium gave the best results (out of 6 studied media) both in terms of biomass growth and AgNPs yield

2. Materials and methods

The fungal strain *T. reesei* PF, used in the present study, was received from the collection of the Biotechnology Department, Faculty of Biology, Sofia University "St. Kliment Ohridski". Studies were carried out with seven days old culture of *T. reesei* cultivated on potato-dextrose agar at 29±1°C.

To study the influence of growth medium on the AgNPs synthesis, the fungal biomass was harvested after inoculating the strain in 100 mL of five different growth media. The main medium contained inorganic salts and glucose as carbon source. It is referred to as medium 1, with the following composition: glucose – 2 %, NH_4Cl 0.1 %, $CO(NH_2)_2$ - 0.3 %, $(NH_4)_2SO_4$ – 0.1 %, KH_2PO_4 - 0.2 %, $(NH_4)_2SO_4$ - 0.14 %, $MgSO_4 . 7H_2O$ - 0.03 %, $CaCl_2 . 2H_2O$ - 0.04 %. The other media were medium 1 to which different additives were added. Medium 2 was 1 + 0.1% yeast extract, medium 3 was medium 1 + 0.1% corn steep liquor, medium 4 was medium 1 + 0.1% peptone, and medium 5 was medium 1 + 0.1% casamino acids. All reagents used were p.a., purchased from Sigma Aldrigh. Sterile distilled water was used as diluent, extractant and for fungi washing. The fungal culture was incubated in 500 cm^3 flasks for 72 h at 29±1 °C and under continuous shaking at 220 rpm. Then the fungus biomass was separated from the culture broth by using sterile Whatman No. 1 filter paper and washed more than twenty times with distilled water. Further, 20 g of the wet biomass was suspended in 200 mL sterile distilled H_2O and it was extracted for 24 h, under continuous shaking at 150 rpm. Then the utilized fungus biomass was separated by filtration and the obtained CFE (containing a mixture of fungus metabolites) was used in the further experiments on extracellular biotransformation of Ag^+ to AgNPs. For this purpose $AgNO_3$ (p.a.) was dissolved in the produced different CFE to obtain a concentration of 10 mM $AgNO_3$. The extracellular fungi cell-free biosynthesis of AgNPs was carried out in 500 cm^3 flasks, at 29±1°C, under continuous shaking at 150 rpm, in dark condition.

The process of AgNPs formation was monitored from 0 h till 144 h by medium sampling for conducting UV-Vis absorption spectroscopy by using BOECO S-220 UV/VIS spectrophotometer at the wavelength from 200 to 600 nm. Samples of CFE, containing the produced AgNPs, were taken at every 24 hours and their absorbance was recorded. Further, the synthesized AgNPs were isolated (by centrifugation - 30 min, 13300 rpm), washed several times with distilled water, re-suspended and their spectra were also recorded. The amount of bio-transformed Ag^+ ions was measured by Inductively-Coupled Plasma-Atomic Emission Spectroscopy (ICP-AES) after dissolving the carefully washed AgNPs that were separated from known volume of CFE, containing the prepared AgNPs.

The synthesized and isolated nanoparticles were characterized by Transmission Electron Microscopy (TEM) and infrared spectroscopy (FTIR-ATR). For the TEM analysis the washed AgNPs were re-suspended in distilled water. Further a drop of this suspension was casted on carbon-coated copper grids and dried under ambient conditions prior the measurement. JEOL model JEM 2100, 200 kV analytical electron microscope was used. For the IR analysis, biosynthesized AgNPs were washed with distilled water several times. FTIR spectra were recorded on Bruker Tensor 27 spectrometer by direct deposition of the samples on a diamond attenuated total reflectance (ATR) accessory (INFRAMAT Equipment - DO1-284/17.12.2019). The spectra were collected in the frequency region 4000 – 600 cm^{-1} with 64 scanning at resolution of 2 cm^{-1}. Air spectrum was used as a background.

3. Results and discussion

Under the same cultivation conditions different media gave different yield of *T. reseei biomass*, that generally could be presented by the order: medium 2 ≅ medium 3 ≅ medium 4 > medium 1 >> medium 5. For the period of the Ag^+ ions reduction experiments the pale yellow color of the all CFEs changed to dark yellow and then through yellowish-brown to brown color. The color appearance is an obvious sign of the formation of silver nanoparticles in the reaction mixture. The excitation of surface plasmon vibrations in the AgNPs is the cause for the color change of the solution [9, 17]. The reaction was followed by recording the UV–Vis absorption spectra of the five studied media with biosynthesized AgNPs. Figure 1 depicts the recorded spectra.

The spectra of all the samples (figure 1a – 1e) exhibited a pronounced peak around 280 nm in the beginning of the experiments. This peak is assigned to the superposition of the absorption bands of proteins and $AgNO_3$. As it can be seen out of media containing different additives, the highest peak is observed for the CFE obtained from fungi grown on medium 3 (with addition of corn steep liquor). This could be related to the extraction and availability in the CFE of higher amount of proteins. Interestingly, this peak is the highest for the CFE from fungi cultivated on medium 1, however this has not resulted in higher production of AgNPs (see figure 1f). The intensity of this peak drops off with time implying conversion of the components of the reaction medium. This effect can be clearly seen at 24th h for CFE obtained from fungi grown on media 5, 4, 3 and 2 (the most pronounced for medium 5 and the least – for medium 2) and after 72 h – for the CFE obtained from fungi grown on medium 1. At the same time a broad peak appears, starting from around 350 nm. This indicates the presence of AgNPs. For the mixtures of the CFE and synthesized AgNPs the peak appearance was observed at 24th hour for CFE from fungi grown on media 4 and 5, at 48th h for CFE from fungi grown on media 3, at 72th h - for CFE from fungi grown on media 2, and at 96th h - for CFE from fungi grown on media 1. Since it is recognized that metal nanoparticles synthesis by fungi is a bio-reduction process on the basis of the above mentioned observations it could be assumed, that either the CFE from fungi grown on media 4 and 5 contain bio-active molecules in higher concentration or contain some molecules with more pronounced catalytic activity compared to the CFE from fungi grown on the other media. The intensity of the peak that is characteristic for the AgNPs increases with the time for all media which is indicative for the continued reduction of the Ag^+ ions and an increase in concentration of AgNPs. However, at 144th h the highest relative increase in the absorption was observed for the mixture of AgNPs and CFE from fungi grown on media 3, implying the higher total reduction ability of this CFE. The peak height of isolated AgNPs (figure 1f) hints to the same idea.

The peak indicative for the AgNPs is broad for the mixtures CFE + AgNPs, as well as for the isolated AgNPs from all media, except of medium 3, where it is relatively sharp – figure 1f. The latter finding is a sign for the formation of mono-disperse particles. The peaks recorded in the mixtures AgNPs + CFE from fungi grown on media 3, 4, and 5 are symmetric, which points at the stability of the synthesized nanoparticles [9]. The absorbance peak of AgNPs synthesized with CFE from fungi grown on media 4 and 5 are observed at a longer wavelength which shows the presence of larger nanoparticles [18], compared to those formed in CFE from fungi grown on media 3. The small blue and red shifts in the wavelength of the absorbance peak observed for AgNPs synthesized with CFE from fungi grown on media 4 and 5 (more pronounced for medium 5) could be related to obtaining AgNPs that are different in shape and size [9].

Based on the data presented in figure 1 it could be stated that the highest yield of AgNPs is achieved when the medium for *T. reesei* cultivation contained corn steep liquor (medium 3). This is confirmed by the data from ICP analysis – table 2. Our yield of AgNPs is in the range of the obtained by Ag^+ ions extracellular reduction by the metabolites of the *Fusarium oxysporum* [9].

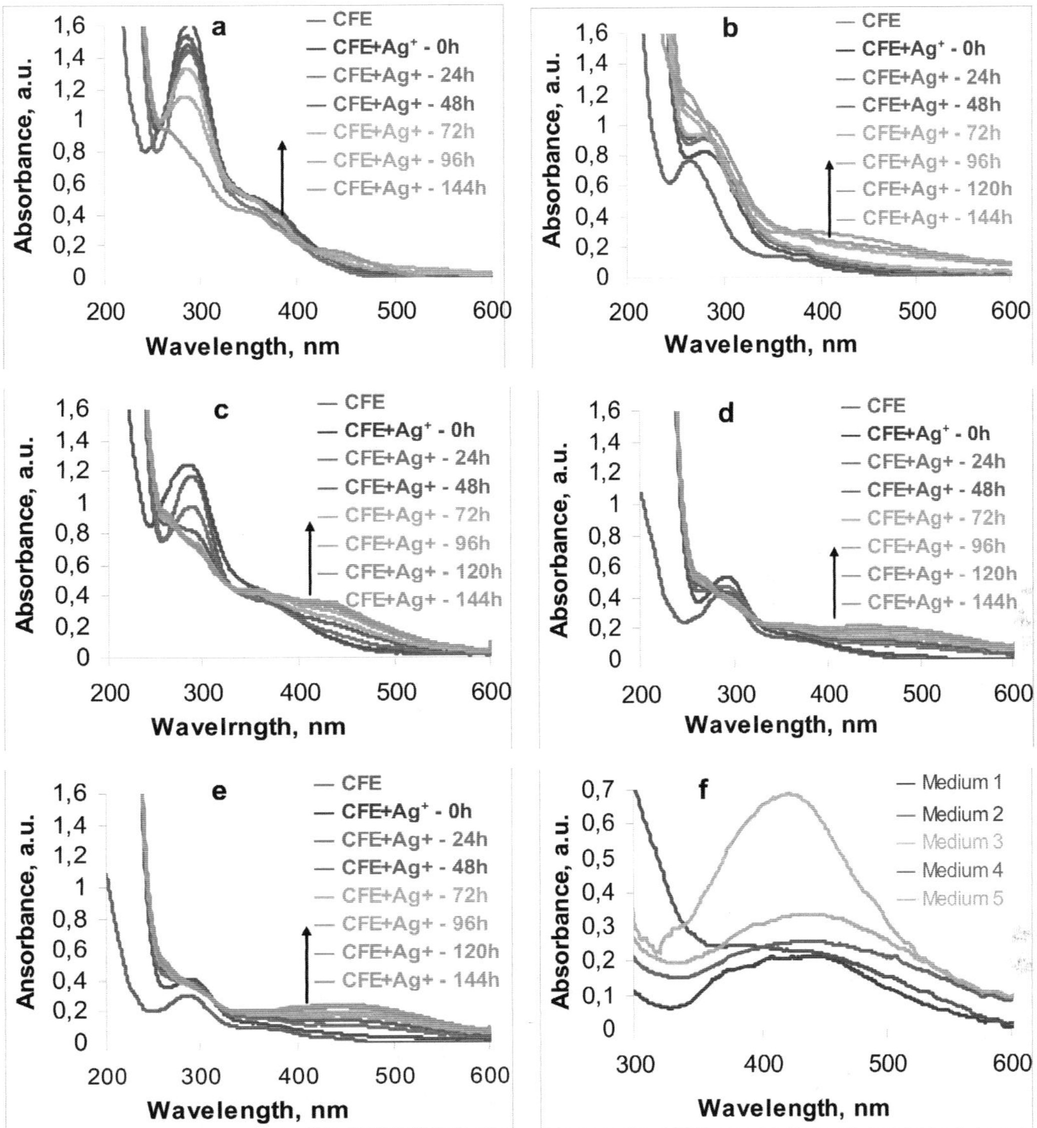

Figure. 1 UV–Vis absorption spectra of CFEs containing AgNPs synthesized by *T. reesei* metabolites at various time and initial concentration 10 mM $AgNO_3$: synthesized AgNPs + CFE of fungi cultivated in: medium 1 - a), medium 2 - b) medium 3 - c) medium 4 - d), medium 5- e), Spectra of the isolated, washed and re-suspended AgNPs at 144 h - f). Arrows point the peak development in time.

Table 2. Degree of biotransformation of Ag^+ to AgNPs by CFE from *T. reesei.*

Medium for *T. reseei* cultivation	No 1	No 2	No 3	No 4	No 5
Biotransformation degree, %	2.4	4.4	6.1	5.2	5.4

Figure 2 depicts TEM micrographs of biosynthesized AgNPs obtained with CFE from fungi grown on media giving the higher degree of biotransformation, namely media 3, 4 and 5.

As it can be seen in the TEM micrographs the AgNPs formed were mostly spherical (till spheroid shape), with diameters in the range 2 - 6 nm – for AgNPs produced from CFE from fungi grown on media 3; and 4 - 11 nm – for AgNPs produced from CFE from fungi grown on the other media.

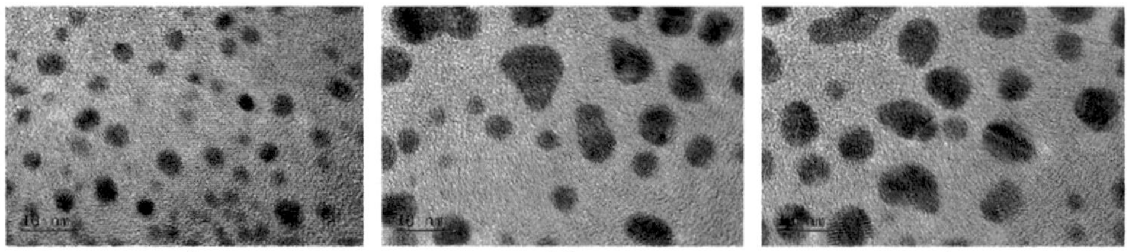

a) b) c)

Figure. 2 TEM micrograph of biosynthesized AgNPs by reduction with a cell-free extract of *T. reesei* biomass for 144 hours: a) medium 3, b) medium 4, c) medium 5.

Besides that are smaller compared to the AgNPs formed in CFE from fungi grown on the other media with comparable yield, i.e. media 4 and 5, the AgNPs obtained in CFE from fungi grown on media 3 are nearly mono-disperse - figure 2. This suggests that the composition of culture media is important both for the yield of AgNPs synthesis and for the particles size. However, even bigger synthesized nanoparticles are with relatively low-size. This finding implies that the enzymes extracted from the biomass act not only as reducing but also as capping reagents. As it can be seen in figure 2, the synthesized particles were not in direct contact, indicating good stability. The latter may be assigned to the functional groups of the protein molecules wrapping AgNPs and acting as capping reagents. They were identified using FTIR-ATR analysis - figure 3.

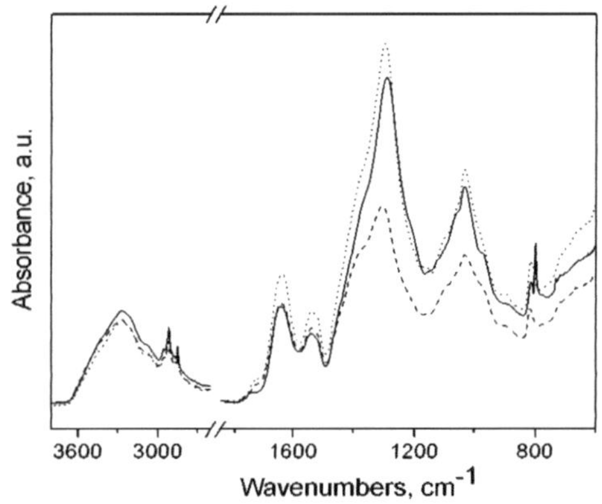

Figure. 3 FTIR-ATR spectra of AgNPs synthesized by using CFE obtained from *T. reesei* cultivated on medium 3 (solid line), medium 4 (dotted line) and medium 5 (dots).

The IR bands of isolated and carefully washed biocoated AgNPs practically appeared at the same cm^{-1} for the three studied media. The ATR-FTIR analysis showed the presence of IR bands at the typical positions for protein vibrations such as N-H stretching νN-H at app. 3270 cm^{-1}, amide I bands (carbonyl stretching) at app. 1638 cm^{-1} and N-H deformation vibration at app. 1535 cm^{-1}. Besides, weaker bands appeared at app. 1730 cm^{-1} for medium 5 which correspond to the carbonyl stretching vibration of ester groups. There is an evidence that some lipids were also deposited on the AgNPs along with the protein molecules. The C-H stretching vibrations of both protein and lipid molecules were found within the IR interval 3000-2800 cm^{-1}. The IR spectra of the three samples showed very strong absorptions at app. 1300 cm^{-1} for P-O vibrations and 1030 cm^{-1} for C-O vibrations, pointing out to the presence of phospholipids. The IR spectra of the three samples have minor differences – mainly in the low-frequency region below 900 cm^{-1}, where medium 3, shows an additional band at 801-809 cm^{-1}. The characteristic infrared bands for biocoated AgNPs in the three samples and the corresponding vibrational mode assignments are shown in table 3.

The CFE is the only source of proteins and phospholipids available on the AgNPs. The biomolecules and/or their parts that present in filtrate are responsible for synthesis and stabilization of AgNPs by their capping. In addition to ensuring stability of the produced AgNPs the proteins can act in the anchoring of drugs to transport them into cells [6].

Table 3. Characteristic infrared bands and vibrational mode assignments for bio-coated AgNPs.

Characteristic IR band positions			Assignments
Medium 3	Medium 4	Medium 5	
3267	3277	3272	N-H stretching
2956		2955	assym. C-H stretching of $-CH_3$ groups
2918	2918	2917	assym. C-H stretching of $-CH_2-$ groups
2850	2850	2850	sym. C-H stretching of $-CH_2-$ groups
		1730	C=O stretching of ester groups
1638	1639	1638	C=O stretching of amide groups
1541	1536	1536	N-H deformation of amide groups
1291	1307	1297	P=O asymmetric stretching vibration
1032	1032	1032	C-O stretching vibrations

4. Conclusions

As a result of the studies carried out the following conclusions can be drawn:

AgNPs can be synthesized by using CFE from the non-toxic and easy to cultivate in high amounts fungus *Trichoderma reesei,* grown on medium containing inorganic salts and glucose as carbon source, in presence of different nutrient additives (yeast extract, corn steep liquor, peptone or casamino acids).

Having in mind (i) the yield of *T. reesei* cultivated on the different media, (ii) the biotransformation degree of Ag^+ ion by the CFE from *T. reesei* cultivated on the different media, and (iii) the size and size dispersion of the synthesized AgNPs, it can be concluded that the media containing as additive corn steep liquor is the most suitable (among the studied media) for the cultivation of *T. reesei* whose aqueous CFE can be used for the biosynthesis of AgNPs.

The metabolites released in sterile distilled water by the cultivated *T. reesei* act both as reducing and capping reagents for obtaining nearly monodispersed AgNPs with size in the range of 2-11 nm.

5. References

[1] Elgorban A, Al-Rahmah A, Sayed S, Hirad A, Mostafa A and Bahkali A 2016 *Biotechnology & Biotechnological Equipment* **30** 299-304
[2] Saxena J, Sharma P, Sharma M and Singh A 2016 *SpringerPlus* **5** 861
[3] Moustafa M 2017 *Water Science* **31** 164–76
[4] Shelar G and Chavan A 2015 *Biolife* **3** 109-13
[5] EL-Moslamy S, Elkady M, Rezk A and Abdel-Fattah Y 2017 *Scientific Reports* **7** 45297
[6] Guilger-Casagrande M and Lima R 2019 *Frontiers in Bioengineering and Biotechnology* **7** 287
[7] Müller A, Behsnilian D, Walz E, Gräf V, Hogekamp L and Greiner R 2016 *Biocatalysis and Agricultural Biotechnology* **6** 107–15
[8] Baltazar-Encarnación E, Escárcega-González C, Vasto-Anzaldo X, Cantú-Cárdenas M and Morones-Ramírez J 2019 *Hindawi Journal of Nanomaterials* ID 4637325
[9] Birla S, Gaikwad S, Gade A and Rai M 2013 *The Scientific World Journal* ID 796018.
[10] Luo K, Jung S, Park K-H and Kim Y-R 2018 *Journal of Agricultural* and *Food Chemistry* **66** 957-62
[11] Vahabi K and Dorcheh S 2014 *Biotechnology and Biology of Trichoderma,* ed. V Gupta, M. Schmoll, A Herrera-Estrella, R Upadhyay, I Druzhinina and M Tuohy (Poland: Elsevier B.V.)
[12] Oksanen T, Pere J, Paavilainen L, Buchert J and Viikari L 2000 *Journal of Biotechnology* **78** 39–48
[13] Saravanakumar K and Wang M-H 2018 *Microbial Pathogenesis* **114** 269–273
[14] Hussein M 2016 *European Journal* of *Biological Research* **6** 28-35
[15] Fayaz M, Tiwary C, Kalaichelvan P and Venkatesan R 2010 *Colloids* and *Surfaces B* **75** 175–8
[16] Ahluwalia V, Kumar J, Sisodia R, Shakil N and Walia S 2014 *Industrial Crops and Products* **55** 202–6
[17] Vahabi K, Mansoori G and Karimi S 2011 *Insciences Journal* **1**, 65-79

IOP Conf. Series: Materials Science and Engineering 1117 (2021) 012007 doi:10.1088/1757-899X/1117/1/012007

[18] Elamawi R, Al-Harbi R and Hendi A 2018 *Egyptian Journal of Biological Pest Control* **28** 28

Effects of hybrid flame-retardant fillers on fire-resistive and mechanical properties of solvent-borne intumescent coatings

*,1Ming Chian Yew, [1]Guo Jie Ooi, [1]Ming Kun Yew, [1]Lip Huat Saw, [1]Jing Han Beh, [1]Tan Ching Ng, [1]Wei Hong Yeo

*,1Lee Kong Chian Faculty of Engineering and Science, University of Tunku Abdul Rahman, Jalan Sungai Long, Bandar Sungai Long, Cheras, 43000 Kajang, Selangor, Malaysia.

*Corresponding author, E-mail address: yewmc@utar.edu.my; yewmingchian@gmail.com

Abstract. The aim of this research project was to investigate the effects of hybrid flame-retardant fillers on the fire protection and mechanical properties of solvent-borne intumescent coatings. Formulations of intumescent coatings with different combinations of flame-retardant fillers were developed and investigated through Bunsen burner test, adhesion strength test and Scanning Electron Microscopy (SEM). It was found that the formulation of intumescent coating was optimized with four combinations of flame-retardant fillers in proper compositions and had improved the fire protection performance and mechanical properties of the coating. In overall, filler compositions of coating sample D (1 wt.% nano CES/ 2 wt.% expandable graphite/ 1 wt.% zinc borate/ 4 wt.% calcium silicate) had contributed to positive fire protection performance (equilibrium temperature of 190 °C) and adhesion strength (2.62 MPa). Significantly, the incorporation of an appropriate combination of hybrid nano CES bio-filler, expandable graphite, zinc borate and calcium silicate had led to a better fire protection performance and mechanical properties of the solvent-borne intumescent coating.

1. Introduction

Fire protection has become one of the most important key considerations especially in building designs around the world due to the risks of fire. Fire can bring the risks of taking human's life and loss of property, at the same time will cause the environmental effects, which are the air pollution problems resulting from the emission of toxic gases and smokes during the combustion process. Generally, the fire protection system is divided into two main types, which are active and passive systems. The examples of passive systems mainly consist of fire doors and fire protective coatings [1].

Intumescent coatings which act as passive fire protection (PFP) have been widely used in steel structures, offshore oil rigs, air crafts and building components to slow down the spread of fire to ensure the fire safety by following the building regulations practice in many countries [2]. It is considered as one of the most economical, easiest, and effective methods to prevent the rapid spread of flame during the event of fire to protect the substrates and building structures [3]. In addition, intumescent fire protective coatings consist of the flame-retardant materials which have obtained broad acceptance for fire protection application ~~in~~ worldwide [4]. Application of intumescent fire protective coatings have been highly recommended as passive fire protection (PFP) in buildings or infrastructure projects

IOP Conf. Series: Materials Science and Engineering 1117 (2021) 012008 doi:10.1088/1757-899X/1117/1/012008

especially commercial buildings as they can reduce the destructing cost of fire in terms of both human life risks and damage of property.

The advantages of intumescent coatings are light in weight, easily to be coated, have a good decorative effect and do not change the intrinsic properties of steel [5]. Normally, intumescent coatings are composed of three flame-retardant additives, which are an acid source namely ammonium polyphosphate phase II (APP), a carbon source namely pentaerythritol (PER) and melamine (MEL) which acts as blowing agent combined with various flame-retardant fillers and a polymer binder. The coating will expand once exposed to adequately high temperature by releasing non-combustible gases to form a protection barrier against oxygen, thereby protecting the substrate from heat or flame. Moreover, incorporation of any flame-retardant filler has a significant effect on the end use properties of intumescent coatings, especially the fire-retardant performance [6]. This research project highlights a beneficial bio-filler extracted from the chicken eggshells (CES) bio-waste and its potential role in the fire protection coatings industry. Recently, CES has accumulated attention due to its reclamation potential. The chemical composition and availability of CES has made it to become a potential source of compatible bio-filler to improve the thermal and mechanical properties to the intumescent coatings [7,8].

2. Experimental program

2.1. Materials and sample preparation

In this research project, different formulations of intumescent fire protective coatings were prepared with various combinations of flame-retardant fillers with fixed ingredient of flame-retardant additives and polymer binder. Table 1 shows the compositions of solvent-based intumescent coatings.

Table 1. Compositions of Intumescent Coatings

Materials	*Coatings (grams)*			
	A	B	C	D
Flame-retardant additives				
APP	20	20	20	20
MEL	10	10	10	10
PER	10	10	10	10
Binder				
Acrylic Resin	50	50	50	50
Pigment				
TiO_2	2	2	2	2
Flame-retardant fillers				
Nano CES	-	-	-	1
Expandable graphite	-	4	2	2
Zinc borate	8	4	2	1
Calcium silicate	-	-	4	4

Bunsen burner test

The purpose of Bunsen burner test is to investigate the performance in fire protection of intumescent coating during fire and to determine the thickness of the protective char layer formed after exposed to fire. Each coating sample prepared had applied by painting method onto a steel plate, each with a dimension of 100 mm × 100 mm × 3 mm. This coating process was repeated for several times until the coating thickness of 2 mm ± 0.2 mm was achieved. In this research project, the critical temperature of 400 °C was chosen and set for the steel plate to ensure a higher level of safety. Basically, the critical temperature of 400 °C was chosen due to the reason that steel will lose its properties at around 550 °C. During the burning process (about 1000 °C), the temperature profile of the coated steel plate was mounted vertically and recorded using a digital handheld thermometer, which connected to a thermocouple plate located behind the coated steel plate for 60 minutes as shown in Figure 1. The gas consumption of the Bunsen burner was set at about 160 g/h. The distance between the coated steel plate and fire source was fixed at about 8 cm. The response of intumescent coating during the burning process was observed. The temperature changes of the steel plate were recorded for the first 6 minutes (at 1-minute time interval) and every increment of 3 minutes after the first 6 minutes onwards. The thickness of the char layer formed by each sample after the test was measured and recorded using the steel ruler.

Figure. 1 Bunsen burner test setup: (1) digital thermometer; (2) retort stands; (3) Bunsen burner; (4) coated sample; (5) Galvanized sheet.

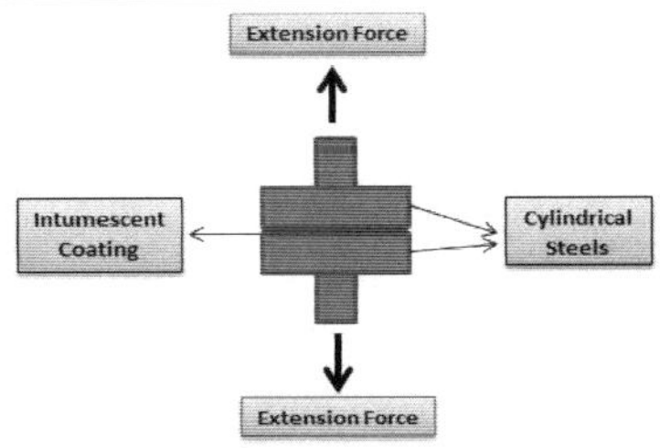

Figure. 2 Method of adhesion strength test

2.2. Adhesion strength test

The purpose of adhesion strength test is to determine the bonding strength of the intumescent coatings prepared. The adhesion strength of each coating sample was investigated with the aid of the Instron Micro Tester. The two attached cylindrical steels coated with sample coating and epoxy resin were fixed on the Instron Micro Tester machine and then drawn apart vertically and continuously in tensile mode at constant rate of 1 mm per minute, as shown in Figure 2, until the intumescent coating on the cylindrical steel cracks. Meanwhile, the adhesion strength, f_b (N/m^2) for each coating sample was calculated using the equation below:

$$f_b = F/A \quad \text{(equation 1)}$$

where

f_b = the adhesion strength (N/m^2);
F = the crack charge (N);
A = sticking area of the intumescent coating on the cylindrical steel surface (m^2).

2.3. Scanning Electron Microscopy (SEM)

Scanning electron microscopy, which is commonly known as SEM, was carried out to investigate the surface morphology of the char layers formed by intumescent coating. The cracking parts and porosity of the char layers formed by different intumescent coating samples can be observed and analysed by using SEM. In this research project, SEM was operated at low beam energy of 1 kV for observation, in order to minimize the thermal damage's possibility to the char layer. In addition, SEM was operated under two levels of magnification, which are 1000 and 8000 magnification.

3. Results and discussion

3.1. Bunsen burner test

There were four kinds of coating samples (marked A, B, C and D) that were conducted through Bunsen burner test. The evolution of temperatures at the back of each coated steel plate and the thickness of char layer formed were recorded and compared. The temperature profiles for coating samples were shown in Figure 3, while the thickness of char layer formed after the 60 minutes burning process was presented in Figure 4.

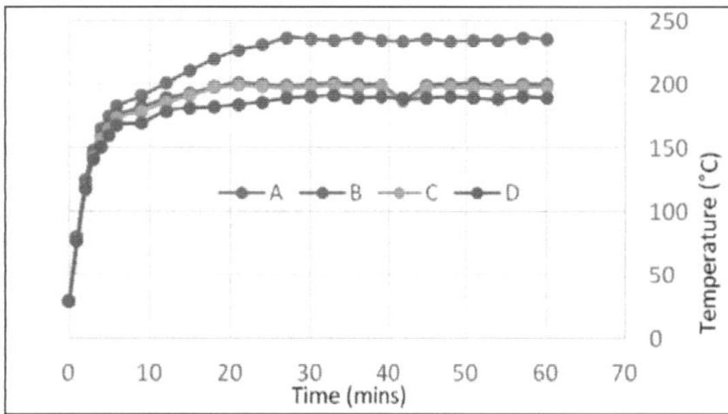

Figure. 3 Temperature profiles of coating samples

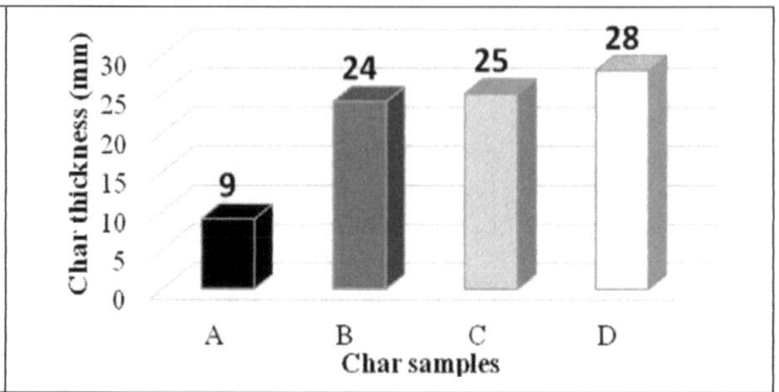

Figure. 4 Thickness of char layer formed by coating samples

The fire protection results showed that the equilibrium temperature and thickness of char layer formed after 60 minutes burning process of coatings A, B, C and D were about 237 °C (9 mm), 201 °C (24 mm), 198 °C (25 mm) and 190 °C (28 mm), respectively. All coatings showed a similar rise in temperature at the first five minutes resulting from thermal degradation of the coatings and the formation of the char layer due to the physical and chemical reactions from the coating ingredients [9]. Coating A showed the worst fire protection performance with the highest equilibrium temperature of about 237 °C due to the char layer hard to expand with the addition of 8 wt.% zinc borate. A high dosage of zinc borate promotes the formation of a strong char layer, but it did not show a strong synergic effect with flame-retardant additives and polymer binder on the formation of multicellular char layer in protecting the steel. Coating D showed the best fire protection performance with a lower equilibrium temperature of about 190 °C, while coatings B and C had a slightly higher equilibrium temperature of about 200 °C. The difference in equilibrium temperatures of these two coatings was about 10 °C. Both coating formulations were added with same flame-retardant fillers (expandable graphite and zinc borate), but only differ in the amount of calcium silicate. The results showed that coating D which incorporated an ~~with~~ additional 1 wt.% of nano has promoted a better fire protection performance by having a lower equilibrium temperature. Since both coatings B and C had the almost same thickness of multi-cellular char layer formed (about 25 mm). The reason is that denser, uniform and less porous multi-cellular char layers promote a better efficiency in inhibiting the heat transfer to the underlying steel plate, thus resulting in lower equilibrium temperature. As a conclusion, the coating D (thickest char layer of 28 mm) shows a positive effect on fire protection performance to the underlying steel plate, with optimized formulation by adding proper combinations of flame-retardant fillers (nano CES, expandable graphite, zinc borate and calcium silicate). This indicates that there is a significant improvement in fire protection performance by incorporating hybrid flame-retardant fillers into the coating formulation.

3.2. Surface morphologies of char layers

Surface morphologies of the char structure formed by intumescent fire protective coating can be observed under high magnification surface micrographs through SEM. The char layers formed by the coatings A-D were conducted through SEM to observe the surface morphologies of their char structures. Figure 5 shows the surface morphologies of coatings A-D under magnification of 8000.

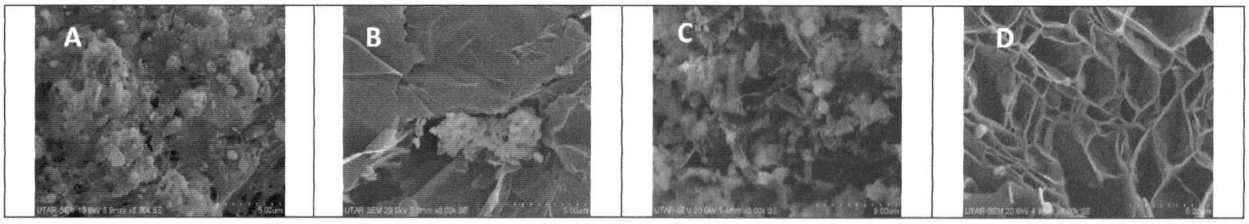

Figure. 5 Surface morphologies of char layers A, B, C and D

From the observation, it was found that the foam structure of coating A had a porous and non-uniform which was unable the char layer to insulate the heat effectively from reaching the underlying steel plate. In addition, there are some holes in its foam structure which led to the poor performance in fire protection due to the heat might transfer through the holes easily. The foam structure of char layers B and C was significantly improved by incorporating expandable graphite (which produced a denser and more uniform foam structure) compared to coating A which incorporated with only zinc borate flame-retardant filler (which produced porous, non-uniform and tiny foam structure). One contributing factor is that char layers formed by coatings B and C had a uniform and dense foam structure, which was able to isolate the underlying steel plate from heat or fire by offering a better fire protection performance. From the observation, it was found that the foam structure of coating D was significantly improved with an additional of 1 wt.% nano CES bio-filler content (which produced a denser and more uniform tiny foam structure) compared to other coatings. One contributing factor is that the foam structure of coating D demonstrated more even expansion cells burst which led to heat reduction by lessening heat transfusion rate and thus enhancing its fire protection performance.

3.3. Adhesion strength

The values of adhesion strength of coating samples A, B, C and D were 1249.89, 1620.31, 1979.86 and 2066.50, respectively. Coating D had the highest adhesion strength (about 2.62 MPa), while coating A had the lowest adhesion strength (about 1.58 MPa). One contributing factor is that coating D formulation has extra 1 wt.% of nano CES as compared to coatings A, B and C. The results indicated that appropriate combination of amount of nano CES, zinc borate, expandable graphite and calcium silicate would enhance the adhesion strength of the coating due to metal–polymer composites adhesive interfaces. The difference between the adhesion strength of both coatings (A and D) were about 1.34 MPa (about 39.69 % difference). Figure 6 shows the appearance of the bonding surface area for the coatings A (lowest), B, C and D (highest) adhesion strength after undergoing the pull off test.

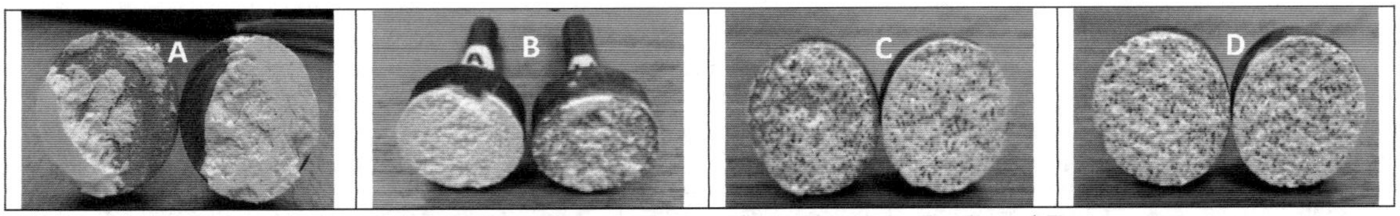

Figure. 6 Bonding surfaces of coatings A, B, C and D

As comparing the adhesion strength test results obtained, the bonding strength of the coatings was improved with proper combination of flame-retardant fillers in the coating formulation. The improvement in adhesion strength was about 40 % increment, by comparing the coating samples with the lowest adhesion strength. This indicated that the coating formulation to achieve the best adhesion strength for the coating has been optimized, whereby an appropriate amount of the combination of nano CES, expandable graphite, zinc borate and calcium silicate flame-retardant fillers with polymer matrix has contributed to the best adhesion strength.

4. Conclusions

Influence of flame-retardant filler on the fire protection and mechanical properties of solvent-borne intumescent fire protective coatings were analysed and investigated. Based on the overall results obtained, Coating D showed the best formulation in terms of fire protection performance and mechanical properties. The fire protection performance of the Coating D formulation (1 wt.% nano CES/ 2 wt.% expandable graphite/ 1 wt.% zinc borate/ 4 wt.% calcium silicate) had significantly improved with an appropriate combination of nano CES, expandable graphite, zinc borate and calcium silicate fillers, which contributed to the lowest equilibrium temperature performance of 190 °C and the highest adhesion strength of 2.62 MPa. Another contribution is that the Coating D has promoted a good expansion of char layer by forming the greatest thickness of char layer, which is able to insulate the heat effectively. Furthermore, the incorporation of nano CES bio-filler had reduced the ingredient cost of the coating, at the same time preserve the environment. Hence, the finding from this research project reveals that an appropriate combination of hybrid flame-retardant fillers is important to optimize the formulation in terms of fire protection and mechanical properties of solvent-borne intumescent coatings.

Acknowledgments
Authors gratefully acknowledge the financial support from Lee Kong Chian, Faculty of Engineering and Science, Department of Mechanical and Material Engineering, University Tunku Abdul Rahman (UTAR).

References
[1] Yew, M.C., Ramli Sulong, N.H., Yew, M.K., Amalina, M.A. and Johan, M.R., 2014. Investigation on solvent-borne intumescent flame-retardant coatings for steel. *Materials Research Innovations,* 18(6), pp. 1-2.
[2] Liu, Z., Dai, M., Zhang, Y., Gao, X. and Zhang, Q., 2016. Preparation and performances of novel waterborne intumescent fire retardant coatings. *Progress in Organic Coatings,* 95 (100-106), pp. 1 - 2.
[3] Yew, M.C., Yew, M.K., Saw, L.H, Ng, T.C, Rajkumar, D. Beh, J.H., 2018. Influences of nano bio-filler on the fire-resistive and mechanical properties of water-based intumescent coatings. *Progress in Organic Coatings,* 124, pp. 33-40.
[4] Horrocks, A.R. and Price, D., 2001. *Fire Retardant Materials.* Cambridge, England: Woodhead Publishing Ltd.
[5] Yew, M.C., Ramli Sulong, N.H., Yew, M.K., Amalina, M.A. and Johan, M.R., 2015. Eggshells: A novel bio-filler for intumescent flame-retardant coatings. *Progress in Organic Coatings,* 2015 (116-124), pp. 1-2.
[6] Faiz, A, Sami, U, Nurul H.H.M, Eugenio, O, Abdullah, G. A, Guan H.Y, 2019, An investigation on thermal performance of wollastonite and bentonite reinforced intumescent fire-retardant coating for steel structures, *Construction and Building Materials,* 228, pp.116734.
[7] Yew, M.C., Ramli Sulong, N.H., Yew, M.K., Amalina, M.A. and Johan, M.R, 2014. Fire propagation performance of intumescent fire protective coatings using eggshells as a novel bio-filler, *Scientific World Journal,* 2014, pp. 9.
[8] Beh, J.H.; Yew, M.C.; Yew, M.K.; Saw, L.H., 2019. Fire protection performance and thermal behavior of thin film intumescent coatings. *Coatings,* 9, pp. 483.
[9] Aziz, H. Ahmad, F, 2016. Effects from nano-titanium oxide on the thermal resistance of an intumescent fire retardant coating for structural applications. *Progress in Organic Coatings,* 101, pp. 431 - 439.

Influences of macro polypropylene fibre-reinforced lightweight concrete incorporating recycled crushed LECA aggregate

Ming Kun Yew*, [a], Ming Chian Yew [a], Jing Han Beh [a], Lip Huat Saw [a], Siong Kang Lim [a], Foo Wei Lee[a], Yee Ling Lee[a]

[a]Lee Kong Chian Faculty of Engineering & Science, UTAR, Cheras 43000 Kajang, Malaysia. *Corresponding author: Ming Kun, Yew

E-mail address: yewmk@utar.edu.my (M.K. Yew)

Abstract. The huge amount disposal of light expanded clay aggregate (LECA) had become a serious environmental issue across the globe which represents an immediate threat to the environment. One of the potential solutions is to incorporate these wastes into construction materials, mainly concrete. These waste materials are abundant and easily available. The addition of recycled crushed LECA creates void space in the concrete which helps in decreasing the density of concrete, and reducing the dead loads needed to be sustained for buildings. In this study, different volume fraction (0, 0.15, 0.3 and 0.45%) of macro polypropylene (MPP) fibre have been used to enhance the strength properties of lightweight concrete (LWC). The combination of lightweight LECA and MPP fibre have been used to achieve compressive strength between 25 to 40 MPa with a density ranged between 1800 to 2100 kg/m^3. It is found that the addition of MPP fibres at an optimum volume fraction concrete enhances the compressive strength. Hence, the findings of this research revealed that it can be used as an alternative solution for concrete industry.

1. Introduction

Concrete is the most used building material since the 20th century. With the rapid growth of urbanisation, the amount of concrete used also increasing. Concrete had been the 2nd most consumed material in the world, and its production consumed an enormous amount of natural resources such as aggregate, cement and water. With the current trends of concrete consumption, it has negative impacts toward the concrete industry. The natural resources, aggregate, are extracts from the seabed, riverbeds and other water sources. These activities cause environment problem after prolonged extractions and it is slowly depleting. Thus, a minor amount of recycled materials comes in to replace natural aggregate will result in saving natural resources, and more sustainable. This pursues researchers to use diverse recycled material in concrete in the early 2000s [1].

Among all types of lightweight aggregate concretes (LWAs), it can be divided into three (3) types, which are industrial waste lightweight aggregate, natural aggregate, and artificial lightweight aggregate. One type of industrial waste lightweight aggregate is known as recycled lightweight expanded clay aggregate. LWA are great in lightweight construction applications such as lightweight concrete wall, building facades, and architectural units. Lightweight construction means constructing the structure with minimal weight possible yet keeps the structure safe. As civil construction consumes 50~60% of resources worldwide, it is vital to study its behavior and apply lightweight construction in building structure [2]. In this research, the LWA in the form of crushed LECA will be used to replace portions of coarse aggregate to investigate its effect on the engineering properties of the LWAC.

Structural lightweight concrete (LWC) plays a crucial role in the lightweight construction. One of the current method to achieve this structural LWC is using the lightweight aggregate (LWA). It has been reported by Aslam *et al.* (2017) [3] showed that the used of blended oil palm boiler clinker (OPBC) to substitute both coarse and fine aggregate in concrete able to create a LWC with a 28 days compressive strength range from 17 MPa to 47 MPa with density in range of 1440 kg/m^3 to 1850 kg/m^3. The application of this blended LWA produced a high strength LWC which is suitable for structure design.

Ironically, the increase the concrete strength results in an increment of its brittleness [4]. These brittle properties cause concrete more susceptible to cracking. Initiation and propagation of cracks cause detriment effect on structural health and its lifetime, which will eventually affect the serviceability of the concrete structure. Instead of prevent cracking, the development of fibre-reinforced concrete (FRC) controls the cracking by restraining the cracking openings [5]. In current research, non-experimental researches have been performed by incorporating macro polypropylene fibre with the combination of crushed lightweight expanded clay aggregate (LECA) method to investigate its influences on the strength properties of the lightweight aggregate concrete (LWAC).

2. Materials and Methods

2.1. Materials

2.1.1 Ordinary Portland Cement (OPC)

The cement used in the mixing was Ordinary Portland Cement (OPC) Type 1 which meets the Malaysian Standard MS 522: Part 1: 2003 and certified by MS ISO 14001. It is a product from one of the largest manufactures in Malaysia, Tasik Cement Sdn. Bhd. The density and fineness of Tasik OPC are 3150 kg/m^3 and 3170 cm^2/g respectively.

2.1.2 Water and superplasticizer (SP)

Potable water with pH value of 6 was used for both mixing and curing. The water to cement ratio of 0.33 was used for all the mixes. The high range water reducing admixture namely superplasticizer used in this study is the MasterGlenium SKY 8808 supplied by BASF.

2.1.3 Fine and coarse aggregate

Fine aggregate utilized in this research is namely localized river sand from LABTECH Laboratory Supplies company. To ensure moisture is completely driven out of the pores, sand is oven-dried to constant weight at temperature of about (105±5) °C for 24 hours. The sand is required to sieve through 4.75 mm sieve. The sand available having a specific gravity, fineness modulus, water absorption and maximum grain size of 2.65, 2.70, 0.96% and 4.75 mm, respectively.

Yew et al. [6] reported that crushed oil palm shell (OPS) aggregate significantly increased the compressive strength compared to original OPS aggregate. Therefore, crushed LECA aggregate was used as the coarse aggregate in this study. These LECAs are obtained from Weng Thye Horticulture Sdn. Bhd. located at Kota Damansara, Selangor. Furthermore, all these coarse aggregates must have size able to retain on 4.75 mm sieve. In this study, crushed LECA was utilized, as shown in **Figure 1 (a)**.

Figure. 1 (a) Crushed lightweight expanded clay aggregate and (b) Macro polypropylene fibre

2.1.4 Fibres

A photograph of the macro polypropylene (MPP) fibre is shown in **Figure 1 (b)**

2.2. Mix proportions

A total of four (4) mixes were prepared. The mix proportions of all the LECA lightweight aggregate concrete (LLWAC) mixes used in this study are presented in **Table 1**.

Table 1. Mix Proportions of LLWAC-50-SPP

Specimen	SPP (%)	Material (kg/m^3)						
		Cement	Sand	Water	Granite	LECA	SP	SPP
LLWAC-50-MPP-0%[1]	0.00	420	875	138.6	232.5	240	7.175	0.000
LLWAC-50-MPP-0.15%	0.15	420	875	138.6	232.5	240	7.175	1.365
LLWAC-50-MPP-0.30%	0.30	420	875	138.6	232.5	240	7.175	2.730
LLWAC-50-MPP-0.45%	0.45	420	875	138.6	232.5	240	7.175	4.095

Note:[1]LLAC-50-PP-0% = LLWAC-50 with 0% of macro polypropylene fibre inclusion

2.3. Testing methods

The procedure adopted for mixing the crushed LECA lightweight aggregate fibre-reinforced concrete (LLWAFRC) with different volume fraction involves the following steps. Firstly, the sand and aggregate were poured into a concrete mixer and dry mixed for 60 seconds. Secondly, the cement was spread and dry mixed for 60 seconds, after which the specified amount of fibres is distributed and mixed for 180 seconds in the mix. This is followed by the addition of water and superplasticizer with a mixing time of 300 seconds. The concrete specimens were cast in 100-mm cube steel oiled moulds, and a poker vibrator was used to eliminate air bubbles in the mixture. The specimens were demoulded approximately 24 +/- 1 hours after casting and were cured in water at 25 ± 2 °C until age of testing.

3. Results and discussion

3.1. Properties of fresh concrete (Workability)

The workability of concrete determines the ease and homogeneity, with which it can be mixed, placed, consolidated, and finished. It is a way to understand the behaviour of concrete and to recognize the requirements of workability on site. In this study, the slump tests were carried out to determine the consistency of fresh concrete. The workability of LLWAC-50 with different volume percentage of polypropylene fibre is represented by the normal slump value as indicated in **Figure 2**.

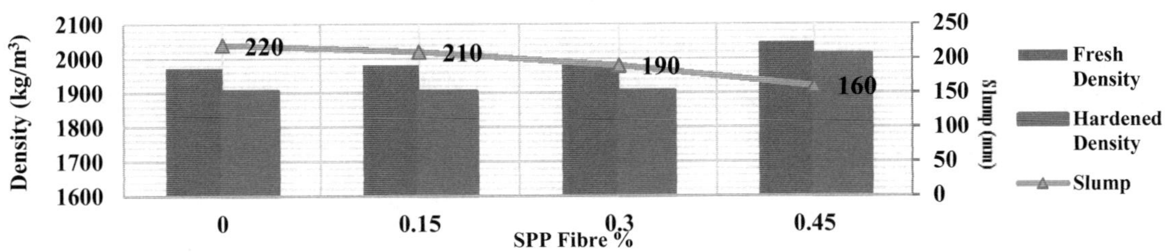

Figure. 2 Relationship of fresh density, hardened density and slump with various % of polypropylene fibre

The use of fibres is well known to affect the workability and flowability of plain concrete intrinsically [7]. In this study, the quantity of water and SP were kept constant for all mixes in order to access the effects of different content of SPP fibre on the workability of LLWAC. From Figure 3, it could be seen that the workability of fresh LLWAC decreases due to an increase of SPP fibre content. The addition of MPP fibre in LLWAC-70 has a negative effect on the workability. Slump value reduces remarkably with the increase of MPP fibre percent. The slump is declined gradually by 4.55%, 13.64% and 27.27% at 0.15%, 0.30% and 0.45% with the inclusion of SPP fibre respectively.

Addition of fibre decreases the workability of concrete in a way which bridge and hold the cement matrix, forming a network structure in concrete. This structure hence promotes cohesion and adhesion among the matrices. It can be ascertained that the fibres will absorb more of the cement paste in order to 'wrap around' due to the high content and large surface area of the fibres, and the increase in viscosity of the mixture promotes a decrease in workability [6].

It can be concluded that the inclusion of SSP fibre up to 0.45% V_f had significantly reduced the workability of LLWAC. Furthermore, the geometry of PP fibres through the surface area had an impact on workability. It has been reported that the fibrillated fibres produced higher slumps as compared to multi-filament fibres is mainly due to the fibre geometry [8]. However, LLWAC-70 inclusion of MPP fibre from 0%-0.45% had achieved the high workability with the slump value between 160- 220 mm.

3.2. Density

Three types of density, namely, fresh density (FD), demoulded density (DD) and oven-dry density (ODD) were measured for all mixes. Structural lightweight concrete (SLWC) is typically defined as concrete with ODD of not greater than 2000 kg/m³. It can be observed that all LLWAFRC density having an ODD and DD within the range of 1905 – 1982 kg/m³ and 1963 – 1993 kg/m³ which fulfilled the requirement of structural LWC for all mixes.

From **Table 2**, all densities are observed to have a slight increment as the inclusion of polypropylene fibre increased, except for the oven-dry density of LLWAC-50-SPP0.15%. This might be caused by the non-uniform of the dispersion of PP fibre in each specimen, thereby affected the average density obtained consequently. Besides, the fresh, hardened and oven-dry density of LLWAC-50-SPP0.45% are noticed to be escalated sharply. In general, density increases as the inclusion of fibre increase.

Even though the incorporation of SPP fibre with relatively low unit weight might displace certain amount of cement, however, the effect of fibre in stiffening the mixture capable to overcome the former, thus result in the increment of density. Similar outcome is evidenced by Bagherzadeh et al. [9]. This can be further explained by the packing density theory, where SPP fibre holds the cement matrix close to one another, causing a packing effect. As a result, increase the material occupied within a unit volume, thus increase overall density.

Table 2. Fresh and hardened properties of LLWAC-50 with various volume fraction of polypropylene fibre

Specimen	**Density (kg/m³)**				**Stability**	**Consistency**
	Target	Fresh	Hardened	Oven-Dry		
LLWAC-50-MPP0%	1896.9	1971.8	1962.8	1906.1	1.005	1.039
LLWAC-50-MPP0.15%	1898.2	1980.2	1981.2	1904.6	1.000	1.043
LLWAC-50-MPP0.30%	1899.6	1988.3	1982.2	1906.4	1.003	1.047
LLWAC-50-MPP0.45%	1901.0	2048.0	1993.4	1982.7	0.983	1.077

3.3. Compressive strength

3.3.1. Continuous moist curing

The results of the compressive strength test for all concrete mixes subject to moist curing up to 28 days are presented. **Figure 3** shows the compressive strength development for LLWAC containing SPP fibre with different percentage of V_f. As can be seen, there is significant difference between LLWAC-50-SPP0% and LLWAC-50-PP0.45%. It can be found that a positive development of compressive strength can be noticed with the increase of PP fibre percentage at both curing age. As the curing period increase, the compressive strength increased by 11.75%, 6.71%, 14.89% and 13.55% respectively from 0% to 0.45% inclusion of fibre. The highest early strength and later strength are given by LLWAC-50-SPP0.45% with strength of 32.32 MPa and 36.70 MPa respectively. In comparison among various percentage of fibre, LLWAC-50-PP0.15% possessed a slight escalation of strength. Furthermore, LLWAC-50-PP0.45% highlighted a dramatic development of compressive strength about 28.30% at 7 days and 30.37% at 28 days respectively.

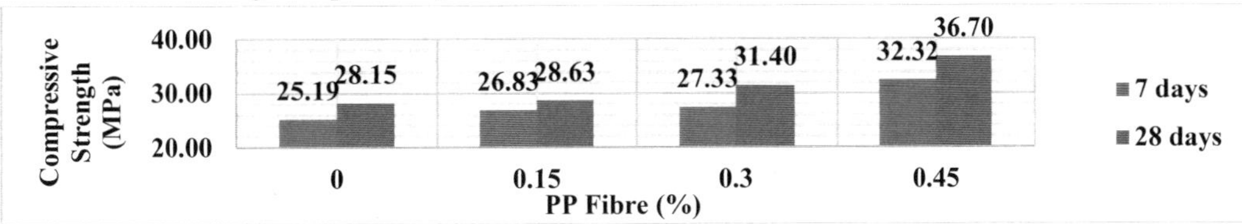

Figure. 3 Compressive strength of LLWAC-70 with different percent of SPP Fibre at curing age of 7 days and 28 days

This positive trend in the development of compressive strength have subverted the statements reported that addition of SPP fibre at different quantities has no effect on the compressive strength. This incremental of strength is mainly influenced by the fibre's stiffness and energy absorption induced by fibre. The improvement came principally from the fibres interacting with the advancing cracks and showed the higher values of compressive strength compared to LLWAC-50-SPP0% concrete. Therefore, improve the capability of concrete to capture larger loadings, resulting in the extension of failure mode that prone to ductile mode, hence give rise to higher compressive strength. This process is attributed to the ability of the fibres to arrest cracks or bridging effect in concrete.

4. Conclusion

Based on the experimental results of this study, the following conclusions can be drawn:

(1) The presence of synthetic polypropylene (SPP) fibre has a minor effect on density. There is slight increment in density as the SPP fibre percentage increased. Density is increased from 1906.1 kg/m^3 to 1982.7 kg/m^3 with addition of 0.45% SPP fibre.

(2) The presence of SPP fibre in LLWAC had significantly reduced the workability, where the rate of slump drop accelerated as the fibre content increased. The inclusion of 0.45% of SPP fibre has reduced the workability from 220 mm to 160 mm.

(3) Compressive strength increased as the volume percentage of SPP fibre in LLWAC increased. The compressive strength at 7 days and 28 days has increased by 28.30% and 30.37% respectively for 0.45% volume fraction of SPP fibre.

Acknowledgment

The authors gratefully acknowledge the financial support from Universiti Tunku Abdul Rahman for providing the financial support for this work under the Universiti Tunku Abdul Rahman Research Fund (UTARRF), Grant No.: IPSR/RMC/UTARRF/2020-C2/Y02.

References

[1] Mendis, ASM, Al-Deen S & Ashraf M. (2017). Effect of rubber particles on the flexural behaviour of reinforced crumbed rubber concrete beams, *Construction and Building Materials*, **154**, 644-57.

[2] Sobek W. (2016). Ultra-lightweight construction. *International Journal of Space Structures*, **31(1)**, 74–80.

[3] Aslam M, Shafigh P, Alizadeh Nomeli M & Zamin Jumaat M. (2017). Manufacturing of high-strength lightweight aggregate concrete using blended coarse lightweight aggregates. *Journal of Building Engineering*, **13(2)**, 53–62.

[4] Turatsinze A & Garros M. (2008). On the modulus of elasticity and strain capacity of self-compacting concrete incorporating rubber aggregates. *Resources, Conservation and Recycling*, **52(10)**, 1209–15.

[5] Shafigh P, Jumaat MZ & Mahmud H. (2011). Effect of steel fibre on the mechanical properties of oil palm shell lightweight concrete. *Materials & Design*, **32**, 3926-32.

[6] Yew MK, Mahmud H, Ang BC & Yew MC. (2016). Effects of polypropylene twisted bundle fibres on the mechanical properties of high-strength oil palm shell lightweight concrete. *Materials & Structures*, **49(4)**, 1221-33.
[7] Yew MK, Yew MC, Beh JH, Saw LH, Lee FW & Ng TC. (2018). Influence of high-performance polypropylene fibre and heat-treated dura oil palm shell on durability properties of lightweight concrete. *European Journal of Environmental and Civil Engineering*, 1-20.
[8] Yap SP, Alengaram UJ, Mo KH & Jumaat MZ. (2017). Ductility behaviours of oil palm shell steel fibre-reinforced concrete beams under flexural loading. European *Journal of Environmental and Civil Engineering*, 1-13.
[9] Bagherzadeh R, Pakravan HR, Sadeghi AH, Latifi M & Merati AA. (2012). An investigation on adding polypropylene fibers to reinforce lightweight cement composites (LWC). *Journal of Engineered Fibers and Fabrics*, **7(4)**, 13–21.

The Effects of Sintering Additives on the Sintering of 3Y-TZP Ceramic

S Theenagaran[1], C H Ting [1*], K L Chin[1], W H Yeo[1], C K Ng[2], S Ramesh[3] and W D Teng[4]

[1]Department of Mechanical and Materials & Manufacturing Engineering, Universiti, Tunku Abdul Rahman, Jalan Sungai Long, Bandar Sungai Long, Cheras, Kajang, 43000, Malaysia
[2]Department of Materials Engineering, Faculty of Engineering & Technology, Tunku Abdul Rahman University College, Jalan Genting Kelang, Setapak, Kuala Lumpur 53300, Malaysia
[3]Department of Mechanical Engineering, Faculty of Engineering, University of Malaya, Kuala Lumpur 50603, Malaysia
[4]Ceramics Technology Group, SIRIM Berhad, Shah Alam 40911, Malaysia

E-mail: tingch@utar.edu.my

Abstract. The influence of 0.5 wt.% of manganese oxide (MnO2) addition as dopant into 3 mol% yttria stabilized tetragonal zirconia polycrystals (3Y-TZP) were studied in comparison to undoped 3Y-TZP using different sintering profiles. Samples were sintered at 1200℃, 1300℃ and 1400℃ with holding time of 1 min, 12 min, 2 hours and 20 hours all for single step sintering. The optimal sintering profiles are A4 and C2 where 3Y-TZP doped with 0.5% wt. MnO2 sintered at 1400℃ for 12 minutes and 1200 ℃ for 20 hours respectively. At these sintering parameters, the MnO2 doped 3Y-TZP sample exhibited relative density of ~97%, Vicker's Hardness value of ~13 GPa, Fracture toughness value of ~4 MPam1/2 and Young's modulus of ~210 GPa.

1. Introduction

Zirconia is known as an oxide ceramic that is present with crystal structure transition properties when its subjected to increasing temperature. Zirconia at room temperature adopts monoclinic crystal structure and slowly transitions to tetragonal and cubic phase at temperature up to 1170℃ and 2370℃ respectively. Tetragonal phase structure particularly possesses great properties such as high flexural strength, excellent chemical and thermal stability and great biocompatibility. In the early seventies, medical industry had the necessity for metal free material to be used as replacement for titanium steel that were widely used for hip head replacement. In 20th century, a research work identified a mechanism in zirconia ceramics known as transformation toughening where the material transforms from tetragonal phase (t-phase) to monoclinic phase (m-phase) in the presence of external stress which in return contributes to toughening effect. [1]

One of the major drawbacks of this material is that this material is unstable and has the tendency to alter from its tetragonal (t) structure that is also a metastable phase which has great properties to monoclinic (m) phase that is stable but with weaker properties. Through research it was found that by adding Yttria, particularly 3 mol% helps to alleviate the phase by retaining tetragonal phase structure at normal room temperature. Doping manganese oxide into yttria stabilized tetragonal zirconia polycrystals, Y-TZP, had been substantial in assisting pressureless sintering process. [2] This is as undoped 3Y-TZP requires longer dwell time to achieve desired mechanical properties and hydrothermal aging behaviour. Where, it was concluded that the addition of >0.3 wt.% of MnO_2 was effective in improving densification, material's matrix stiffness and also the hardness when they are single step sintered at temperature below 1350 ℃ as compared to undoped Y-TZP samples. Particularly it was found 0.5 wt.% of manganese oxide has improved the mechanical properties when sintered below 1350℃ using the conventional sintering method. This is where the tetragonal structure phase stability in zirconia ceramic's matrix remains undisrupted at that dopant weight percentage addition. [2] [5] There are many works in past that have discussed the effects of MnO_2 addition as dopants in enhancing the properties of 3Y-TZP. [6] [7]

However, the optimal sintering parameter using 0.5 wt.% of manganese oxide as dopant produced for structural application by enhancing the densification while not losing its mechanical properties of 3Y-TZP were not proposed. Thus, the objective of this study is to identify the optimal sintering profile to fabricate the 3Y-TZP doped 0.5 wt.% of manganese oxide with relative density >95% at varying sintering parameters whilst comparing it to undoped 3Y-TZP samples.

2. Experimental

2.1. *Sample Preparation*

The precursor powder used was 3Y-TZP (TOSOH) Japan also known as tetragonal zirconia polycrystals stabilized with 3 mol % of yttria, along with manganese oxide (MnO_2) as dopant. Both the material, 3Y-TZP and MnO_2 were then mixed through ultra-sonification process for 45 minutes using an ultrasound probe that remits pulse of 29-36k Hz. The mixture then continued with ball milling process with 200ml of ethanol for 60 minutes at 450rpm. 2mm diameter zirconia milling balls were added to the mixture to assist milling process. The slurry media prepared was separated from the zirconia milling balls and was poured evenly on a tray before drying them in a pre heated drying oven at 70℃ overnight to ensure all ethanol evaporates. Once dried, the residue is sieved using a 250 μm stainless steel mesh sieve and the mixed powder was collected. Two different batches of powder prepared, undoped 3Y-TZP and 0.5 wt.% of MnO_2 doped 3Y-TZP powder samples. All of the samples from different batches were uniaxially pressed at 3 MPa to rectangular bar and disc shape with dimensions, 30x10x10 mm weighing 3 grams and 20mm diameter, 5mm thickness weighing 2.5 grams, respectively. All pressed green samples then subjected to cold isostatic pressing (CIP) at 210 MPa.

2.2. *Sintering Method*

Single step sintering method deployed for all samples. Referring to the Table 1, three temperature range sintering carried out starting from 1200℃, 1300℃ and 1400℃ at four dwell periods of 1 min, 12 min, 2 hours, and 20 hours. The single step sintering (SSS) process was carried out at with a constant heating rate of 10℃/min using carbolite furnace from room temperature and once dwell period was completed it was cooled down with similar cooling rate of 10℃/min.

Table 1: Listed are the sintering profiles and additives.

Sintering Profile	Sintering Time	Sintering Temperature (℃)	Sintering Profile Dopants
a1	1 min	1200	
a2	12 mins		
a3	2 hours		
a4	20 hours		
b1	1 min	1300	

b2	12 mins		Undoped
b3	2 hours		
b4	20 hours		
c1	1 min	1400	
c2	12 min		
c3	2 hours		
c4	20 hours		
A1	1 min	1200	3Y-TZP Doped 0.5% MnO_2
A2	12 mins		
A3	2 hours		
A4	20 hours		
B1	1 min	1300	
B2	12 mins		
B3	2 hours		
B4	20 hours		
C1	1 min	1400	
C2	12 mins		
C3	2 hours		
C4	20 hours		

3. Result and Discussion

3.1. Relative Density

Figure 1 represents relative density of varying sintering profiles in comparison to doped and undoped 3Y-TZP samples. All doped samples achieve relative density of >95% when sintered with dwell time of above 2 hours at 1200℃, 1300℃ and 1400℃. Undoped samples however achieve relative density >95% when sintered at 1200℃ at longer dwell time of 20 hours. Relative density of 3Y-TZP that are more than 95% are accepted as a criterion for structural engineering ceramics applications. Whereby, this aligns with a research done in 2013 where the addition of >0.3 wt.% manganese and sintered at low temperature helps the consolidation process of 3Y-TZP that improves the densification of the material.

At lower sintering temperature, longer dwell time needed to achieve densification >95% for undoped samples a4. Whereas for 0.5 wt.% manganese oxide doped A3 samples, shorter dwell period of 2 hours at 1200℃ sintering temperature enables it to achieve densification >95%. Moreover, at temperature 1300℃, doped B1 sample achieves relative density of >95% when held for 1 minute, where undoped sample needed 12-minute dwell time. Nevertheless the additive has no apparent densification effect with longer sintering holding time at sintering temperature above 1400 ℃.

3.2. Vicker's hardness

The effects of varying sintering profiles are shown in Figure 2. The hardness of undoped samples increases with increasing dwell time especially at lower sintering temperature of 1200℃. For 0.5 wt.% manganese oxide doped samples, when sintered at 1200℃, the hardness value reaches 13 GPa range only when sintered with dwell time of >2 hours. However, when it was sintered at temperature 1300℃ and 1400℃, hardness was hovering around 13 GPa at all dwell time before reducing to 12.88 GPa when sintered at 1400℃, for 20 hours.

The peak hardness value attained through undoped sample that was sintered at 1200℃ for 20 hours, recording value at 15.26 GPa, while the lowest value was 3.13 GPa that of sintered at 1200℃ for 1 minute. Through the conventional single step sintering process, it shows that longer sintering time required when sintered at temperature 1200℃, and similar hardness value can be achieved when samples are sintered at temperature >1300℃, for both doped and undoped samples.

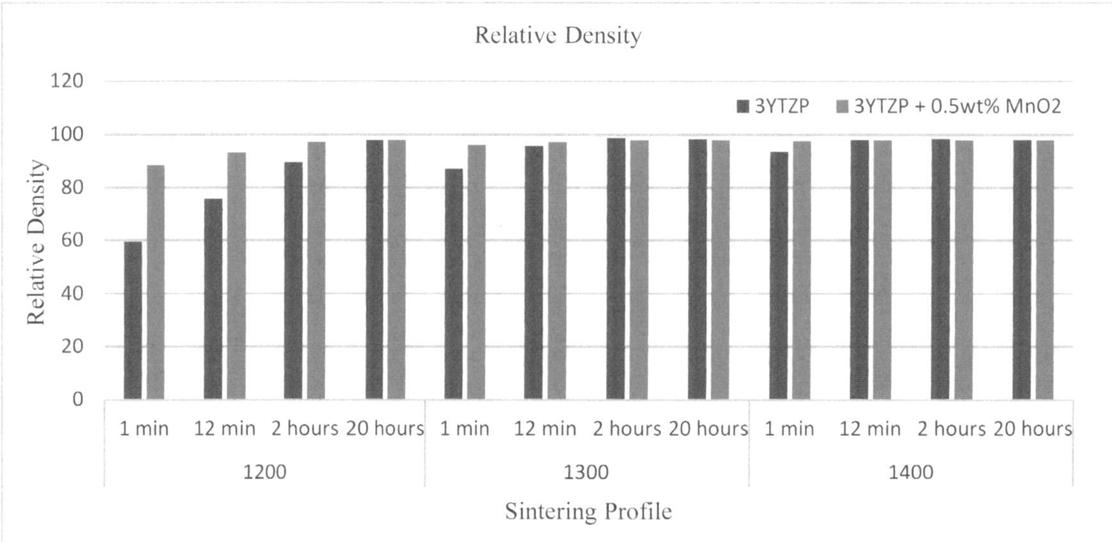

Figure. 1 Bar Chart of Relative Density with varying sintering profile

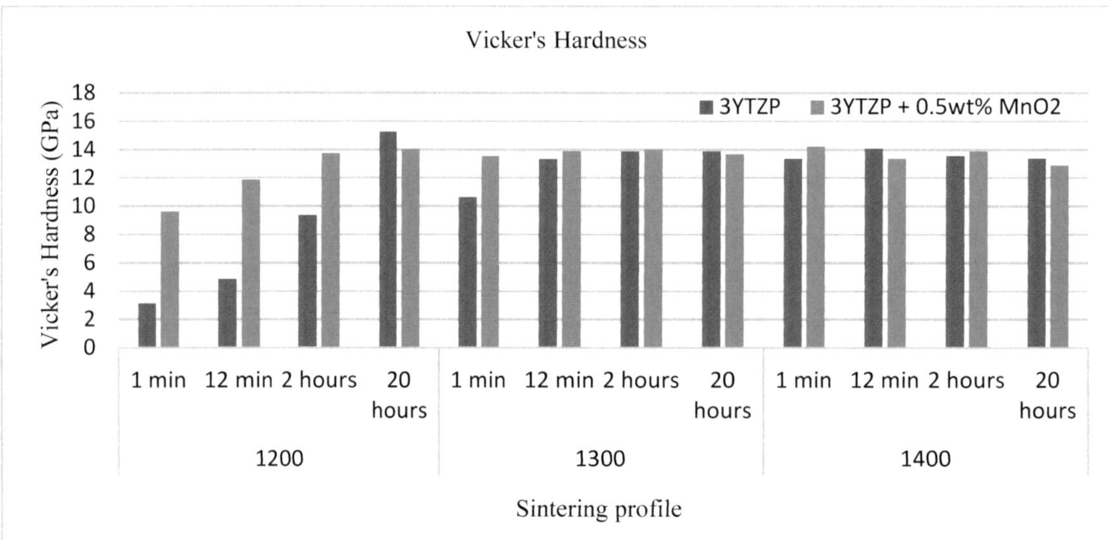

Figure. 2 Bar Chart of Vicker's Hardness with varying sintering profiles

3.3. Young's Modulus

The varying Young's Modulus (E) for doped 3Y-TZP and undoped 3Y-TZP are depicted as bar chart in Figure 3. The doped samples are in correlation with the relative density where the dopants addition have improved the tetragonal zirconia matrix stiffness when sintered at temperature above 1200℃ at dwell time of more than 2 hours, 1300℃ at dwell time more than 12 min and 1400℃ at dwell time more than 1 minute. All these samples at these parameters are able to achieve Young's modulus of >200GPa. The addition of dopant has been significant in enhancing the elastic modulus of 3Y-TZP as compared to the undoped samples. The highest E value of 207.1 GPa obtained through 0.5 wt.% MnO_2 doped 3Y-TZP sintered at 1300℃ for 20 hours followed by a slight drop to 206.6 GPa for sample sintered at 1400℃ for 20 hours.

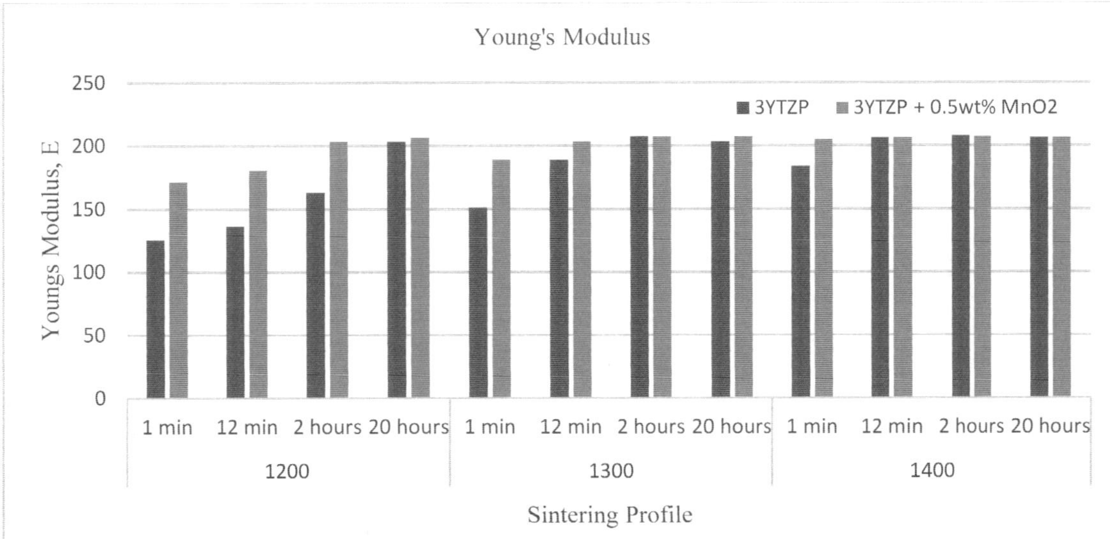

Figure. 3 Bar Chart of Young's Modulus with varying sintering profiles

4. Conclusion

In conclusion, MnO_2 as dopant addition to 3Y-TZP are important in improving the matrix stiffness of 3Y-TZP to achieve increased densification at low sintering temperature or short holding time. Through conventional pressureless sintering method, the mechanical properties can be enhanced with higher temperature at shorter holding period. Sample C2, sintered at 1400℃ and 12 minutes dwell time achieves Vicker's hardness value of ~13 GPa, with relative density of >97% and Young's Modulus of ~206 GPa. On the other hand, similar properties are achieved for sample A4 with 20 hours of holding time at 1200℃. Both samples A4 and C2 comply within the requirement of structural application engineering ceramics material.

5. References

[1] Yuri M. Solonin, 1995. Powder Metalurgy and Metal Ceramics. In: *Theory and Technology of Sintering Process, Thermal and Thermochemical Treatment. Kiev*: Plenum Publishing Corporation, pp. 43-52
[2] Chong F.D., Tan C.Y., Ramesh S., Muchtar A., 2016. *Effects of manganese oxide on the sinterability of 8 mol% yttria stabilized zirconia*. Material Characterization, pp. 331-336
[3] Golieskardi M., Satgunam M., Ragurajan D., 2017. *Effects of holding time on the mechanical properties and microstructural evaluation of 3Y-TZP ceramics*. Australian Ceramic Society
[4] Kwa S.M., Ramesh S., 2015. *Effects of sintering holding time on the properties and low temperature degradation behaviour of manganese oxide-doped Y-TZP ceramics*. Journal of Ceramic Processing Research, pp. 193-198
[5] Ramesh S., Amiriyan M., Meenaloshini S., Tolouei R., Hamdi M., Pruboloksono J., Teng W.D., 2011. *Densification behaviour and properties of manganese oxide doped Y-TZP ceramics*. Ceramics International , 37(8), pp. 3583-3590.
[6] Ramesh S., Kelvin Chew W.J., Tan C.Y., 2013. *Influence of Manganese on the Sintering Properties of Tetragonal Zirconia*. Ceramics Silikaty, 1(57), pp. 28-32.
[7] Khan, M., Ramesh S., Bang L.T., Wong Y.H., Ubenthiran S., Tan C.Y., Purbolaksono J., Misran H., 2014. *Effect of Copper Oxide and Manganese Oxide on Properties and Low Temperature Degradation of Sintered Y-TZP*. Journal of Materials Engineering and Performance, pp. 4328-4335.

Acknowledgments
This research was supported under the UTAR Research Fund (Project No:IPSR/RMC/UTARRF/2018-C2/T04).

Research on the dissociation of yucca to the organic solubility of wastewater

Lu Sun, Xin-Da Wu*, Ze-Sheng Qiu, Tong-Huai Zheng, Yu-Hong Wen, Min Zhu
School of Chemistry and Materials Engineering, HuiZhou University, China

Corresponding author E-mail:13478511396@139.com, wutomp@hzu.edu.cn

Abstract. With the development of society, the pollution of wastewater has become extremely serious, and the amount of wastewater is also gradually increasing. The water quality discharge does not meet the standards, which has a serious impact on the water quality of our country, it also has a very important impact on people's daily water use. Yucca has a wide variety of plants, and the main components of its extract are steroidal saponins, free radical saponins and carbohydrate complexes. It has a special physiological structure, macromolecules and sequence molecules are directly adsorbed, and it has a strong adsorption capacity for harmful gases, which can improve the odours and is widely used, But it is less used in wastewater treatment, this project uses yucca extract to conduct preliminary analysis and exploration on the dissociation of organic solubility in wastewater, to improve the solubility and treatment of free ions and organic dissolved matter in water, preliminary results found that adding 200ppm-20% and 500ppm-10% concentration of yucca is the most suitable range for water quality treatment. It will be a bigger breakthrough when used in wastewater treatment, and my country's water pollution will be in great improvement.
Keywords: wastewater pollution; yucca extract; organic solubility.

1. Introduction

Nowadays, with the development of society, no matter in industry or agriculture, wastewater pollution is extremely serious. In the process of economic development, my country's factories are gradually increasing. The amount of waste water is also gradually increasing, and the water quality discharge does not meet the standards, which has a serious impact on the water quality of our country, and also has an impact on the people's daily water use [1-4]. As we all know, water is an indispensable substance in our daily lives, but nowadays, the ubiquitous waste water ponds can't help but make people worry. Rural sewage research shows that the rural sewage processing and treatment status, backward rural equipment infrastructure, poor water pollution control and other issues. Rural domestic sewage is directly dumped out of farmland and ditches through daily washing and household use, which has a serious impact on water quality [5-7]. In order to comply with the current environmental protection concept of green and sustainable development, there is an urgent need for a process to treat and prevent wastewater. Yucca, also known as soft-leaf yucca, can be found in many places in our country. It is an evergreen shrub for many years and has a wide variety of plants in the genus Yucca. The main components of the extract are steroidal saponins, free radical saponins and carbohydrate complexes. It has a special physiological structure, macromolecules and sequence molecules are directly adsorbed, and it has a strong adsorption capacity for harmful gases, which can improve the smell. At the same time, it can reduce animal ammonia emissions and improve the breeding environment of livestock and poultry. Yucca saponins are a natural detergent or surfactant, which can change the morphology of the

IOP Conf. Series: Materials Science and Engineering 1117 (2021) 012011 doi:10.1088/1757-899X/1117/1/012011

epithelial cell membrane of the digestive tract and reduce the surface tension of the cell membrane. It can combine with cholesterol on the cell membrane to form an irreversible complex and excrete from the body, thereby enhancing the body's immunity and improve animal performance. It also has the functions of regulating the intestinal microenvironment and promoting cell nutrition. It has been widely studied and used in animal production. Studies have also shown that it has a certain promoting effect on improving the water quality of aquaculture water bodies [8-12]. Combining with the current background of "five water co-treatment" and conforming to the needs of social development, this project explores the initial evaluation and exploration of the dissociation of yucca extract in the organic solubility of wastewater and the extent of the impact of different concentrations of yucca on water quality improvement. It further illustrates the problem of yucca improving the dissociation of the organic solubility of wastewater. Due to the problem of wastewater treatment in the plant area, the research team used natural yucca extract to treat the water quality to improve the treatment of free ions and organic dissolved matter in the water, especially to achieve the reusable water quality and the environmental problems of the plant water treatment at the first phase research discussion.

2. Experiment process

In this study, the conductivity of the yucca extracts in a series of percentages and concentrations of the plant solution changes at different times. The experimental process is shown in Figure 1.

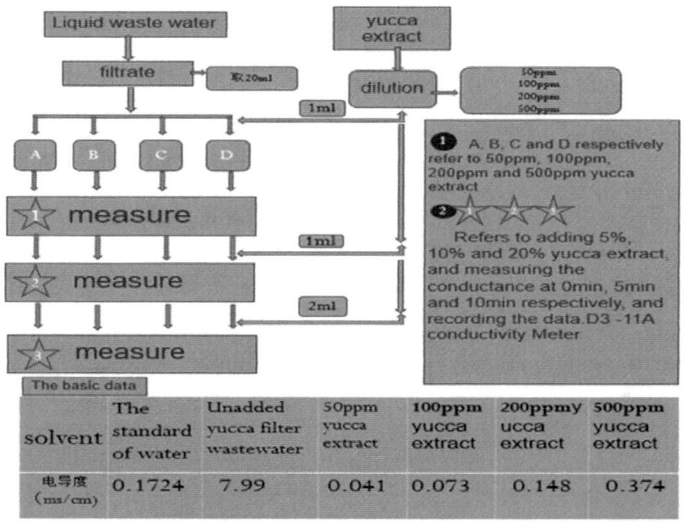

solvent	The standard of water	Unadded yucca filter wastewater	50ppm yucca extract	100ppm yucca extract	200ppmy ucca extract	500ppm yucca extract
电导度 (ms/cm)	0.1724	7.99	0.041	0.073	0.148	0.374

The sample processing
1.Prepare 20ml yucca 50ppm, 100ppm, 200ppm and 500ppm respectively
2.Prepare a 500ml bottle of industrial wastewater

Figure. 1 Experimental flowchart

This research shows the experimental flow chart of this project as shown in Figure 1. The waste water comes from the waste water that the project members personally visit the physical factory and personally treat the waste water, and then bring it into the laboratory for the yucca treatment experiment. After filtering, the waste water is tested to 50ppm, 100ppm, 200ppm and 500ppm yucca were studied. Experimental process steps:

1） Conductivity instrument——DDS-11A conductivity mete, put its electrode in deionized water first.
2） The conductivity instrument measures and records the conductivity of the standard water and filtered wastewater without yucca.
3） Dilute the 10000ppm yucca extract into 50ppm, 100ppm, 200ppm, and 500ppm yucca extracts. Take 10ml of it and put it into four clean beakers, and use a conductivity instrument to measure

the conductivity of the yucca extract at this concentration.

4) Take out 100ml of factory liquid wastewater in a beaker and filter it into another beaker with a funnel. Pour 80ml of the filtered plant solution into four clean beakers, labeled as A.B.C.D.

5) Pour 1ml of yucca extract of different concentrations into a beaker A.B.C.D and stir with a glass rod. Measure and record the conductivity immediately. Re-measure and record with conductivity after standing for 5min and 10min.

6) Follow the fifth test to take the yucca extract and take 1ml into the beaker A.B.C.D. Stir with a glass rod, measure and record. Re-measure and record with conductivity after standing for 5min and 10min

7) Follow the sixth test to take the yucca extract and take 2ml into the beaker A.B.C.D and stir with a glass rod. Immediately use conductivity to measure and record. After standing for 5 min and 10 min, the conductivity is measured and recorded again.

3. Study the flow chart of water improvement

The water improvement process in this experiment is shown in Figure 2. The Figure 2 and 3 illustrate the process of improving the water body. After the wastewater is removed from the factory, it is pre-treated. The experiment is preliminary filtration. After the filter residue is filtered, it will be further studied, and then then use different concentrations of yucca to drip the wastewater and improve water saturation of organic phenadoxones and water. This process will be returned to the factory for use as raw materials for industrial production and the next industrial production, such as shown in Figure 4.

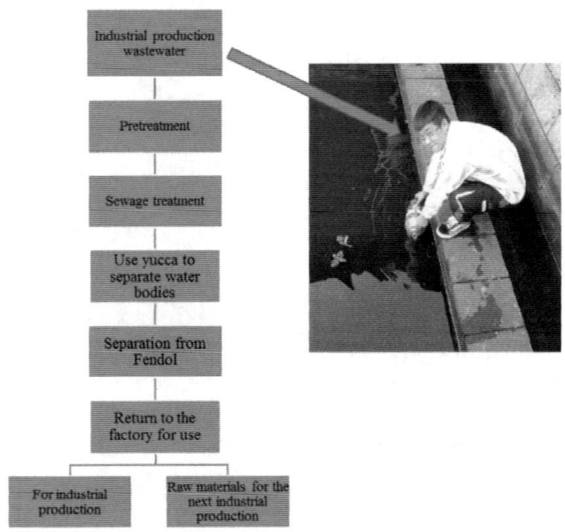

Figure. 2 Take samples for analysis

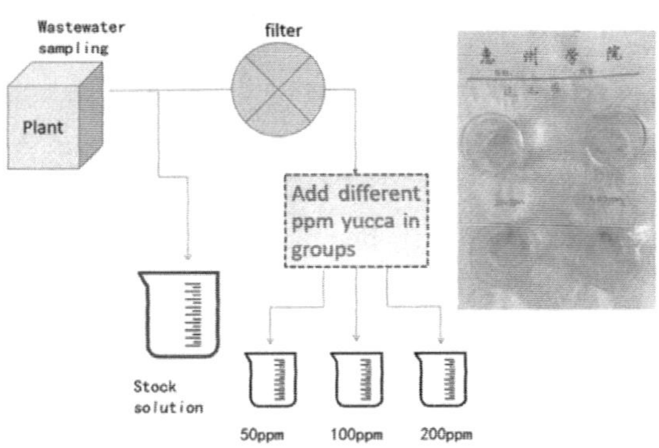

Figure. 3 Sample preparations for analysis

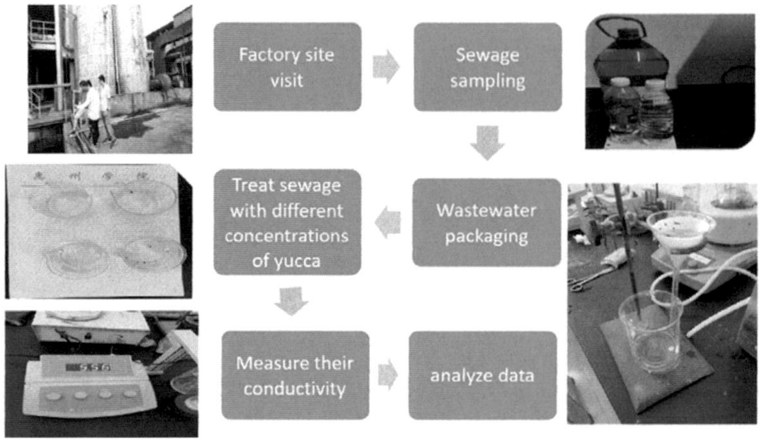

Figure. 4 Experimental process

4. Initial results and discussion

In this project, standard water, filtered wastewater without yucca added and conductivity of 50ppm, 100ppm, 200ppm, 500ppm are measured, as shown in Table 1.

Table. 1 Original conductivity of stock solution and filler

Solvent	Standard water	Filtered wastewater without yucca	50ppm yucca	100ppm yucca	200ppm yucca	500ppm yucca
Conductivity (ms/cm)	0.172	7.990	0.041	0.073	0.148	0.374

Test the conductivity of standard water and filtered wastewater without yucca as a standard and comparison, and then measure the conductivity of 50 ppm, 100 ppm, 200 ppm, and 500 ppm respectively. The table 2 shows that the greater the concentration of yucca extract, the higher the conductivity of the solution. The difference in conductivity caused by adding different concentrations of different doses of yucca additives to wastewater is shown in Table 2.

Table. 2 Different additions and changes in water

Yucca addition amount /ppm Conductivity /(ms/cm)	50ppm	100ppm	200ppm	500ppm
Wastewater + 1mlAdditives (5%)	5.41	4.99	4.82	4.75
Wastewater + 2mlAdditives (10%)	5.16	4.78	4.75	4.64
Wastewater + 4mlAdditives (20%)	4.79	4.46	4.38	4.58

Wastewater contains a lot of wood particles, ss, and part of wood organic matter. After filtering, some suspended matter is removed, and the remaining organic matter that is difficult to filter causes the conductivity of the solution to increase. The yucca extract contains a fat-soluble and water-soluble surfactant-yucca saponins, which can effectively remove the organic matter contained in the wastewater and greatly reduce the conductivity of the wastewater. As shown in Figure 5-7

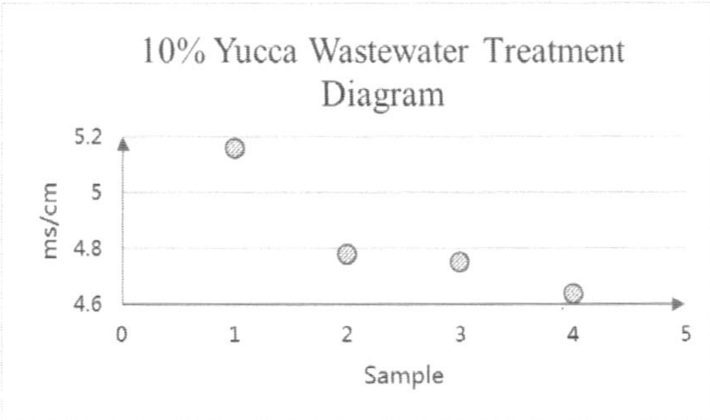

Figure. 5 10% Yucca wastewater treatment diagram

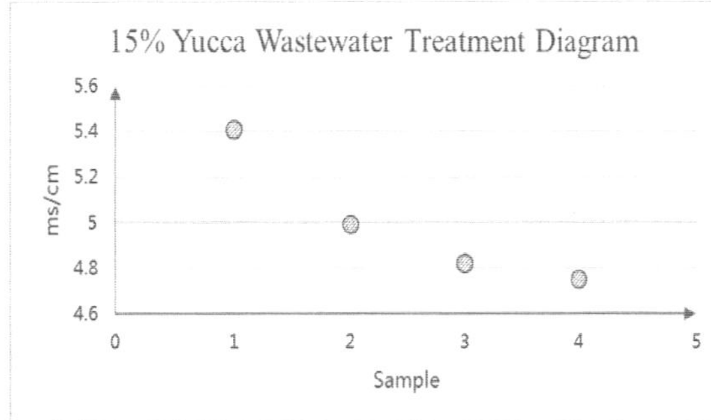

Figure. 6 15% Yucca wastewater treatment diagram

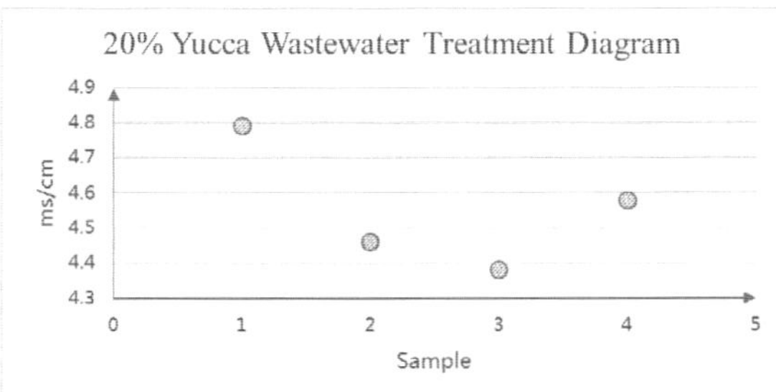

Figure. 7 20% Yucca Wastewater Treatment Diagram

It can be seen from Table 5-7 that when 4ml of yucca extract (20%) is added, the conductivity of wastewater with 500ppm yucca extract is greater than that of 200ppm silk extract. Therefore, the conductivity of the solution when adding 4ml of 500ppm yucca extract is higher than adding 4ml of 200ppm. The reason for the conductivity of the yucca extract is preliminary hypothesis: the reaction is complete when 500ppm yucca extract is added, and the residual yucca organic matter is greater than the residual organic matter of the solution when 200ppm is added, making the solution conductivity increase.

5. Conclusion

It can be seen from the entire experimental process that we initially explored the influence of different concentrations of yucca on the conductivity of wastewater. According to the experimental data, the addition of yucca can effectively reduce the conductivity of the solution, indicating that yucca can effectively reduce the solubility of the solution, improve the size of the solution molecules in the solution, and can effectively promote the dissociation of the organic solubility of wastewater. And under different concentrations (except 20% concentration), the conductivity of 10min<5min<the conductivity of yucca is added immediately, indicating that a certain period of time can effectively promote the organic dissolution of yucca in wastewater, which can be effective in a certain period of time to improve water quality. At the same time, as the concentration of yucca increases, the conductivity of the solution gradually increases. At a concentration of 500ppm, the conductivity table showed an increase in 10 minutes, indicating that the addition of 200ppm-20% and 500ppm-10% concentration of yucca should be within the most appropriate range for water quality treatment. In summary, this project effectively proves that yucca can effectively treat wastewater, and within a certain time and concentration range, it can achieve the best results. Yucca is easily available and affordable and it will be a bigger breakthrough when used in wastewater treatment. My country's water pollution will be greatly improved. Our country will be greener and environmentally friendly and become a greener country.

6. Acknowledgements
The study special thanks, supported by The Professorial and Doctoral Scientific Research Foundation of Huizhou University (2020JB034), FERRY READING CAFE and Staff canteen provide space for this discussion. Thanks to Huizhou Minyi Industry Co., Ltd. and Guangxi Yanding New Energy Materials Co., Ltd for providing the experimental site. Thanks to Huizhou Zhongkai High-tech Zone ZONGHUA Battery science and technology Institute for Project support.

References
[1]Jiang Yeli, "Analysis and countermeasures of common problems in environmental engineering industrial sewage treatment", china academic journal electronic publishing house, DOI: 10.16647/j.cnki.cn15-1369/X.2020.10.026, HUANJINGYUFAZHAN p45-46
[2]CAO Lei, LU Ji-lai, Jiang Hai, LI Xiao-lu, TU Yong, JIANG Yong-wei, WU Yun-bo, "A Study on Rural Ddomestic Sewage Maragement Measures in Taihu Basin of Jiangsu Province", Pollution Control Technology, 2014,27(06),35-40
[3]HE Yong, "Quality Improvement and Implementation of Sewage Treatment System at River Outlet of Panzhihua Steel", Metallurgical Power, 2020, (10),51-54
[4]Shinn-Dar Wu, Sui-Qun He, Yu-Lin Yang, Yea-Chyi Lin, Ting-Yu Chang, Ching-Yu Peng and Meng-Ting Hsieh, "Research on Environmentally Friendly Chemical Technology for Green Reusable and Sustainable Water Metal Copper Ions", Earth and Environmental Science, 555 (2020) 012013.
[5]WANG Junneng, ZHAO Xuetao, CAI Nan, CHEN Minmin, WANG Hao, "Pollution Discharge and Environmental Treatment Efficiency of Rural Domestic Sewage in China", Research of Environmental Sciences, DOI： 10.13198/j.issn.1001-6929.2020.11.08
[6]SHENG Xiaolin, QIU Songkai, XU Feng, Management of rural domestic wastewater in a city of Yangtze delta region:performance and remaining challenges[J].Bioresource Technology Reports,2020,11:100507.
[7]CHENG Hong, ZHANG Mingmei. Occurrence and removal of antibiotic resistance genes in municipal wastewater and rural domestic sewage treatment systems in eastern China[J]. Environment International,2013,55:9-14.
[8]ZHOU Jia, ZHAO Xiaolu, TANG Chao, "The effect of Yucca Plant extract on form, odor and ammonium nitrogen content in dog fecal", Guangdong Journal of Animal and Veterinary Science, 2018,43(04),35-38
[9]KANG Xing, "Research progress on the application of Yucca Shidigera extract in animal production", Feed Research, 2020,43(04),126-129
[10] CHEN DANDie, HUANG Yanling, XI Lin, "Research progress of yucca extract in animal feed", China Feed, 2020, (03),46-50
[11]Wang Jinping, Zhang Mi, "Study on the Pet Odor and Application of Deodorizers", China Cleaning Industry, 2019, (12),67-71
[12]LU Huijie, CHEN Weidong, TIAN Zhimei, MA Xianyong, "Physiological Function of Yucca Extract and Its Applications in Husbandry Production", Chinese Journal of Animal Nutrition, 2019,31(05),1990-2001

The Study on penetration variation of nickel ion in wastewater by laser light technology

Dan Lin, Fang Xie, Xin-Da Wu*, Hao- Lin Zhou

School of Chemistry and Materials Engineering, HuiZhou University, China
Corresponding author E-mail:13478511396@139.com and wutomp@hzu.edu.cn

Abstract. The rapid development of industry in the world has brought serious negative effects, so pollution control and prevention is an important issue for the future environment. The discharge and treatment of metal ions in water environment are becoming more and more serious. Common heavy metal analysis methods need the support of large-scale instruments. In the study, the differently control of wavelength, temperature and concentration. Explore theoretical analysis solutions based on laser absorption and scattering, Design of metal ion analysis of laser light penetration. The influence of medium change of metal ions on the light dispersion path is measured by direct light. Preliminary results found that the 532nm wavelength laser light has a deviation in the determination of metal ions. Laser wave was analysed by water diffusion chromatography. In order to facilitate the change from large equipment to portable metal ion detection technology.
Keywords: Metal ion analysis, Nickel ion, Laser beams, Direct light.

1. Introduction

In recent years, international pollution control has been paid more attention. The most important issue is the development of science and technology, such as nuclear energy, chemical production, industrial waste water and many other pending solutions. This study explores the environmental problems caused by heavy metal pollution in the water environment, to address the serious negative impact of rapid industrial development brings. Water resources are vital to the ecosystem. With the rapid development of industrialization, the discharge and treatment of metal ions in the environment are becoming more and more serious. Therefore, it is a very important task to do a good job in waste liquid discharge and metal ion detection. So, do the disposal of emissions and metal ion detection is a very important job. Shinn-Dar Wu et[1-2] proposed to solve the recovery and treatment of copper metal in 2020, and also to research and solve the pollution problem of greening, especially for low-concentration Cu^{2+} separation as shown in Fig 1 battery recycling technology. Especially in the process of battery waste treatment, water treatment is necessary in every stage. Study the extraction of waste liquid from different processed products. As the test fluid for experimental research, discuss whether there are any metal ions remaining in the treatment process (M^{+}), and whether the battery waste liquid meets the emission standard after treatment. The battery waste liquid treatment process is as shown in Fig1. In this study, from the filtrate of three finished products, conduct a polarization experiment to detect the transmitted light point. G. R. Strbac [3] use Raman scattering and density functional theory to calculate the optical properties and structure of As−Sb calcium salt glass. Key results of optical measurements are set forth refractive index, and the effect of the absorption edge of the optical band gap and other parameters As-Sb atoms. YUAN Min et. [4] made a summary and review of the research progress of heavy metal detection methods in water. Analyze and discuss the advantages and disadvantages of various detection methods in actual environmental detection. Common heavy metal analysis methods are: ultraviolet spectrophotometry (UV), atomic absorption method (AAS), atomic fluorescence method (AFS), inductively coupled plasma method (ICP), X fluorescence spectroscopy (XRF), inductively coupled plasma mass spectrometry Method (ICP-MS). However, the above measuring method requires large equipment support. Its high cost and time-consuming. The detection of heavy metals in the environment is a long-term and urgent task. Various detection methods are required to develop in the direction of higher sensitivity, higher selectivity, and more convenient and fast. Due to the sampling and analysis of aqueous solution, it takes a long time to go to the laboratory for instrument testing. According to the standard limits of metal ions stipulated in China's surface water environmental quality standard GB3838-2002[5] and the United States CWA[6] regulates the discharge of pollutants into American waters and surface water quality standards. In order to strengthen the operability of my country's water environmental standards, to improve measurement efficiency, as soon as possible to reduce environmental harm metal. Previous studies on the transmission properties of neutral salts in solutions have been carried out. Control different wavelengths, temperatures and concentrations, explore theoretical analysis solutions based on laser absorption and scattering. Due to the scattering of laser by solution medium, some inherent optical properties of different metal ions are considered. The physical and chemical interactions of media in water are studied by laser optics. Minimize the impact of the beam incident angle of the direct laser displacement sensor on the measurement accuracy [7-11]. Therefore, a qualitative measuring instrument was designed and developed to excite metal ions in the solution by laser. Finally, an instrument for more convenient monitoring of metal ions contained in environmental water resources. It may be implemented anywhere in the detection of metal ions.

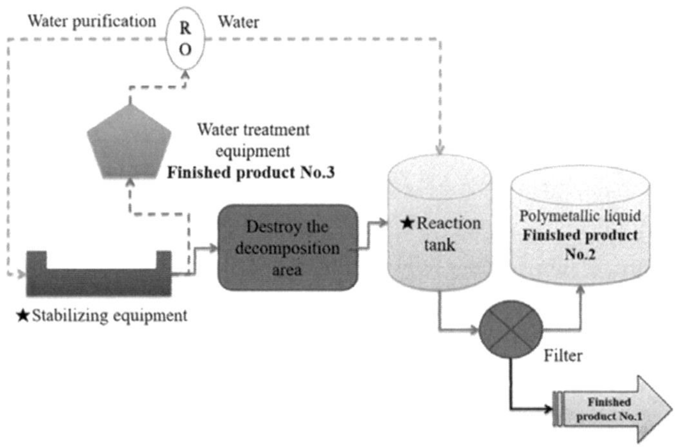

Figure. 1 Technical process of recovery

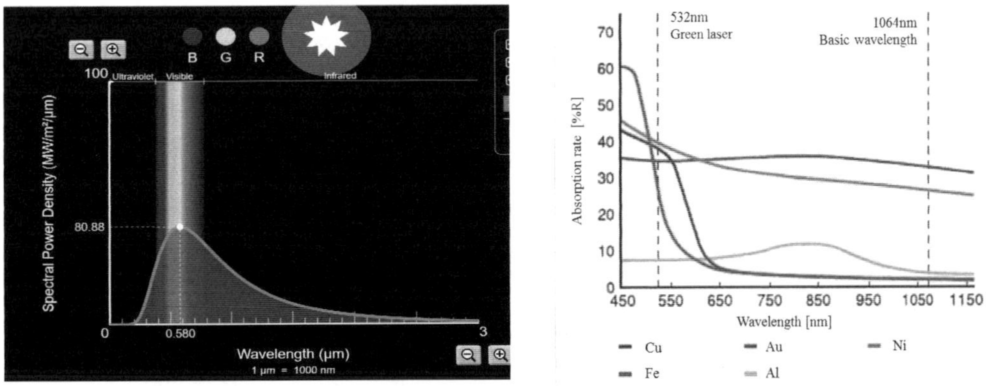

(a). Spectral power density of Different wavelength

(b). Laser absorptivity of metals

Figure. 2 Optical wavelength diagram [3]

When the light propagates in the medium, the laser wave path interacts with the medium. Its propagation characteristics also vary with the wavelength of light waves, at the same time, the transmission law also follows guided wave optics, and the Principles of fiber optics and micro optics. The wavelength of the SHG laser is half of the standard wavelength (1064 nm) as shown in Figure 2(b). Located at 532 nm in the visible spectrum appear green. The test solution in this experiment is 20% nickel solution， different metal ions can have different absorption rates for the same wavelength of light, to analyse whether a residue of the metal ion. The transmission characteristics of light are related to the energy of light waves. When the light wave energy (mainly energy density) is large to a certain extent, produce various nonlinear scattering effects. Both are the research fields of nonlinear optics and polarization optics. Through optical principles, optical detection and analysis of samples is an aspect discussed in this study.

2. Experimental design

The experimental process of this research aims at the separation of metal ions from battery recycling waste liquid. Analysis of residual metal ions (M^+) in water samples. Experimental Discussion on the First Phase of the Research on Nickel, as Fig 3 experimental process:

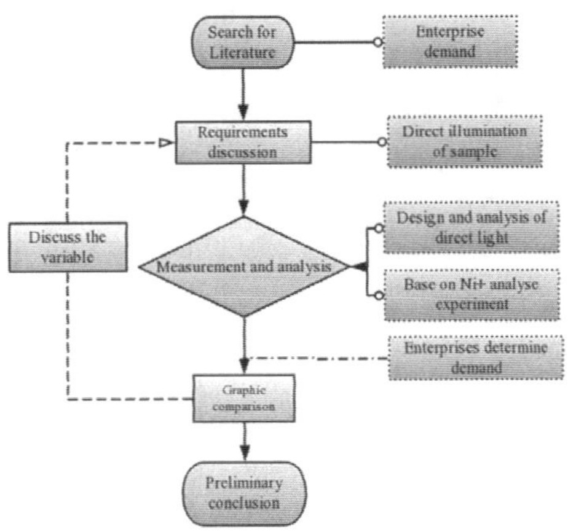

Figure. 3 Experimental procedure

As shown in Fig 2, optical analysis of physical properties of metal ions in sewage system. Because for the enterprise to waste battery treatment of sewage discharge will result in environmental problems

IOP Conf. Series: Materials Science and Engineering 1117 (2021) 012012 doi:10.1088/1757-899X/1117/1/012012

demand. Therefore, preliminary estimates of different ions contained in the waste battery discussions. The properties of the metal ion of the metal waste battery as a main sample analysis, and refer to the known experimental results in the past literature, designed this experiment. In this experiment, a green laser (power of 100mW, wavelength of 532nm) is used as the sample to do the actual measurement of direct light. In the early stage, the main experimental analysis and discussion object was the scattering of Ni^{+} characteristic metal ions, as shown in the present study Fig 3. By 532 nm laser, the transition is from excited ions. At this time radiates light corresponding to transition energies. This phenomenon is called spontaneous emission. The light emitted by the laser collides with other atoms that are also in an excited state and will also stimulate the same transition.

2.1. Measuring platform design and measuring method
This research is shown in Fig4, in different waste liquid treatment processes. Detect the change of the different light dispersion path of the metal ion medium by direct light:

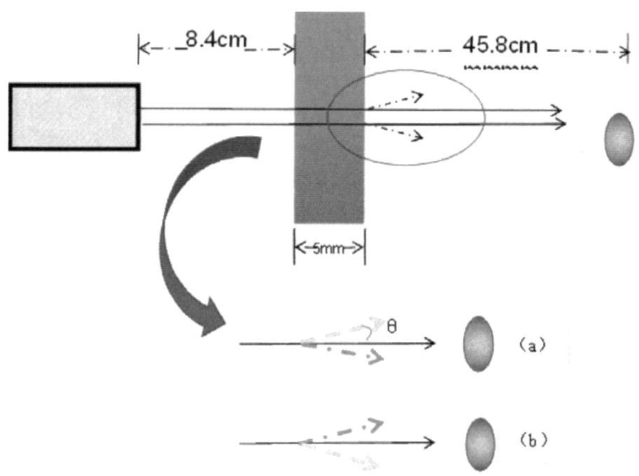

Figure. 4 Diagram of the sample with direct light

Due to the influence of M^{+} on the beam, the propagation direction of the original light wave is deviated due to the ion characteristics. Therefore, the degree of deviation of light waves has different effects on different propagation media. The path of direct light detection is discussed in Fig 4. The deviation of direct light passing through the medium causes the difference in the scattering phenomenon of the final light spot. Fig4 (a) and (b) indicates that the light spots and angles produced by different media are deviated. Lead to light penetration point state measurement. And Xiao tao Mi's[12] point of view is consistent. The type of scattered ions (refractive index, conductivity, uniformity and orientation of ions) is also one of the factors that cause light scattering. Fixed light source and sample to be tested, and the istance of light spot receiving surface, set up two control groups, explore the difference in the propagation of light waves in different media in the analysis sample. Conclusion By comparing the area of the light spot on the receiving surface, to make sure that this discussion is established.

3. The light amount measurement data analysis
Design experiments with reference to theoretical analysis. According to the experimental phenomenon and experimental results, the conclusion is drawn:

IOP Conf. Series: Materials Science and Engineering 1117 (2021) 012012 doi:10.1088/1757-899X/1117/1/012012

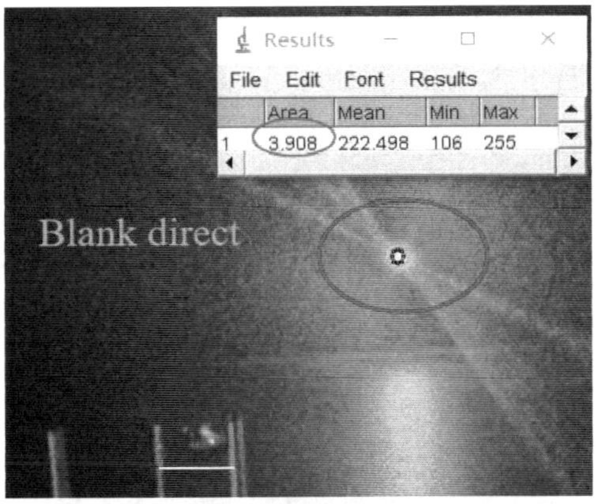

(a) Blank direct

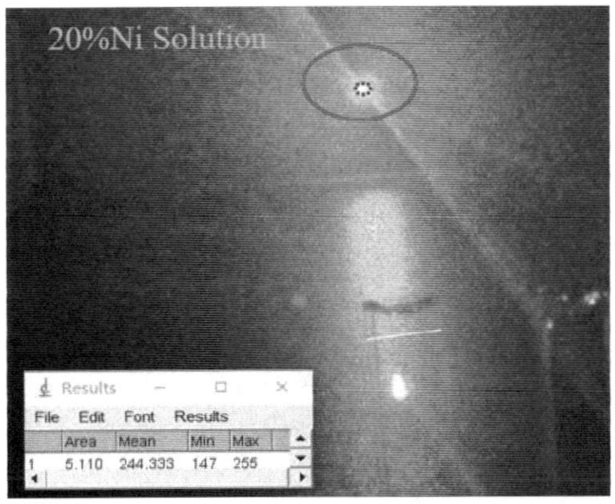

(b) 20%Ni Solution

Table 1. Calculation and analysis

Sample solution	**Blank direct**	**20%Ni Solution**	**Distilled Water**
Length (x,y)	1, 1	1.100, 1.158	0.700, 0.7
Area	3.908	5.110	1.328
Angle of deviation (tanθ)	0	1.2×10^{-3}	7.64×10^{-4}

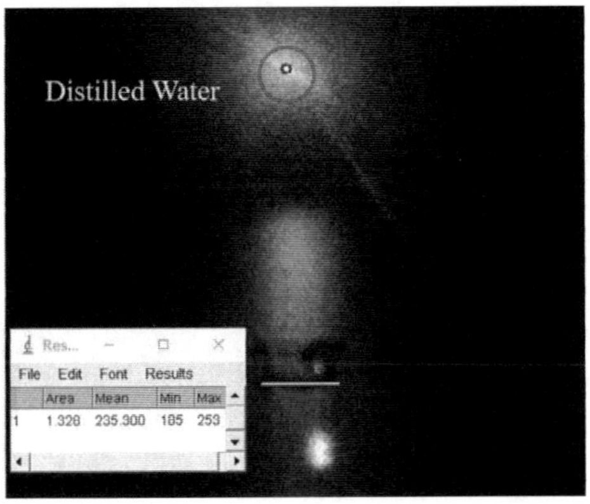

(c)Distilled Water

Figure. 5 actual measurement and data recording

The Fig 5 is the Actual measurement diagram during experiment. The experimental results of 20% Ni solution and distilled water blank control were analysed. Measuring the area of the spot of light in different media generated. Because the halo forms a spot on the receiving surface, when the strong light is projected on the receiving surface, the surface reflection will be caused and the image will be blurred. Thus, when calculating the area of the light spot , use the largest ellipse area as the experimental record data, get the analysis results of Fig 5 (a) Blank direct, (b) 20%Ni Solution, and (c) Distilled Water.

Through calculation and analysis, the description in Table 1 is completed. The direct light detection light scattering route has the strongest scattering for nickel liquid. The spot area is 5.110mm^2; In distilled water medium is the weakest area is 1.328mm^2; In the blank comparison, the medium is air. The spot area is 3.908mm^2. For the blank control 20% Ni Solution, the deviation angle tan value is: 1.2x10-3 mm^2 and the Distilled Water is 7.64x10-4 mm^2.

Preliminary experimental demonstration results, for the difference between the different pairs of metal ions contained in the solution to be tested and the optical dispersion route of direct light. This phenomenon can be used to analyse the metal ions contained in the battery waste liquid. Through the physical properties of different metal ions, to analyse whether there are residual metal ions in the waste liquid, so as to strengthen the control of waste liquid discharge and achieve green environmental protection.

4. Conclusion

In this study, 532nm wavelength laser as the light source, to discuss the scattering image of metal ions (M^{+}) in water according to the direct light dispersion of the laser beam penetrating metal ions in water. In the first stage, 20% nickel liquid (industrial pollution liquid) and distilled water are used for preliminary discussion of this paper. In the future, we will observe the direct radiation of different metal ions, conduct a comprehensive discussion. Experiment with Table1 and Fig 5, comparison of direct light deflection angle of nickel liquid and distilled water: Calculate the x and y axis dimensions based on air, the air medium is x 1y1 as the standard value, Calculate area with x y axis, 20% nickel liquid: x1.100, y1.158, distilled water: x0.700 y0.776, by the experiment to determine that there is a difference in deflection angle, Refer to fig 6. Future research will modify the light intensity to discuss the variation in the color interval of waviness. Future research, Calculated bias angle with different amount of metal ions will be derived, intensity light waves to analyse diffusion chromatography. It is beneficial to change from large equipment to portable metal ion detection equipment.

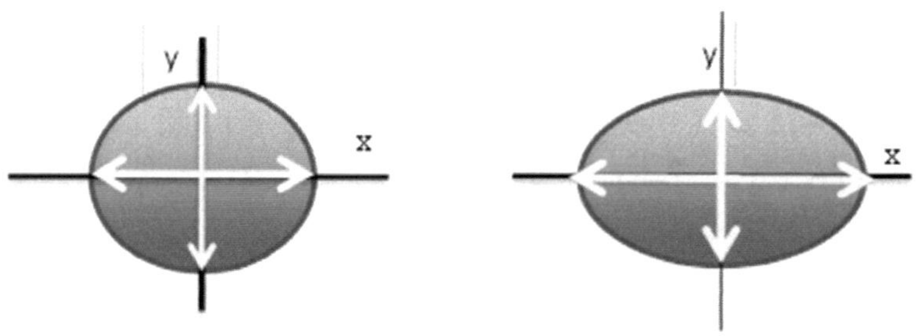

Figure. 6 diagram of the deviation angle

5. Acknowledgements

The study special thanks, supported by The Professorial and Doctoral Scientific Research Foundation of Huizhou University (2020JB034), FERRY READING CAFE and Staff canteen provide space for this discussion. Thanks to Guangdong CHANGSI Sustainability Technology Co., Ltd. and Foshan XINDA New Material Technology Co., Ltd. for Project support. Thanks to Huizhou Zhongkai High-tech Zone ZONGHUA Battery science and technology Institute for Project support.

References

[1]Shinn-Dar Wu, Sui-Qun He, Yu-Lin Yang, Yea-Chyi Lin, Ting-Yu Chang, Ching-Yu Peng and Meng-Ting Hsieh, "Research on Environmentally Friendly Chemical Technology for Green Reusable and Sustainable Water Metal Copper Ions", Earth and Environmental Science, 555 (2020) 012013.

[2]Wen-Na Hu, Shinn-Dar Wu, Zhaoying Cai, Yea-Chyi Lin and Xiong-Dong Sun, "Study on the Recycling Technology of Street Tree Waste Recycling and New Lead Carbon Alloy Processing Technology", Materials Science and Engineering, 774 (2020) 012136.

[3]Štrbac G R, Lukić-Petrović S R, Strbac D D, Benekou V, Chrissanthopoulos A, Yannopoulos S N, "Optical Properties and Structure of As-Sb Chalcohalide Glasses by Raman Scattering and Density Functional Theory Calculations"[J], The journal of physical chemistry, B, 124, 2950–2960: 10.1021.

[4]YUAN Min, WU Jian-chao, YU Jin-song, CAO Hui, XU Fei, "Progress on detection methods of heavy metals in water", Applied Chemical Industry, 2015, 44(04):10.16581.

[5]TITLE 33—NAVIGATION AND NAVIGABLE WATERS, AUTHENTICATED U.S. GOVERNMENT, INFORMATION

[6]BOOK: TECHNISCHER LEITFADEN FOR LASER.

[7]ZhangHua, Xu Jiaw en, Wang Jim ing, Song Qiang, "Research on the Properties of Laser Propagation in Neutral Solution", APPLIED LASER, 2008(04):301-305.

[8]Smith R C, Baker K S., "Optical properties of the clearest natural waters (200-800 nm)"[J] Applied optics, 1981,20(2).

[9]Martin Von Allmen, "Laser-Beam Interactions with Ma-terials Physical Principles and Applications" [M], Berlin:Springer-Verlag, 1987: 6- 8.

[10]Ronald A.J. Litjens, Terence I. Quickenden, Colin G.Freeman, "Visible and near-ultraviolet absorption spec-trum of liquid water"[J], Applied Optics, 1999, 38(7):1216- 1223

[11]Claudia Schneeweiss, Jürgen Eichler,Martin Brose, Daniela Weiskopf, "Physikalische Grundlagen der Lasertechnik", Leitfaden für Fachkundige im Laserschutz,2020 (21-60)10.1007.

[12]Xiaotao Mi, Shanwen Zhang, Xiangdong Qi, Haili Yu, Hongzhu Yu, Yu Lin, Xuefeng Yao, Sibo Jiang, Jingxuan Zhou, "Effect of the changes in refractive index of air on grating diffraction wavefront", Optics Communications, 2020, 457.

The study of Laser 532nm wavelength in $Ni^{+}\cdot H_20$ for reflective effect

Hao-Lin Zhou, Xin-Da Wu *, Dan Lin, Li-Qiong Zhang

School of Chemistry and Materials Engineering, HuiZhou University, China
Corresponding author E-mail:13478511396@139.com and wutomp@hzu.edu.cn

Abstract. The rapid development of science and technology in china, treatment of pollutants is also relatively the biggest problem in developed countries, and heavy metal pollution in many water pollution accounted for a larger proportion. The traditional analytical sampling methods include ultraviolet spectrophotometry (UV), atomic absorption spectrometry (AAS) and atomic fluorescence spectrometry (AFS), to improve timeliness, in this study, laser light method was used to preliminarily analyse the water body, using the difference of reflected facula analysis may contain metal ions in the sample. It was found that after the laser with the wavelength of 532nm was refracted in the nickel solution and distilled water, the facula reflected by the outer glass surface had a significant deviation, from the preliminary experimental results, the reflect facula area of that nickel liquid was 96.01% with respect to water. Therefore, that lase light can be used as a reference for research on miniaturize measurement equipment in the future.
Keywords: Laser light, Metal ion, Underwater laser propagation, Refractive index, Wastewater ion analysis.

1. Introduction

In recent years, the rapid development of science and technology in china, multi scientific research conditions and international technology on the same track, treatment of pollutants is also relatively the biggest problem in developed countries, the water pollution problem is increasingly serious, and heavy metal pollution in many water pollution accounted for a larger proportion. Because heavy metals are toxic and easily enriched and expanded in biological chains, human as the top of the food chain, there is a greater risk of heavy metal poisoning. Minamata disease in the international is a more famous case in modern times, minamata bay fish and shrimp because of industrial wastewater pollution, and these contaminated fish and shrimp are eaten by animals and humans, to neurasthenia syndrome, acute renal failure and other diseases. Traditional metal ion analysis methods include ultraviolet spectrophotometry (UV), atomic absorption spectrometry (AAS), atomic fluorescence spectrometry (AFS) and other methods. to improve timeliness, this study will break through the traditional analysis methods for polarized light metal ions, can timely understand the condition of ions in water. Reference could be made to the research on recovery of low concentration copper ions and green treatment of abandoned street trees by Shinn-Dar Wu and Wen-Na Hu et al. in 2020. Their study addresses the cost-effective recovery of low-concentration copper solutions and the pollution of air and groundwater by traditional methods of treating street trees, provide solutions for industrial copper wastewater treatment and environmental protection [1,2]. In optical research, the measured data indicate that the refractive index and optical attenuation coefficient of the laser in different salt solutions and liquids with different concentrations are different and have significant regularity [3-7].Studies have used refractive index differences at different concentrations to obtain measurements of solution concentration by analysing the refractive index [8-11]. In the above background, it is very necessary to detect and reuse the wastewater in industry, or to detect the heavy metal ions that may be contained in the drinking water in life. However, since these detection instruments are bulky, expensive and difficult to use, a high level of expertise from the user is required. Usually, water samples need to be sent to professional laboratories and testing institutions for sampling and testing, which is difficult to detect water samples in daily life. In this study, the laser light method was used to preliminarily analyse the water body, and the difference of faculas was used to analyse the metal ions that might be contained in the samples, which can greatly reduce the size, operation difficulty, cost and other aspects, has a wide application prospect in the future.

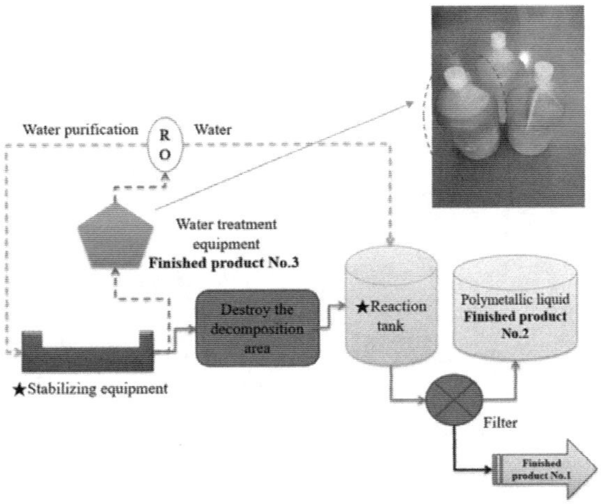

Figure. 1 Recovery butter technology process

In this study, the metal ions in the battery waste liquid before and after the treatment of finished product No.3 in Fig1 water treatment were qualitatively detected, judge whether there are a lot of recyclable metal ions in the waste liquid before and after recovery.

2. Experimental design

The experimental process of this study is for battery recycling waste liquid, residual analysis of water sample after separation of metal ions (M^+), as shown in Fig 2.

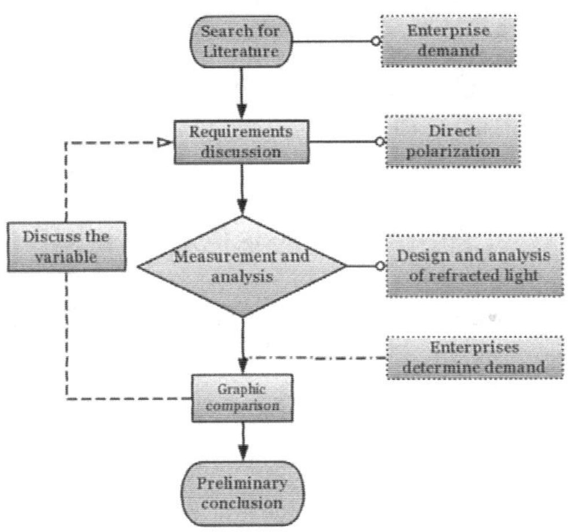

Figure. 2 Experimental process

Such as Fig1 sewage system metal ion refraction characteristic analysis experimental process, according to the battery waste sampling metal ions as the main analysis, and consider the results of physical properties of traditional M^+ polarized light known from previous literature, to design this experiment. In this experiment, the green laser (power 100mW, wavelength 532nm) was used to irradiate the samples, and the actual measurement of the reflected light phenomenon was done, sample group containing distilled water, nickel waste liquid (containing nickel 20%), cuvette (7.5*12.5*45mm) used as a liquid container.

2.1. Direction of light in liquid based on theory
According to theory, the light distribution after refraction in water is shown in Fig3

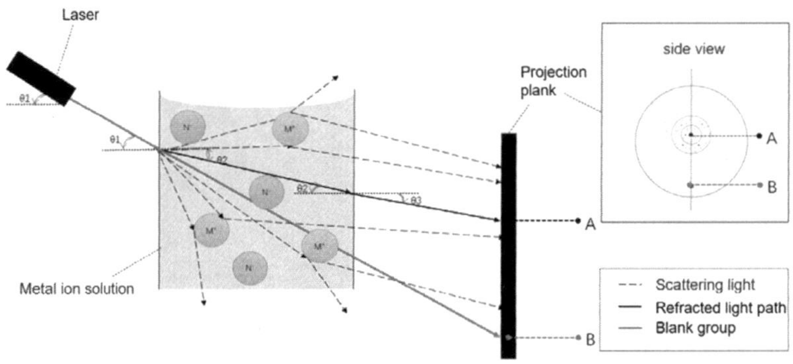

Figure. 3 Theoretical light path map
※Path analysis of light after refraction and scattering through sample

According to the research of LIU Hai-shan et al. the glass wall of a square container is equivalent to a parallel plate, when the light passes through the flat glass, it does not change the propagation direction of the light, nor does it change the angle between the light and the interface of the measured liquid, so the influence of glass on the direction of light propagation can be neglected [8]. Fig 3 shows the difference in refractive index caused by different media. When the light source is used to illuminate the wall directly, such as the red line, it is used as a blank group to discuss the refraction of light in other media. When that light source is inject into 20% nickel liquid at a horizontal angle θ 1, because the refractive index of nickel liquid is different from that of air, the light is deflected. Results the position of the light's landing facula A on the wall and the blank group's landing facula B are different.

3. Experimental analysis and discussion

In this experiment, the physical properties of laser passing through M^+ are discussed by means of actual reflection and reflection measurement, as shown in Fig 4. The light source was placed on the shelf 15cm away from the sample at an included angle of about 40 as shown in Fig4, and the lamp source switch was turned on, the facula of laser light reflected by the sample was recorded in CMOS (honor 9x-imx 582), the differences in the shape and size of the reflected faculas are found and discussed. In view of that effect of the presence of stray light on the reflect facula, the measured facula is not a regular circle, so this experiment in the facula for the largest circle (ellipse), and calculate the area of the circle, the experimental calculation area of the software used for Image J, and finally found that when the light source is refracted by different liquids, the facula area is different, and the specific difference is described by the area ratio, the area ratio is obtained by comparing the facula measured by the sample to the facula reflected without passing through the liquid.

IOP Conf. Series: Materials Science and Engineering 1117 (2021) 012013 doi:10.1088/1757-899X/1117/1/012013

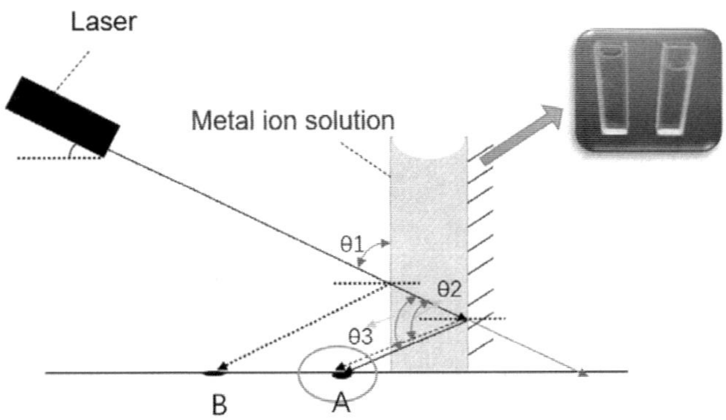

Figure. 4 Schematic diagram of refracted light

The light source shines on 20% nickel solution, and not all the light will pass through the cuvette, which will be reflected back on the glass surfaces on both sides, forming a facula A and a facula B on the desktop. As shown in Fig 4, Analysis of the facula formed after the light source is reflected by the glass surfaces on both sides, facula B is formed after reflection from the outer glass surface. Since it does not pass through the sample solution, it has no practical measurement significance, so do not discuss this. The facula a marked by a red circle in Fig 4 is a facula formed after laser light is reflected by nickel liquid, the position and shape of the measured facula is not the same as the facula formed without passing through the sample due to scattering and loss of light in the liquid.

3.1. Direction of light in liquid based on theory

As shown in Fig 5, an approximate facula area is drawn by measuring the reflected facula, and an area is calculated. The theory is in Fig 3, $\theta 1>\theta 2$, $\theta 2>\theta 3$, resulting in the measured facula A located above the blank group facula B. At the same time, the metal ions in the solution also scatter the incident light source to other different directions. This study refers to the research of Zhang Hua et al. with the increase of solution concentration; the scattering effect of solution on laser is more obvious [3].

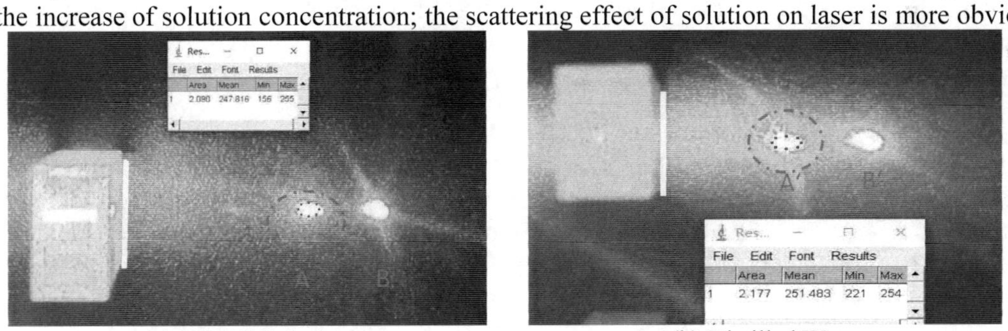

(a) 20%Ni Solution (b) Distilled Water

Figure. 5 Facula screenshot

※Actual reflect light ban of light passing through two set of samples

Table 1. Facula data summary

Sample Solution	20%Ni Solution	Distilled Water	BlankGroup
Area (mm^2)	2.090	2.177	2.281
Area ratio	0.9163	0.9544	1.000

For example, in Fig5 (a) and (b), by comparing the shapes and sizes of the reflect facula when the laser light passes through the nickel solution aednd the distilled water, there are significant differences. In this experiment, an approximate circle is made in the facula to ignore the effect of stray light on the measurement of the actual area of the facula. Image J used in software. In Fig 5 (a) and (b), the length of the cuvette is 12.5mm as a reference, as shown by the yellow line. The largest approximate circle is made within the facula indicated by the red dashed line and the relative area of the circle is calculated. The measurement results are shown in Table 1. In the two groups of experiments, the facula areas after reflection were not equal in size, and the data of the two groups were different.

In this experiment, the area ratio was introduced to compare the two differences, and the test was conducted in the same way. The differences in area ratios of reflected nickel solution and distilled water could be visually compared by comparing their light with those of the blank control group.

4. Conclusion

Based on the known theory and the experimental results, it was concluded that the laser would be refracted and scattered in distilled water and nickel solution, making the measured facula size and shape different from that of the blank group. In this experiment, the specific difference was analysed by measuring the facula are. The area ratio introduced was used to compare the difference between the sample group and the blank group. It was found that the facula area ratio of the nickel solution group was smaller than that of the distilled water group, and the area ratio of the two groups of samples was less than 1. According to the preliminary experimental results, the facula area after reflection with 20% nickel solution is 96.01% of that of distilled water.

More kinds and concentrations of metal ions will be measured and studied in the future, looking for different metal ion liquids, for the characteristics of ions, the difference is calculated by the area of the light reflected by the liquid, through the study of this relationship can be for the measurement of metal ions qualitative and preliminary quantitative, provide the feasibility.

5. Acknowledgements

The study special thanks, supported by The Professorial and Doctoral Scientific Research Foundation of Huizhou University (2020JB034), FERRY READING CAFE and Staff canteen provide space for this discussion. Thanks to Huizhou Zhongkai High-tech Zone ZONGHUA Battery science and technology Institute for Project support.

References

[1]Shinn-Dar Wu, Sui-Qun He, Yu-Lin Yang, Yea-Chyi Lin, Ting-Yu Chang, Ching-Yu Peng and Meng-Ting Hsieh, "Research on Environmentally Friendly Chemical Technology for Green Reusable and Sustainable Water Metal Copper Ions", Earth and Environmental Science, 555 (2020) 012013.

[2]Wen-Na Hu, Shinn-Dar Wu, Zhaoying Cai, Yea-Chyi Lin and Xiong-Dong Sun, "Study on the Recycling Technology of Street Tree Waste Recycling and New Lead Carbon Alloy Processing Technology", Materials Science and Engineering, 774 (2020) 012136.

[3]Zhang Hua, Xu Jia wen, Wang Ji ming, Song Qiang, "Research on the Properties of Laser Propagation in Neutral Solution", APPLIED LASER, 2008(04):301-305.

[4]ZHAI Sheng-cheng, WANG Ya-ge, LU Xiao-ning, "Effect of ternary mixture solution concentration on the refractive index", Journal of nanjing forestry university (natural science edition), 2008 (02) :148-150.

[5]Xinyi Zhao, Yufeng Peng etal, Congcong Zhai, Xueyun Han, and Yi Zhang, "Influence of inorganic salts on the refraction index of water", Applied Mechanics and Materials Vols. 716-717 (2015) 118-121.

[6]Quan X, Fry E S, "Empirical equation for the index of refraction of seawater", Applied optics, 1995, 34(18):3477-80.

[7]Soňa Hříbalová, Willi Pabst, "Light scattering and extinction in polydisperse systems", Journal of the European Ceramic Society, Volume 40, Issue 3. 2020. PP 867-880.

[8]LIU Hai-shan, LI Ming-shi, CHEN Jia-ming, ZHENG Qin-jia, HUANG Jiang, "Measuring the refractive index and the dispersion rate of liquid with vertical prism angle", COLLEGE PHYSICS, 2017(12):62-65.

[9]Jon P. Longtin, Ching-Hua Fan, "PRECISION LASER-BASED CONCENTRATION AND REFRACTIVE INDEX MEASUREMENT OF LIQUIDS", Microscale Thermophysical Engineering, 1998, Volume 2, Issue 4. PP 261-272.

[10]ZHANG Ji-xiong, ZHANG Mei, JIANG Ming-gang, "Concentration Measurment Based on Energy Variety of Reflected Light Near Critical Angle", Journal of Optoelectronics Laser, 2008(04):301-305.

[11]Minghui Chen, Wenyu Jia, Jintao He Martial Geiser, and Gang Zheng, "A miniaturized system for measurement of the refractive index of sub-microliter liquid", CHINESE OPTICS LETTERS, 17(4), 041201(2019).

IOP Conf. Series: Materials Science and Engineering 1117 (2021) 012014 doi:10.1088/1757-899X/1117/1/012014

Comparison of vibrations of high-speed bearings driven by flat layer belts

Z Murcinkova
Faculty of Manufacturing Technology with seat in Presov, Technical university of Kosice, Bayerova 1, 080 01 Presov, Slovak Republic

E-mail: zuzana.murcinkova@tuke.sk

Abstract. The paper presents the results of comparison measuring of absolute vibrations for textile machine driven by flat layered belts with high velocity 50-60 m/s. The dynamic properties of flat belt are important as the flat belt surface is in direct contact with shaft of special textile bearing (115,000-135,000 rpm). Interaction of textile bearing vibration and driven belt may produce significant amplitude and frequency vibration modulation influencing the torsional vibrations and kinematic slip of rolling elements and additional loading of cage. Thus, the bearing life is significantly shortened what is undesirable effect. However, the dynamic properties of driving flat belt are able to influence the vibration of high-speed bearing as it was measured. The two tested flat belts involved one polyester and aramid fabric, respectively, keeping same temperature, belt tension, rotational speed and driven forty bearings. The belt involving the aramid fabric brought the promising results.

1. Introduction

The demand of higher productivity invokes need of high-speed performance of machines and thus invokes demands on the research to ensure reliable and effective operation and production process. The researchers are focused on specific elastomeric materials and bring the different internal belt structures and material compositions to improve power and friction transmission, wear, heat and cold resistance and durability.

Modern V or multi-ribbed belts used in automotive applications as well as in conveyor belts are built up of several functional layers [1]. Even flat belts development is in field of specific composition and structure of three basic section of belts: tension and compression sections and cords. The modern belts are belts with sophisticated shape and inner structure that creates the composite hybrid structure involving mainly short fibres and fabrics in inner middle layer/s or on the outside layer. The cords can be recognized as long fibres that also can be with various inner structure.

About 30 and more years ago, the research was focused mainly on development elastomeric material of high and low temperature resistant properties [2,3]. In present and about fifteen years ago, research is focusing on internal structure of V and flat belts along with material composition. The various hybrid composite belt structures were developed containing reinforcing chopped carbon fibres of various dimensions preferably in cushion section, load-carrying and/or tension sections [4], cords across belt width [5], short fibres such as cotton, polyamide, vinylon, rayon, aramid, etc., with the lengths of the fibres aligned generally widthwise of the belt [6], three rubber layers different by shape and composition with randomly distributed short fibres in tension section and with short fibres implemented on all of the rib surfaces as a solid lubricant [7]. There are other belt embodiments with elastomeric compositions of power transmissions belt containing reinforcement of short fibres of different materials even natural

IOP Conf. Series: Materials Science and Engineering 1117 (2021) 012014 doi:10.1088/1757-899X/1117/1/012014

fibres as kenaf fibre [8] and sophisticated orientation and location of short fibres [9,10], reinforcing fabrics [11,12] and specific porous structure rubber at rib surfaces [13].

2. Measurement description

We performed measurements of the absolute vibrations of two flat multilayer driving belts. We observed the influence of the tested flat belts on the dynamic properties of a spinning unit including a bearing as part of a textile machine. The greatest emphasis in the performed measurements and analysis was placed on the evaluation of the influence of selected two flat driving belts on the operation of driven high-speed bearings. The following were measured:

- absolute vibrations of the spinning unit,
- frequency modulation and variation of bearing speed.

During the measurement, the temperature was constant, i.e. 21°C, and the belt tension was the same.

Measurements of absolute vibrations of the spinning unit were performed at the test spinning section of textile machine with 40 spinning units of textile machine. The frequency modulation measurements were performed on a test bench. All measurements were in the range of 70,000 -130,000 rpm. Test bench and spinning section of textile machine differ in number of spinning units (Figure 1(a)). Test bench involves 4 spinning units (Figure 1(b)) consisting of bearing housing and bearing itself.

We focused on high-speed driving flat belts designed for the highest circumferential speeds, i.e. in our case 50-60m/s. The spinning units are designed for speeds up to 115,000 rpm. The goal of the research is to increase the speed while maintaining the quality of the spinning. The increase in speed itself is not a problem, but in the different interactions of the individual components, it can have undesirable effects, i.e. large amplitude and frequency modulation of dynamic signal that lead to the shortened bearing life. In paper, we study the way of decrease the vibration through the material and structure of flat belt. The direct interaction (contact) of a driving flat belt and driven bearing is at the shaft end (Figure 2).

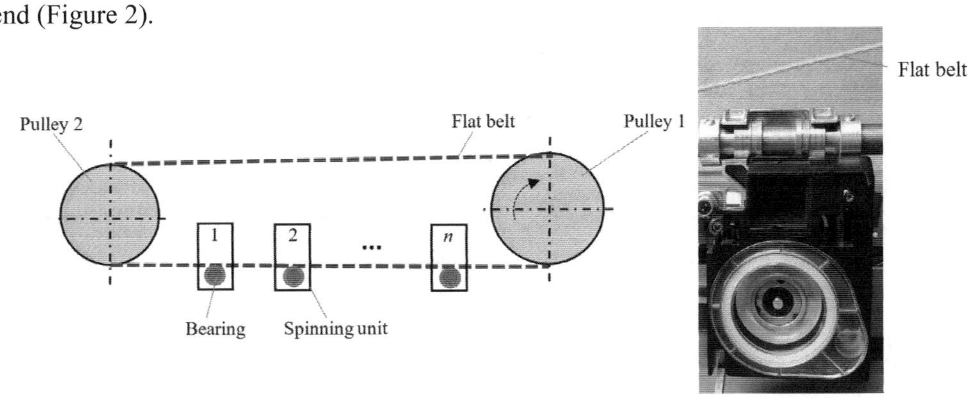

(a) (b)

Figure. 1 Schema of test bench (n=4) and spinning section of textile machine (n=40) (a) and spinning unit (b); n is number of spinning units.

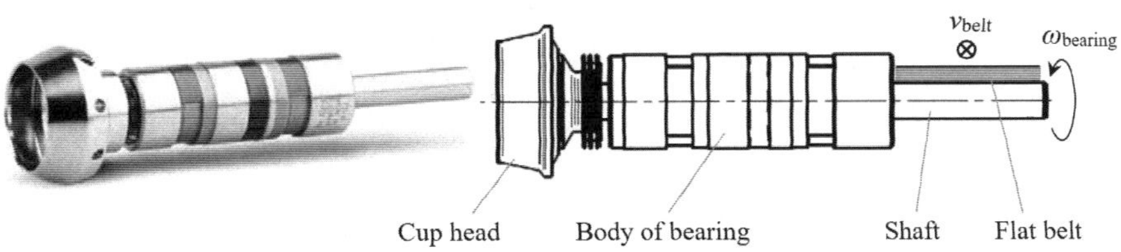

(a) (b)

Figure. 2 Bearing (a) and its parts with belt (b) v_{belt} is velocity of belt, ω_{belt} is angular frequency of shaft.

Absolute vibrations were measured by measuring chain involving:

- NI PXI 4426 Sound and Vibration Module (sampling frequency up to 200 kHz) and software LabView Sound and Vibration Toolkit,
- SKF Microlog analyser GX series, amplitude analyst and measuring method Enveloping Acceleration in frequency range 0.5 – 10 kHz,
- Accelerometer PCB Piezotronics model 352A60 in frequency range up to 65 kHz and sensitivity 10mV/g.

2.1. Description of belts

We tested layer flat belts with one polyester (flat belt 1) and aramid (flat belt 2) fabric layer (Figure 3, Table 1) as representatives of commonly used flat belt and a new constructional and material design of flat belt. Table 1 provides some data of both belts.

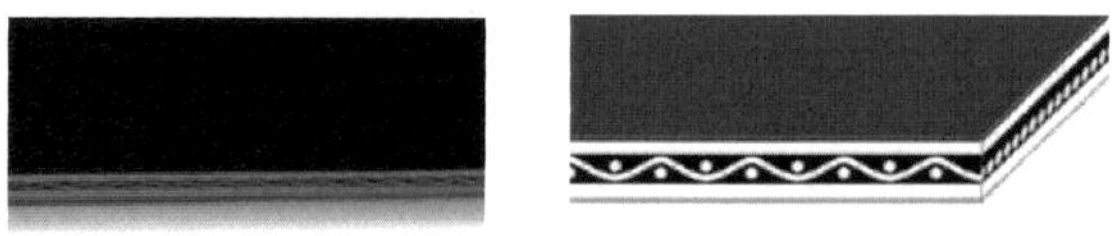

(a) (b)

Figure. 3 Flat belt – photo (a) and schema (b).

Table 1. Data of flat belt 1 and 2.

	Flat belt 1	Flat belt 2
Pulley side material	NBR[a]	NBR
Pulley side surface	Rough	Rough
Traction layer material	*Polyester*	*Aramid fabric*
Number of fabrics	1	1
Opposite side material	NBR	NBR
Opposite side surface	Fine	Rough

[a]Acrylonitrile-Butadiene-Rubber

3. Results

3.1. Amplitude and frequency modulation and speed variability of textile bearings

Each spinning unit involves the bearing (Figures 1 and 2). The character of high-frequency absolute vibrations of bearing is various. The example of undesirable and very significant amplitude and frequency modulation is in Figure 4. Such significant modulation considerably decreasing the bearing life.

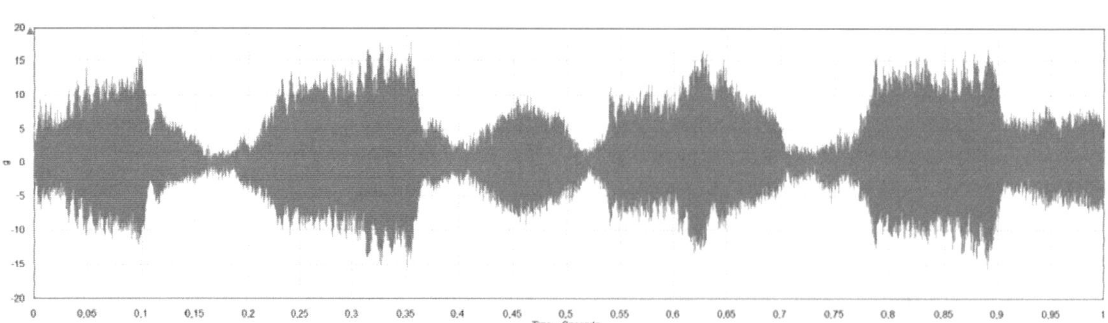

Figure. 4 Significant amplitude and frequency modulation.

Sources of amplitude and frequency modulation in presented case of driving flat belt are following:

- variable driving belt pressure on the shaft of bearing,
- vibrations of belt due to insufficient belt tension,
- variable kinematic slip of rolling elements,
- variable contact angle in angular contact ball bearing.

As the mentioned factors are variable, the frictional moment is not constant. The variable contact belt-shaft causes the changing of bearing revolutions and the torsional vibration of bearing appears. Such vibrations cause the additional load of bearing cage and decrease of its life.

3.2. Comparison of absolute vibrations

Vibrations of bearings are in range up to 10 kHz. The absolute vibrations are evaluated by method Envelope Acceleration (Figure 5). Each column is for one spinning unit of test spinning section of textile machine. Values 5gE and 7gE are limit values for Warning and Danger. Bearing operation up to 5gE is allowed Safe operation. The vibrations were measured for flat belt 1, then disassembled and flat belt 2 was assembled. The environment conditions of both measurements are the same. The bearings are the same for both measurements. Spinning units 11, 16, 31, 33-35 and 38 are in Danger regime.

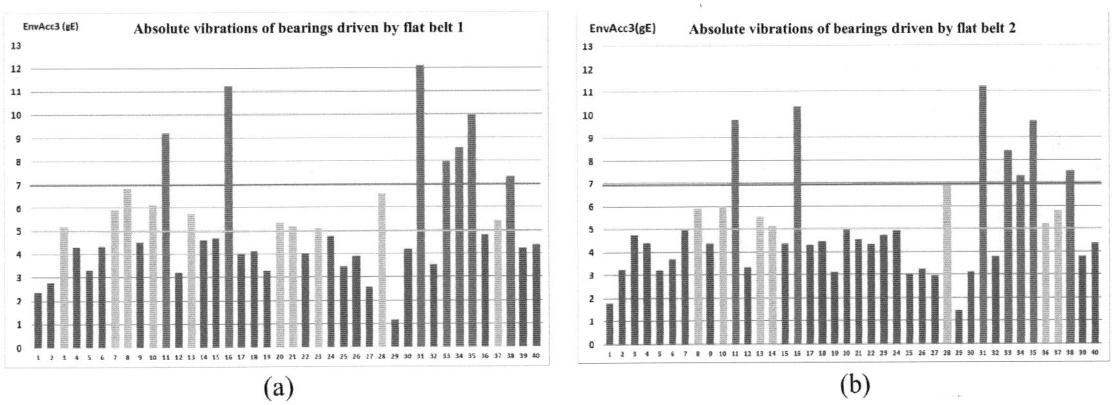

Figure. 5 Absolute vibrations of bearings driven by flat belt 1 (a) and 2 (b).

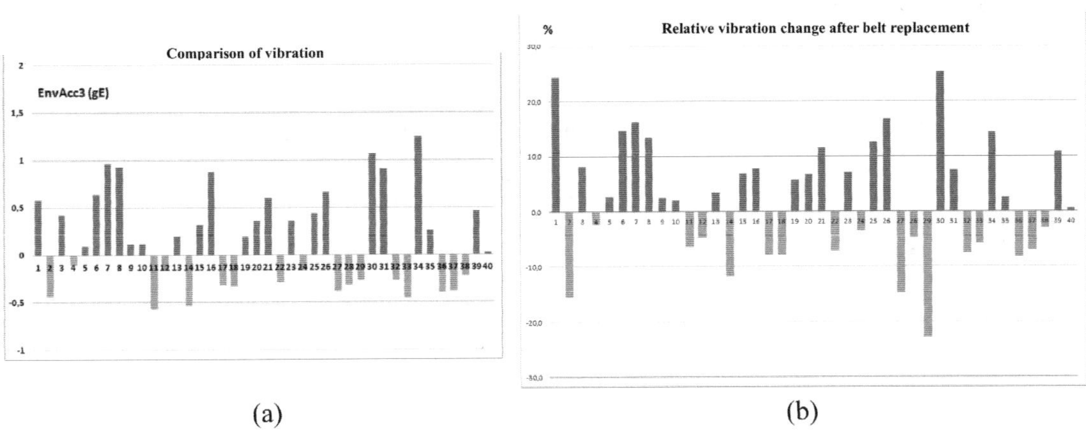

Figure. 6 Comparison of vibration values (a) and expressed in percentage (b).

Comparison of absolute vibrations is made in Figure 6. The comparison of the changes in the bearing vibrations is realized by quantifying the difference between the vibrations values of belt 1 (B1) and belt 2 (B2): (B1 – B2), Figure 6(a). If the vibrations were larger for belt 1, then they are plotted as values with + (above horizontal axis), i.e. it means the flat belt 2 is better as it generated lower vibrations.

If the vibrations were larger for belt 2, then the values are with minus (under horizontal axis), i.e. belt 1 is better. Flat belt 1 with polyester fabric in middle layer generated lower vibrations for 17 positions of spinning units in the sum 137%. Flat belt 2 with aramid fabric had lower vibrations for 23 positions in the sum 220%. Figure 6(b) expressed the same as Figure 6 (a) in percentage evaluated as:

$$\text{Relative change} = \frac{\text{Vibrations of B1} - \text{Vibrations of B2}}{\text{Vibrations of B2}} .100\% \qquad (1)$$

4. Conclusions

Paper presents one way of decreasing the vibrations of bearing driven by flat belt through the use of dynamic properties of flat belt materials. The both flat belts are of the same structure, the main difference is in material of fabric in the middle layer of flat belt. The all other factors as temperature, belt tension and 40 bearings are the same. At same time, the goal to decrease vibration causes the more constant pressure of driving belt on the shaft of bearing and thus the variation of rotational speed (torsional vibration) are lower what enlarging the bearing life. The large torsional vibrations at rotational speed 135,000 rpm are very harmful for bearing and any lowering of that is beneficial. In presented measurement the belt with aramid fabric provides a 60% better behavior regarding generated lower absolute vibrations and thus lower torsional vibrations and lower variation of rotational speed.

Acknowledgment

This work was supported by the Slovak Research and Development Agency under Contract No. APVV-18-0316. The author would like to thank the Agency of Ministry of Education, Science, Research, and Sport of the Slovak Republic for supporting this research through the Grant VEGA 1/0910/17.

References

[1] A. Tonoli, N. Amati, E. Zenerino, “Dynamic modeling of belt drive systems: effects of the shear deformations,” Journal of Vibration and Acoustic, 2006, pp. 555-567.
[2] P.M. Patterson, Gates Rubber Co, “High temperature polyurethane belt,” U.S. Patent 5,112,282, 1992.
[3] S.L. Ciemniecki, R.A. Rorrer, H.D. Visser, H.D., Gates Corp, “High temperature flexible thermoplastic composites for endless belt driving surfaces,” U.S. Patent 6,296,588, 2001.
[4] M.J.W. Gregg, T.G. Burrowes, Goodyear Tire and Rubber Co, “Power transmission belt containing chopped carbon fibers,” U.S. Patent 6,918,849, 2005.
[5] P.S. Knutson, Gates Corp, "Power transmission belt and method," U.S. Patent No. 6,945,891, 2005.
[6] Y. Hineno, T. Nishiyama, T. Takada, H. Takechi, K. Takeda, Mitsuboshi Belting Ltd, “Rubber composition and power transmission belt incorporating the rubber composition,” U.S. Patent 7,485,060, 2009.
[7] M. Mitsutomi, Y. Hineno, S. Shimoo, Mitsuboshi Belting Ltd, “Power transmission belt,” U.S. Patent 9,273,756, 2016.
[8] Y. Feng, Gates Corp, “Kenaf reinforced rubber for power transmission belts,” U.S. Patent 9,255,201, 2016.
[9] S. Uchigashima, M. Yamada, Y. Harada, A. Terasaki, K. Shimamura, Honda Motor Co Ltd and Gates Unitta Asia Co, “Toothed belt,” U.S. Patent 9,353,827, 2016.
[10] Y. Takami, F. Ise, H. Kawahara, H. Enokizono, T. Sawairi, Asahi Kasei Corp, “Short rubber reinforcement fiber, rubber composition containing said short fiber, and power transmission belt,” U.S. Patent 10,215,257, 2019.
[11] T. Yamamoto, T. Nozaki, Nitta Corp, “Flat belt,” U.S. Patent Application 11/393,870, 2006.
[12] H. Okamura, Y. Konishi, M. Ono, N. Nakai, N. Kai, Nitta Corp, “Endless flat belt and method for manufacturing the same,” U.S. Patent 10,570,990, 2020.
[13] T. Okubo, K. Hara, Bando Chemical Industries Ltd, “V-ribbed belt,” U.S. Patent 10,274,045. 2019.

IOP Conf. Series: Materials Science and Engineering 1117 (2021) 012015 doi:10.1088/1757-899X/1117/1/012015

Electrically tunable reflector based on ferroelectric material for millimetre wavelength range

A G Altynnikov[1], R A Platonov[1,*], A M Sosunov[1], A D Ivanov[1] and T K Legkova[1]

[1]Saint Petersburg Electrotechnical University "LETI", Saint Petersburg 197022, Russia

*Corresponding author e-mail: raplatonov@etu.ru

Abstract. The design of a millimeter-wave reflector based on the ferroelectric ceramic was elaborated and considered. Among the advantages of the proposed design are simplicity and cost-effectiveness compared to devices based on active repeaters and reflectarrays. The use of ferroelectric ceramic in the reflector design allows operating in a wide frequency range up to 100 GHz. Parameters of ferroelectric ceramic were experimentally measured and used in the electromagnetic simulation of the proposed reflector. Simulation of the reflector's radiation pattern demonstrated the effective beam scan possibility in a range of 35 deg.

1. Introduction

Due to the explosive demand for high-speed data transmission in wireless communication nowadays, millimeter-wave band carriers have become a wireless system trend. The active development can approve this statement of such innovative infrastructures such as the 5G network [1], WiGig network [2], “smart home,” and it’s a probable extension to the “smart city” [3]. The advantage of millimeter wavelength range use is a significant increase in the transferred information. At the same time, the main drawback is power consumption due to significant attenuation of electromagnetic waves in free space [4]. One of the major hindrances for mm-wave networks' effective coverage is line-of-sight (LoS) blockage. The main solutions to avoid this issue are active repeaters [5] and reflectors [6].

The use of a passive metallic reflector is the simplest way to improving mm-wave propagation in non-line-of-sight (NLoS) areas [5]. This solution can be attractive due to the large life span and low cost. Passive metallic reflectors have been used in past for microwave links for long-distance communications (such as satellite communications) and point-to-point microwave links. However the main drawback of the passive reflectors is the dependence of output radiation on transmitter antenna directivity and its angle of incidence on the reflector, in other words, it is not allowed to rearrange antenna radiation pattern without mechanical steering. In the case of indoor communications, the multifocal reflectors can be used to avoid this issue. But in outdoor communication systems, the use of tunable reflectors with the possibility of beam scan is more attractive. As a rule, such tunable reflectors based on actively phased elements that are operating in reflection mode [7]. In the construction of such reflectors, tunable elements like pin-diodes, varactors, etc. are commonly used [8,9]. This technical design is very useful in the microwave range, but difficult to implement at operating frequencies of K-band and higher [10]. The overcoming of the mentioned limitations is possible by using distributed nonlinear elements like ferroelectric (FE) ceramic plates as the base for tunable reflector design. It is well known that FE materials demonstrate a strong dependence of permittivity on an electric field value, that in combination with the absence of permittivity dispersion in the broad frequency range (up to 100 GHz), allowing to consider FE as a promising material for the use in the electrically controllable reflector.

This work describes the proposed design of an electrically controllable reflector on the base of ferroelectric ceramic plates, its principle of operation, and provides the results of full-wave electromagnetic simulation.

2. Principle of operation

In contrast to mechanical system, where orientation of reflector is variates to source antenna, the ferroelectric (FE) reflector is not moving. The direction of beam reflection electrically controled by the variation of FE permittivity. It is well-known that FE demonstrates dependency of permittivity (ε_{fe}) on the applied electric field (E). Measured dependence shown below, in corresponding section. Since permittivity and thickness of FE plate affects the phase of reflected wave for realization of the deflection of the reflected beam one should provide the spatial distribution of FE permittivity along reflector aperture. The aforementioned spatial distribution is not arbitrary and generally determined by required flatness of the reflected wave front and character of $\varepsilon_{fe}(E)$ dependence. The relation between the phase of outgoing wave and permittivity can be estimated from assumption of electrical equivalent circuit of FE plate.

Schematically the deflection of the reflected wave is presented in figure 1. The wave radiated by the source antenna incidents on FE reflector and reflects by a metalized grounded layer of FE plate. One can see that the permittivity of FE is varied along the reflector's aperture; thus, the wave obtains the different phase shifts at different parts of the FE plate. It should be noted that the variation of ε_{fe} on the electric field is characterized by the tunability factor ($K=\varepsilon_{fe}(0)/\varepsilon_{fe}(E\neq 0)$). The K-factor is more convenient for generality because of the possible specific of $\varepsilon_{fe}(E)$ curves of different ferroelectric materials. One can obtain the phase shift of reflected wave caused by electrical control of FE permittivity:

$$\Delta Phase(K) = \frac{4\pi\sqrt{\varepsilon_{fe}(0)}d_{fe}}{\lambda_0}\left(1 - \frac{1}{\sqrt{K}}\right) \quad (1)$$

This relation allows defining distribution of K-factor along reflector aperture to provide the required wavefront of outgoing wave based on main parameters of FE plate such as thickness and ε_{fe} in the absence of electric field. With specific dependence of permittivity on the electric field, obtained K-factor distribution can be transformed to the distribution of control voltages (U_i in figure 1) applied on each FE plate section.

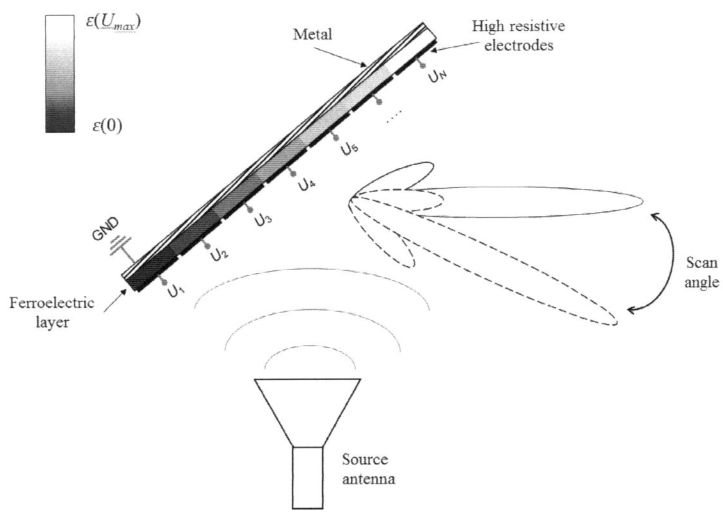

Figure. 1 Schematic of electrically tunable ferroelectric reflector.

To realize the control over FE permittivity, one should provide control electrodes in the lens design. The control electrodes should correspond to several requirements, such as low insertion loss, simple implementation, and minimizing the heating of the FE plate caused by the electric current (Joule heating). It is much more convenient to use radiotransparent electrodes based on a thin film of the high resistive materials instead of metal electrodes were used in [11,12]. The different materials used for the fabrication of thin-film planar resistors also can be a perspective for this purpose in the microwave and millimeter-wave ranges.

The discrete topology of the control electrode consists of the discrete sections of a radiotransparent material was used in the FE reflector design proposed in this article. The electric potential applies to each section of the electrode. In such a configuration, each electrode section is an open-circuit for the control voltage and, therefore, the electric current is not flowing through it. There is the restriction on the width of the electrode section for this type of electrode topology, it should be less than half of the operating wavelength in free space. This restriction was considered in [13] in more detail.

3. Ferroelectric material

As mentioned above, the design and analysis of an electrically tunable FE reflector are better to consider the measured dependence of $\varepsilon_{fe}(E)$ due to this dependence's nonlinear character and possible variation for different FE materials. For the experimental investigation, the ceramic samples of a solid solution of $Ba_xSr_{1-x}TiO_3$ (BSTO) were used. The BSTO is a well-known ferroelectric material. Its properties in the paraelectric state, such as fast time response, low dissipation loss, and the absence of permittivity dispersion, allow considering BSTO a promising material for applications in the millimeter-wave range [14,15]. The characterization of FE nonlinearity is provided by capacitance-voltage CV dependency measurements of FE capacitors under dc control voltages resulting in a tunability factor $K=C(0)/C(U_{dc})$. The FE ceramic capacitors were fabricated with 5×5 mm^2 contact copper metallization deposited on the opposite sides of the ceramic plate. The obtained experimental results of permittivity and tunability dependencies on the electric field are presented in figures 2a and 2b correspondingly. To provide high precision measurements of ceramic tanδ the electrodeless method based on an open resonator (OR) was used. In accordance with the described measure technique, the loss tangent value of the ferroelectric ceramic plate was about 0.015–0.02 at 60 GHz.

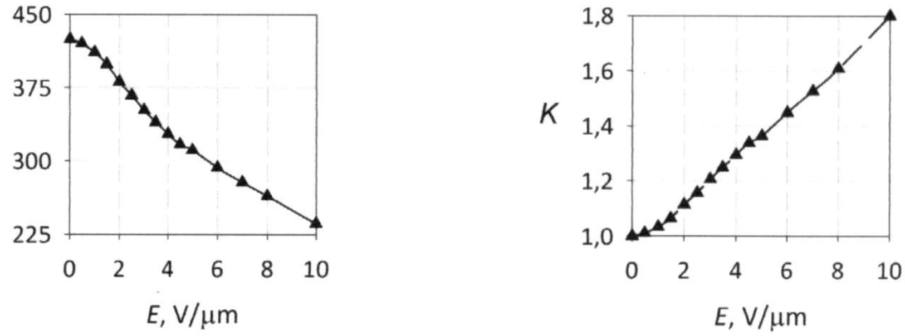

a) b)

Figure. 2 Measured ferroelectric ceramic sample permittivity - a) and tunability – b) dependencies on the electric field

4. Simulation results

The Radio Frequency module (Electromagnetic Waves, Frequency Domain interface) of the COMSOL Multiphysics software was used to perform the 2D electromagnetic simulation of the FE reflector device to prove the proposed beam deflection concept. The region of numerical simulation is presented as an air-filled circle of a radius of 50 wavelengths at 60 GHz with radiation boundaries containing a waveguide flaring to a horn (source antenna) and a ferroelectric plate with impedance matching layer. One side of FE plate was assigned as perfect electric conductor (PEC) boundary to provide reflection of the incoming wave. The horn dimensions were calculated based on the procedure described in [16]. The reflector model consists of ferroelectric plate with quarter wavelength matching layer of linear dielectric. The overall size of the FE lens aperture is $11\lambda_0$ (at 60 GHz) with a total thickness of the FE plates 2 mm. The FE plates were divided into 22 rectangular regions to provide the discrete spatial distribution of phase shifts. The permittivity of each region is determined as ε_{fe}/K_i, where ε_{fe} – ferroelectric permittivity without control electric field ($\varepsilon_{fe} = 440$) and K_i – tunability factor corresponded to the i-th FE region. According to measured tunability of ceramic samples and relation (1) the values of K for each FE plate section was defined to provide plane wave deflection in specified angles.

The simulation results of the wave E-field distribution corresponded beam reflection and deflection are presented in figure 3 correspondingly. The presented E-field distributions clearly demonstrate the flatness of the outgoing wave. The effectiveness of the proposed reflector can be concluded from the simulation results of radiation patterns presented in figure 4. It demonstrates that the achievable scan angle is in range of 35 deg. with side lobes below -14 dB.

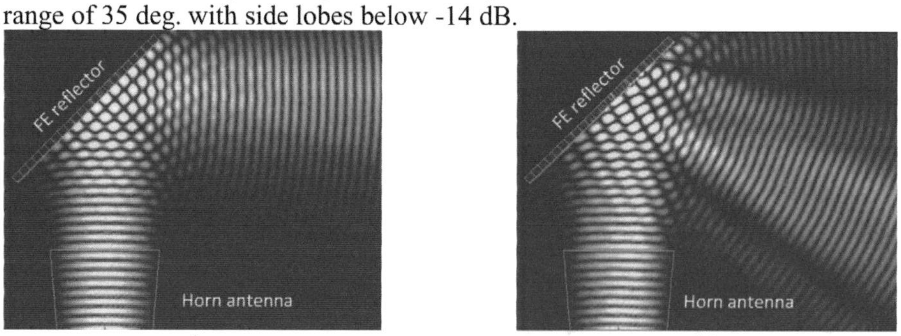

a) b)

Figure. 3 Distribution of reflected wave electric field it the cases of zero control voltage – a) and non-zero control voltage distribution in ferroelectric plate.

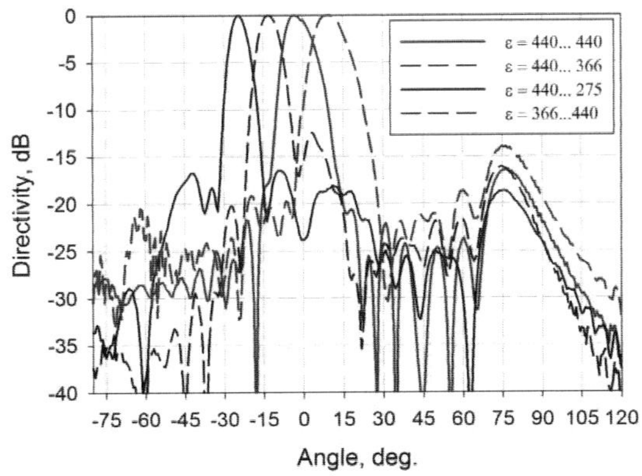

Figure. 4 Simulation results of beam deflection by the FE reflector.

5. Conclusion

The design of an electrically tunable reflector based on ferroelectric ceramic was considered and analyzed. Parameters of reflector design were estimated for 60 GHz operating frequency. Full-wave analysis based on experimental measurement results of ferroelectric ceramic was performed. The obtained results demonstrate the quality of wavefront flatness of the deflected wave. According to simulation results, the beam scan range is about 35 deg. The proposed reflector design can be useful for millimeter-wave wireless communication systems to avoid LoS blockage.

Acknowledgments
The authors gratefully acknowledge the financial support provided by the Russian Science Foundation under grant 18-79-10156.

References
[1] Boccardi F, Heath R W, Lozano A, Marzetta T L, Popovski P 2014 *IEEE Commun. Mag.* **52** 74
[2] Shoaib S, Shoaib N, Khattak R, Shoaib I, Rehman M, Yang X 2019 *Electronics (Switz.)* **8** 1548
[3] Yau K A, Qadir J, Wu C, Imran M A, Ling M H 2018 *IEEE Access* **6** 35072
[4] Marcus M, Pattan B 2005 *IEEE Microw. Mag.* **6** 54
[5] Khawaja W. A. G. et al. 2019 *IEEE Radio and Wireless Symposium (RWS)* 1
[6] Wang L, Saavedra C E 2017 *IEEE 28th Annual International Symposium on Personal, Indoor, and Mobile Radio Communications (PIMRC)* p 1
[7] Jiang Z H, Zhu X W, Hong W 2020 *XXXIIIrd General Assembly and Scientific Symposium of the International Union of Radio Science* p 1
[8] Padilla P, et al. 2016 *Journal of Electromagnetic waves and applications* **30** 12 p 1545
[9] Nie J, et al. 2016 *Progress in Electromagnetic Research Symposium (PIERS)* p 709.
[10] Frank M, Lurz F, Weigel R and Koelpin A. 2019 *IEEE Antennas and Wireless Propagation Letters* **18** 4 p 796
[11] Spencer D B, Fitch J L Electronic Beam Steering Methods and Apparatus. U.S. Patent 4480254, 1984.
[12] Rao J B L . Voltage Controlled Ferroelectric Lens Phased Array. U.S. Patent 5729239, 1998.
[13] Platonov R, Altynnikov A, Kozyrev A 2020 *Coatings* **10** 2 p 180
[14] Nadaud K, Borderon C, Gillard R, Fourn E, Renoud R, Gundel H W 2015 *Thin Solid Films* **591** p 90
[15] Altynnikov A G et al. 2019 *Technical Physics Letters* **45** 6 p 540

[16] Stutzman W L, Thiele G A *Antenna Theory and Design* John Wiley & Sons: Hoboken, NJ, USA, 1981.

IOP Conf. Series: Materials Science and Engineering 1117 (2021) 012016 doi:10.1088/1757-899X/1117/1/012016

Sustainability and life cycle cost analysis of biomass pyrolysis

Yoong Xin Pang [a], Nusrat Sharmin [a], Edward Lester [b], Tao Wu [c,d], Cheng Heng Pang [a,d,e*]

[a] Department of Chemical and Environmental Engineering, The University of Nottingham Ningbo China, Ningbo 315100, PR China
[b] Department of Chemical and Environmental Engineering, University of Nottingham, Nottingham NG7 2RD, U.K.
[c] New Materials Institute, The University of Nottingham Ningbo China, Ningbo 315100, PR China
[d] Key Laboratory for Carbonaceous Wastes Processing and Process Intensification Research of Zhejiang Province, University of Nottingham Ningbo China, Ningbo, 315100, P.R. China
[e] Municipal Key Laboratory of Clean Energy Conversion Technologies, University of Nottingham Ningbo China, Ningbo 315100, China

E-mail address: chengheng.pang@nottingham.edu.cn

Abstract. Pyrolysis is one of the commonly used methods to derive renewable bio-fuels from biomass. A life cycle analysis was performed to evaluate the sustainability and feasibility of such process by utilising simulation results from SuperPro Designer. The process boundary includes biomass pretreatment, main reaction and product collection regimes. Sensitivity analysis was conducted to investigate the influence of cellulose, hemicellulose and lignin on the aspects of economic and environmental impact throughout the life cycle of biomass pyrolysis.

1. Introduction

The progressive depletion of fossil fuels and increasing concern on global warming have resulted in increased efforts in seeking alternative energy sources. In particular, bio-fuels derived from biomass are one of the common alternative fuels used to substitute fossil fuels. Amongst various conversion technologies, pyrolysis is considered as one of the most established method for fuel processing and has been commercialised in the past few decades [1]. Nonetheless, further consideration on life cycle sustainability and energy efficiency of the production system is required for the green production of alternative fuels.

Life cycle analysis (LCA) consist of life cycle environmental impact assessment (LCIA) and cost analysis (LCCA) of a product system over its life cycle chain. LCA collects process outputs i.e. yield, utility demands, emissions; and cost information to conduct environmental and economic assessment in each process step. This study encompasses three life-cycle stages in pyrolysis: (1) biomass pretreatment, (2) feedstock pyrolysis and (3) product collection. The pyrolysis process is first simulated in SuperPro Designer (SPD), a process modelling and optimisation tool, and the obtained results are analysed in LCA. Outcomes from LCA provide comprehensive understanding of production stages that influence costings and potential environmental impacts.

1.1 Methodology
1.1.1. Biomass pyrolysis

A simulation flowsheet as shown in Figure 1 was developed in SuperPro Designer (SPD) v8.5 [2] to simulate biomass pyrolysis in continuous mode. The model was built and improved on the basis reported by Mabrouki [3] and aims to simulate individual lignocellulosic components during biomass pyrolysis. Biomass pyrolysis was simulated in three sections in SPD, namely pretreatment, reaction and product collection regimes. Samples entering the system are shredded and dried at 70 °C before pyrolysis. The primary and secondary pyrolysis reactions were modelled in a combination of a continuous stirred tank reactor (CSTR) and a plug flow reactor (PFR) to achieve the desired conversion with a smaller reactor volume and hence lower costs. Both reactors operate at an actual biomass pyrolysis condition of 550 °C and 1 atm. Solid products (bio-char) from the process are collected from the cyclone whilst bio-oil and bio-gas are separated into different product streams at the condenser. The developed model is useful for any biomass that primarily consists of cellulose, hemicellulose and lignin. Pine wood, a common biomass feedstock for pyrolysis with cellulose, hemicellulose and lignin composition of 59 wt%, 12 wt% and 21 wt%, respectively, is used in this study. Note that apart from the major constituents of cellulose, hemicellulose and lignin [4], the remaining fraction contains constituents such as phenolic substitutes, protein, lipid, starch, and ash [5].

1.2 System boundaries

This analysis aims to access the life cycle of the biomass pyrolysis process. The system boundaries are selected as illustrated in Figure 1, including biomass feedstock shredding, drying, pyrolysis reactions and product collection. Transportation, usage and waste treatment processes are not within the system boundaries.

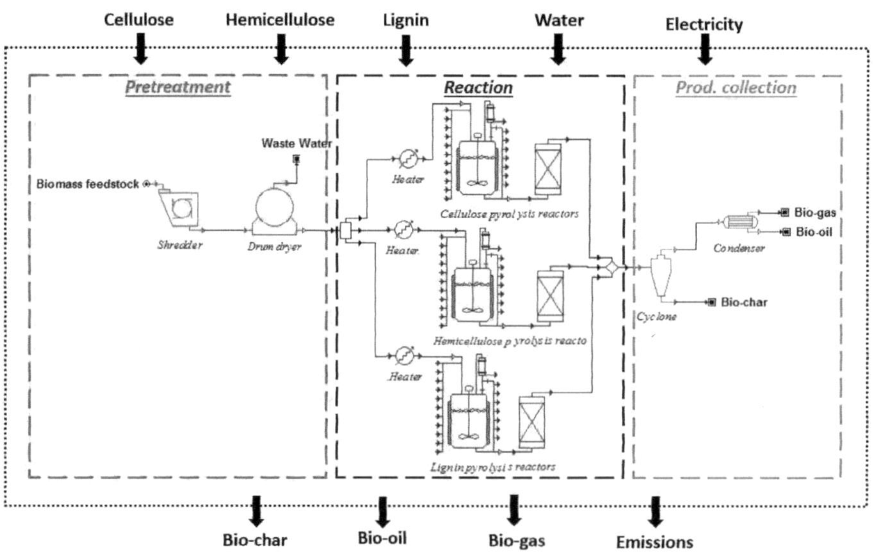

Figure. 1 System boundary and process flowsheet of biomass pyrolysis simulated in SPD.

1.3 Life cycle inventory

Inventory analysis identifies the input and output data in the life cycle of biomass pyrolysis. Biomass feedstock (cellulose, hemicellulose, lignin), water, electricity and pyrolytic products (bio-char, bio-oil, bio-gas) are obtained from SPD. These results and values are then used as input data in life cycle cost analysis and the evaluation of global warming potential based on methods as described in section 1.4.

1.4 Life cycle cost analysis (LCCA) and global warming potential (GWP)

LCCA of biomass pyrolysis is calculated based on equation (1) where LCC is the total cost of the process, I is the investment costs, E is the energy costs, W is the water costs, OM&R is the operating, maintenance and repair costs, and O is other costs incurred in the process.

$$LCC = I + E + W + OM\&R + O \quad (1)$$

Investment costs include direct fixed capital (cost to build a plant) and working capital (start-up costs) estimated based on the evaluation function of SPD. The energy for pyrolysis process is supplied by standard electricity at the cost of \$0.08/kWh [6], whilst water supply costs \$0.4/t in the process [6]. OM&R costs are estimated as 2% of the equipment purchasing cost [7] whilst other costs are assumed to be 10% in addition to the direct fixed capital cost [7].

As the crude pyrolytic products can be purified and upgraded in downstream processing, no direct emission is considered in the process. Thus, the potential environmental impacts are estimated based on global warming potential and greenhouse gas equivalencies by converting kilowatt-hours of electricity to unit of carbon dioxide emission at an emission factor of 0.707kg CO_2/kWh [8].

2. Results and discussion

2.1 LCCA and GWP analyses in pine pyrolysis

Results of LCCA and GWP for the pyrolysis of pine wood based on cellulose, hemicellulose and lignin composition are summarized in Figure 2. Pyrolysis reaction (\$ 29577 /y) is the most expensive stage of production followed by product collection (\$ 17742 /y) and pretreatment (\$ 6738 /y). The relatively high operating temperature of 550 °C in pyrolysis reactors lead to such result trend where pyrolysis process stage consumes the highest electricity (192023 kWh/y), thus equivalently emitting the highest CO_2 amongst all processes. Conversely, product collection stage with a cyclone and a condenser demands the lowest electricity (472 kWh/y) and hence has the lowest impact on GWP. However, LCCA shows that the elevated costs in product collection as compared to pretreatment stage originated from the usage of cooling water for condensation. The extensive cooling water needed (43233 t/y) for bio-gas and bio-oil product separation has increased the cost of product collection. Higher electricity (55912 kWh/y) is consumed by shredder and dryer in biomass pretreatment and therefore reflected in higher GWP as compared to product collection stage.

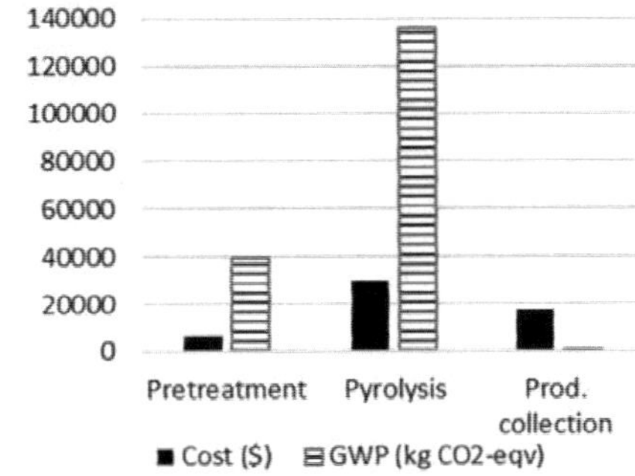

Figure. 2 LCCA and GWP results of pretreatment, pyrolysis reaction and product collection stage in biomass pyrolysis.

2.2 Sensitivity analysis

As the main constituents in biomass samples, sensitivity analysis is performed on lignocellulosic components to study the influence of cellulose, hemicellulose and lignin on LCCA and GWP results. The major parameters that affect LCCA are electricity and water usage; whilst GWP is mainly affiliated with electricity demand. Pine wood is selected as the base case with ±10% change in variable for both the analyses. As apparent in Figure 3, cellulose has the greatest impact on cost and emission, followed by hemicellulose and lignin.

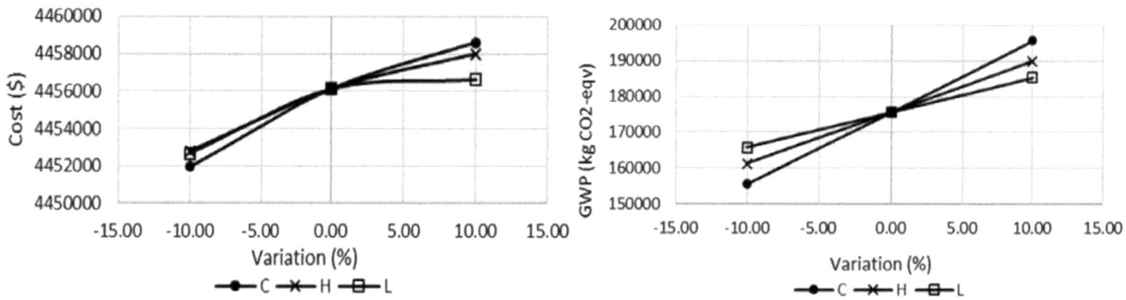

Figure. 3 (LEFT): Sensitivity analysis of pine pyrolysis on LCCA with ±10% variation in cellulose (C), hemicellulose (H) and lignin (L) composition; (RIGHT): Sensitivity analysis of pine pyrolysis on GWP with ±10% variation in cellulose (C), hemicellulose (H) and lignin (L) composition.

Amongst the three lignocellulosic components, hemicellulose commonly decomposes first (~315 °C), followed by cellulose (~400 °C) and the thermally stable lignin (up to 900 °C) during pyrolysis. The higher reaction temperature required by cellulose decomposition is a plausible reason for greater impact on electricity demand as compared to hemicellulose. In addition, the absence of a similar trend in lignin is believed to be a result of the intrinsic properties of lignin which is resilient to decomposition. The reaction rate of thermally stable lignin may be lower at the studied temperature of 550 °C, thus reflected in lower influence on LCCA and GWP results.

Similarly, pyrolytic product yields are closely associated with lignocellulosic compositions, i.e. cellulose produces more bio-oil product, hemicellulose produced higher bio-gas products whilst lignin produces higher bio-char products. Note that condenser is the key unit that consumes water in pyrolysis process for the separation of bio-oil and bio-gas as discussed in section 2.1. As the variation in lignin mainly affects solid products and has minimum impact on water demand, lignin reflects the lowest influence on LCCA and GWP results.

3. Conclusion

Life cycle analysis (LCA) on the aspects of economic and environmental impact was conducted to study the influence of different biomass pyrolysis process stages on costs and global warming potential. The main pyrolysis reaction demands the highest energy supply and thus emit the highest level of CO_2 as compared to other process stages. Conversely, product collection stage utilises the highest volume of cooling water for the separation of bio-oil and bio-gas products, thus reflected in a higher cost than the pretreatment stage. Sensitivity analysis reports cellulose as the component that highly influences LCA results for both the aspects of costing and carbon dioxide emission. Such trend provides insights that are useful for the optimization of pyrolysis process to meet economic and environmental targets.

Acknowledgements
The authors gratefully express gratitude to all parties which have contributed towards the success of this project, both financially and technically, especially the S&T Innovation 2025 Major Special Programme [grant number 2018B10022] and the Ningbo Natural Science Foundation Programme [grant number 2018A610069] funded by the Ningbo Science and Technology Bureau, China, as well as the Industrial Technology Innovation and Industrialization of Science and Technology Project, China [grant number 2014A35001-2] and the UNNC FoSE Faculty Inspiration Grant, China. The Zhejiang Provincial Department of Science and Technology is also acknowledged for this research under its Provincial Key Laboratory Programme (2020E10018).

Reference

[1] Kan, T., Strezov, V., and Evans, T.J., *Lignocellulosic biomass pyrolysis: A review of product properties and effects of pyrolysis parameters.* Renewable and Sustainable Energy Reviews, 2016. **57**: p. 1126-1140.

[2] Intelligen, *SuperPro Designer-User's Guide.* 2014.

[3] Mabrouki, J., Abbassi, M.A., Guedri, K., Omri, A., Jeguirim, M., *Simulation of biofuel production via fast pyrolysis of palm oil residues.* Fuel, 2015. **159**: p. 819-827.

[4] Pang, C.H., Gaddipatti, S., Tucker, G., Lester, E., Wu, T., *Relationship between thermal behaviour of lignocellulosic components and properties of biomass.* Bioresource technology, 2014. **172**: p. 312-320.

[5] Yan, Y., Meng, Y., Tang, L., Kostas, E.T., Lester, E., Wu, T., Pang, C.H., *Ignition and Kinetic Studies: The Influence of Lignin on Biomass Combustion.* Energy & Fuels, 2019. **33**(7): p. 6463-6472.

[6] U.S. Energy Information Administration, *Electric Power Monthly with Data for August* 2019.

[7] Peters, M.S., Timmerhaus, K.D., West, R.E., *Plant design and economics for chemical engineers*. Vol. 4. 1968: McGraw-Hill New York.

[8] U.S. Environmental Protection Agency. *Greenhouse Gases Equivalencies Calculator - Calculations and References*. 2019; Available from: https://www.epa.gov/energy/greenhouse-gases-equivalencies-calculator-calculations-and-references.

IOP Conf. Series: Materials Science and Engineering 1117 (2021) 012017 doi:10.1088/1757-899X/1117/1/012017

Load distribution to contact area analysis of Gecko's limb and feet

K Tjandra[1], R Putra[1] and R Sriwijaya[1*]

[1]Department of Mechanical and Industrial Engineering, Faculty of Engineering, Universitas Gadjah Mada, Grafika Street No. 2, Yogyakarta 55281, Indonesia

*E-mail: sriwijaya@gadjahmada.edu

Abstract. Gecko possesses a distinct feature that enables it to move in lateral and vertical directions due to its adhesive foot structures. In the past few years, the research on Gecko is primarily focused on this adhesive feature. However, Gecko's hind attachment relies heavily on the available contact area between its foot and its attaching medium. In this paper, physical observation and finite element analysis were performed to identify the load distribution pattern and contact area of Gecko's hind. Further investigations were also performed to understand the force in each digit of Gecko's feet. This research will provide readers with a more holistic understanding and complement previous research on Gecko's attachment system.

1. Introduction

Gecko is a unique species that existed mostly in tropical regions. This type of reptile has adhesive foot structures that enable it to move and retain position even along vertical surfaces. It is believed that the adhesion ability of Gecko is possible through van der Waals interaction between millions of fibrillar stalk-like-setae on Gecko's foot and the contact surface [1].

The system that enables Gecko to attach and detach is called hierarchy. Gecko's attachment ability is followed by the fact that it gains full control of its attachment-detachment motion [2], recognized as smart adhesion. While walking, Gecko continually changed its orientation to adjust desired adhesion strength and detached its foot by a peeling motion (pull-off force at 90°).

Extensive research has been performed to study the unusual adhesive behavior on Gecko's feet. [1] explored the hierarchy and structural design of Gecko's setae, including its spatulae. [3] studied the reaction force of vertical climbing geckos and stated that their walking speed does not influence the attachment and detachment times. [4] analyzed the body's lateral bending susceptible to slope degree changes. More recent research performed by Song et al. [5] provided a more specific concern towards distributed control on Gecko's feet, enabling better performance on multiple terrains.

This study aims to measure the magnitude and load distribution pattern on Gecko's feet by observing the movement's movement and its attachment to the contact surface. The results are obtained through Gecko's walking cycle and system hierarchy.

2. Experimental Materials and Methods

2.1. Animals

A Tokay Gecko was obtained from Boggy Sanjaya's Gecko Conservation Facility for this research. The Gecko was 2-years old and weighed 76.9 grams. Its dimension was approximately 290 mm x 35 mm with 121 ml of volume.

2.2. Methodology

The research process began with observations of Gecko's morphology and behavior. Images and videos obtained during the observation were then utilized as references in modeling the Gecko. The models were eventually analyzed using the Finite Element Method to get the magnitude and load distribution pattern on Gecko's feet.

2.2.1. Gecko Observation

Gecko was placed inside a transparent aquarium, and a primary camera (facing upward) was positioned below the aquarium. This camera recorded both videos and images to obtain Gecko's walking pattern and its attachment-detachment cycle. Secondary cameras were also used to capture top-view and side-view photos and videos.

Based on the observation, seven distinct Gecko's steps were obtained in one walking cycle. The photos of each step were compiled in one picture (Fig. 2(a)) for modeling purposes, and the attachment types of each foot (Fig. 2(c)) were recorded.

The process of Gecko's feet attachment (approximately 0.1 s) was far quicker compared to the detachment process (approximately 0.46 s). Therefore, four frames of Gecko's detachment process were captured and compiled next to each other. A mark was given to each of Gecko's feet to identify its attached surface area.

2.2.2. Gecko Modeling

Blender was used for Gecko's initial 3D model design. To generate the 3D model, each view was traced according to the image obtained during the observation process. The most appropriate frames in a video for the tracing process were selected through Filmora, a video editing software. The finished model was exported as .objfile type and converted by SpaceClaim to .scdoc extension, so it was readable by ANSYS software. The final model (Fig.1) was validated by comparing the measured Gecko's volume with the simulation results.

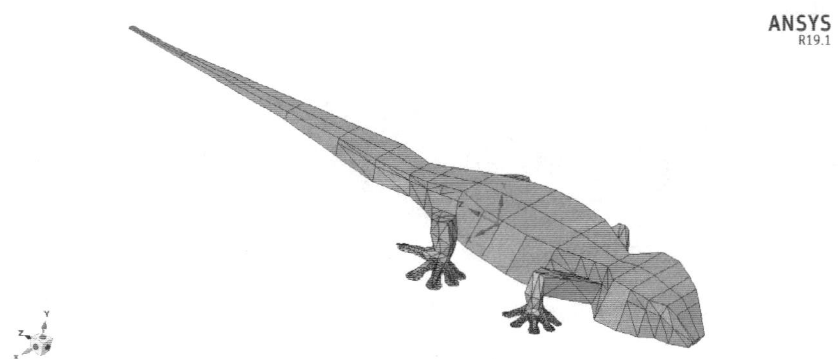

Figure. 1 Gecko's 3D model.

The model shown in Fig.1 was used as the reference to design Gecko's models for one full cycle (Fig. 2(b)). The surface area of Gecko's foot was simplified as a rectangle to aid the force reaction calculation. This surface area was varied to differentiate between fully attached Gecko's limbs and half attached Gecko's limbs (during attachment period).

As for Gecko's feet' modeling, four steps (1 cycle) of Gecko's foot detachment process were modeled in detail. We assumed that the Gecko's left-side feet' geometry were the same as the right side, so we observed (Fig. 3 (a)) and modeled only one fore-foot (Fig. 3 (b)) and one hind-foot for each frame.

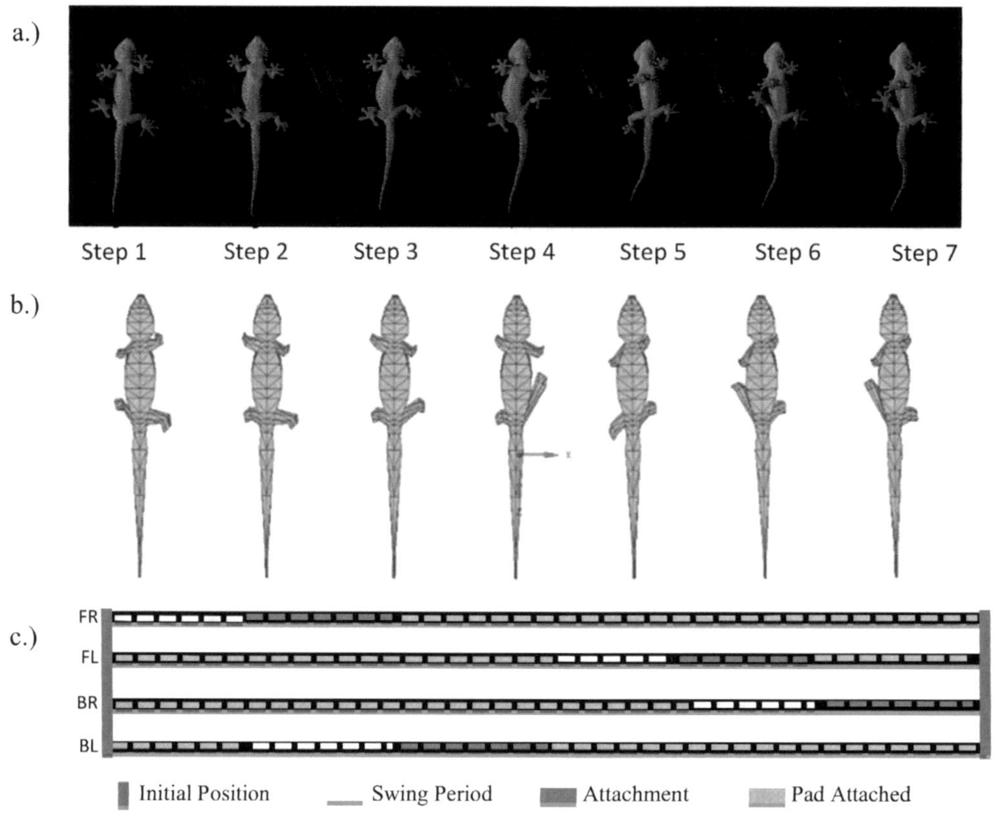

Figure. 2 (a) Gecko's actual full walking cycle, (b) Modelled version of Gecko's walking cycle, and (c) Attachment types of each Gecko's feet.

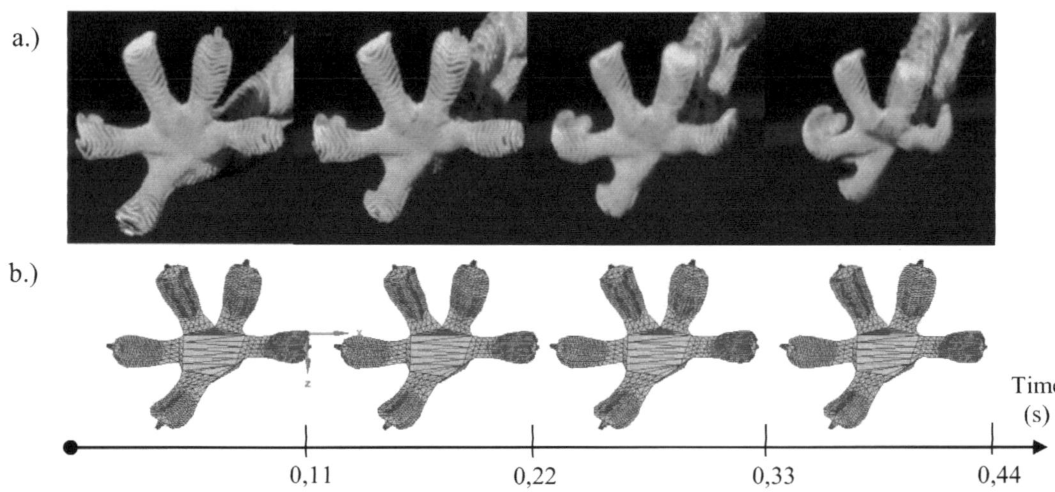

Figure. 3 (a) Gecko's actual detachment process and (b) Modelled version of Gecko's detachment process.

3. Finite Element Analysis

ANSYS R19.1 software is chosen to conduct structural analysis. Gecko's material properties were imported based on the data researched by Gao et al. [1]. Orthogonal and skewness value of 0.75 and 0.25 respectively was determined to produce optimal meshing quality, resulted in 0.002 m of the mesh size.

In analyzing the walking cycle of Gecko, the following parameters are applied to simulate the actual condition: (1) Gravity to replicate the internal load of Gecko's body, (2) fixed support on each of Gecko's foot and bottom part of Gecko's body and tail. Planes were created on each of Gecko's feet to analyze the amount of load applied to each of its feet.

The load distribution analysis on Gecko's foot was performed by highlighting the foot's attached parts at different time frames to obtain the contact area.

4. Findings and Discussion

The first simulation provided us with information regarding load and moment applied to each step of Gecko's walking cycle. This information was plotted in Fig 4.

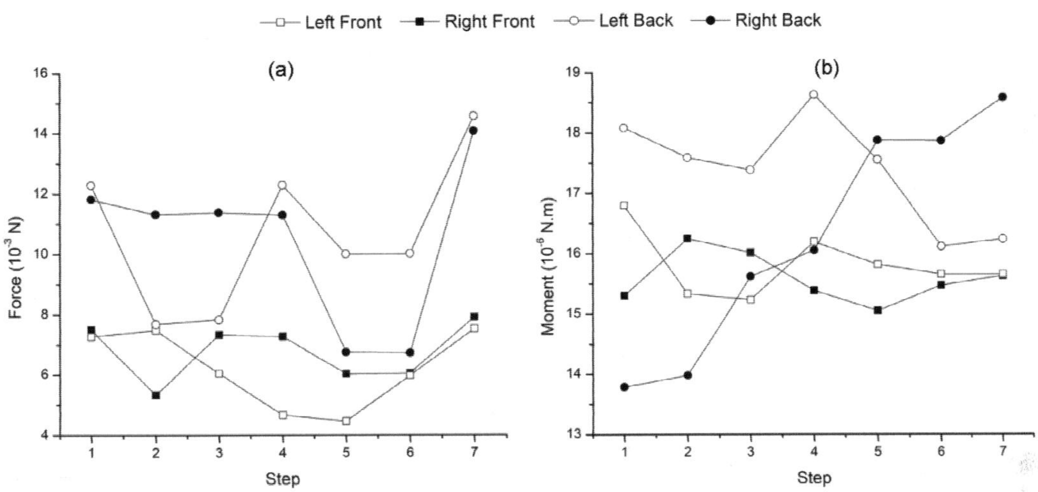

Figure. 4 (a) Force and (b) moment applied on Gecko's limb at each step.

Fig 4(a) showed the load distribution pattern on each Gecko's step. Dissimilar contact areas resulted in different loads applied on each foot. It can be inferred that in one cycle, the increase of load (during the attachment process) is always followed by the decrease of the load while detaching. Furthermore, the force applied to the hind-limb is noticeably higher than on the fore-limb.

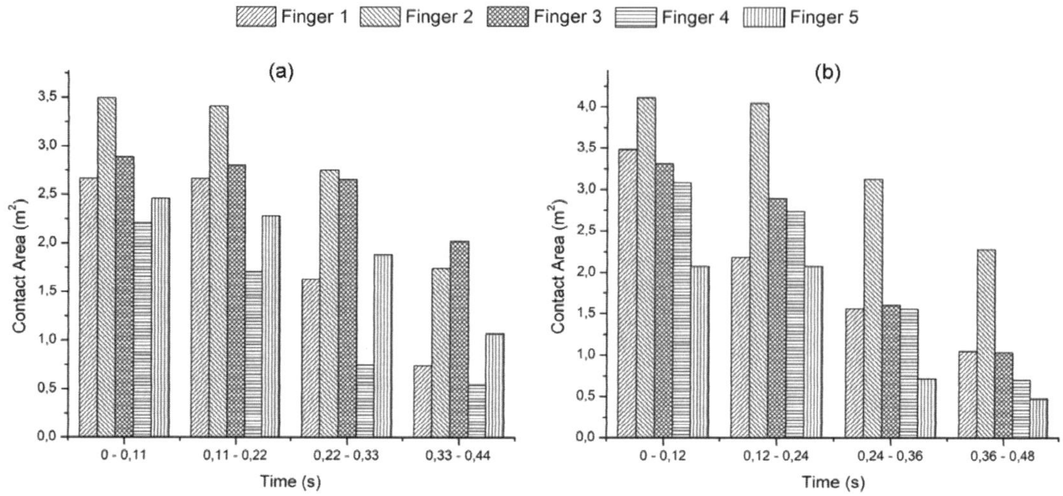

Figure. 5 The contact area of (a) Gecko's fore-digit and (b) Gecko's hind-digit over time.

The moment for each limb was later determined by measuring the reaction force and length of each limb. Based on Fig 4(b), the moment applied on the right hind-limb was increasing over steps due to the increment of its length. The parallel relation between moment and hind length applies to other limbs, but the trends are weaker for fore-hinds due to insignificant limb length differences.

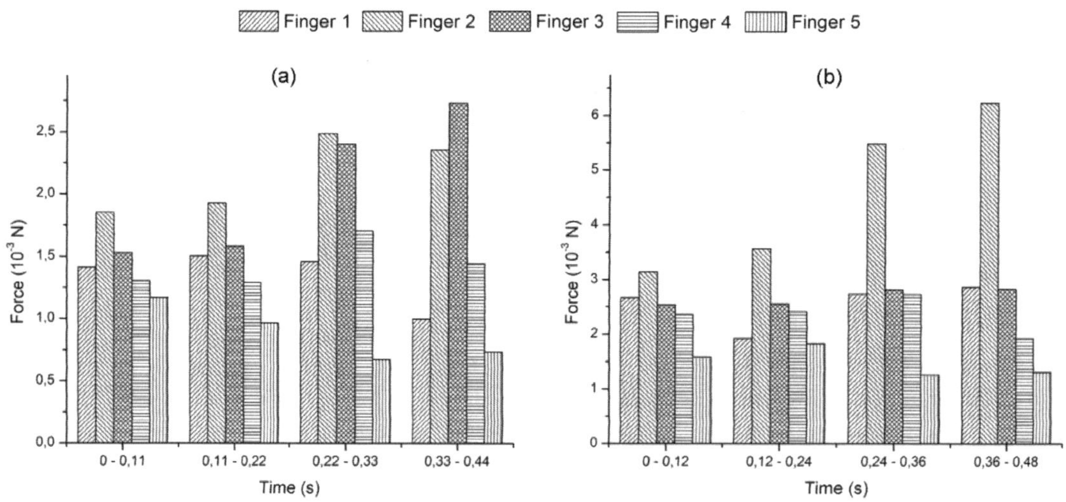

Figure. 6 Force applied on (a) Gecko's fore-digit and (b) Gecko's hind-digit over time.

Gecko's digits are labeled from 1 (far-right digit) to 5 in counter-clockwise order. The contact area measurement results of each digit during Gecko's detachment process were then plotted in Fig. 5. To obtain the force applied on each digit (Fig 6), the area measurement data was converted as a percentage out of the total contact area of the foot, then multiplied by the load applied on each limb (7.27×10^{-3} N for fore-limb and 12.29×10^{-3} N for hind limb).

According to Fig. 5, both fore-limbs and hind-limbs' contact areas decrease over time as they detached. However, contradicting trends are shown in Fig. 6, mainly due to less-even load distribution at the detachment process. The most significant load was applied to digit 2, confirming that digit 2 has the biggest contact area.

Conclusions

In one cycle, the load distribution and the moment on each Gecko's hind forma reciprocating pattern, with peak points achieved at the fully attached stage and stretched limb, respectively. Analysis performed on Gecko's feet proved a linear correlation between the surface area and each digit's force distribution.

Acknowledgments

The research was funded by the Department of Mechanical and Industrial Engineering, Faculty of Engineering Universitas Gadjah Mada through Grant No. 417/UN1/FTK/SK/HK/2020.

References

[1] Gao H, Wang X, Yao H, Gorb S and Arzt E *Mechanics of Materials* 275-85 **37** 2005
[2] Bhushan B , *Journal of Adhesion Science* and *Technology* 1213-58 **21** 2012
[3] Wang Z Y, Wang J T, Ji A H, Zhang Y Y and Dai Z D *Chinese* Science *Bulletin* 573-83 **56** 2011
[4] *Wang* Z Y, CaiL, Li W, Ji A H, Wang W B and Dai Z D *Journal of Bionic Engineering* 238-49 **12** 2015
[5] Song Y, Dai Z D, Wang Z Y and Full R J, *Proceedings of the Royal Society B*: *Biological Sciences* 20200123 **287(126)** 2020

The Effect of Long Term Hygrothermal Aging by Immersion on Carbon/Epoxy Composites Exposed to the Heat Sources for Naval and Marine Applications

SoonKook Hong[1]

[1]Department of Weapon system engineering, Republic of Korea Naval Academy, Korea.

Abstract. Fiber-Reinforced Polymer (FRP) composites have many advantages over conventional materials in terms of high strength-to-weight and stiffness-to-weight ratios, design flexibility and enhanced service life. Due to these advantages, FRP composites are used in many different fields such as Construction, Electric/power generation, Marine, Corrosion/infrastructure, transportation and so on. Most researches of FRP composites were focused on the applications used in ground field. However, in Navy, marine composite materials are increasingly being used in diverse applications: Submarine, Autonomous Underwater vehicle, Unmanned Surface vehicle and Payloads in Naval vessel. It is known that polymer composites are very sensitive and are friendly to water or moisture in any forms. Moisture in polymer composites often causes degradation by swelling and hydrolysis. Especially, if polymer composites will be immersed in water and seawater, faster degradation can be caused by water uptake than any other moisture environments. While general investigations regarding water uptake are concentrated on the composite materials cured in ambient temperature, the immersion effects of polymer composites exposed to elevated temperatures were studied because underwater applications of composite materials can be exposed to the various heat sources. For example, fires on naval vessels and underwater vehicles can be started by any number of causes such as electrical faults and ignition of flammable gases or liquids. Accordingly, immersion effects must be studied on marine composite materials exposed to various temperatures in order to evaluate service life in an underwater environment. This study investigated the water uptake, diffusion coefficient, and Short beam shear (SBS) for Interlaminar Shear Strength(ILSS), delamination between fiber and resin using test specimens exposed to the elevated temperatures. The specimens were immersed for a definite period of time (~one and a half years) in seawater and deionized water (DI water) to compare both degradation mechanisms.

1. Introduction

1.1. Background

FRP composite materials are being used in a variety of underwater applications based on their stiffness, strength, reduced weight, and corrosion-free capabilities. Until recently, the use of FRP composite materials for military applications was limited to aerospace and air force for high-performance applications. Currently, applications of FRP composite materials in the Navy are widely broaden into sonar bow domes, windows, hulls and so on. Moreover, there is a resurgence of interest for the use of composites in military applications including naval vessels, army combat vehicles, underwater robot fish and unmanned underwater vehicles. Beside inherent advantages of FRP composites such as high strength-to weight and stiffness-to-weight ratios, composites materials using the carbon fiber as a reinforcement are particularly useful because they exhibit better mechanical properties than other FRP composites as well as provide the ability for electromagnetic shielding for stealth purpose. The all-composite naval ships are currently operating to perform multifunctional operation with the benefits for FRP composite materials. In early, use of composite materials was constrained to the construction of small patrol boats and landing craft in displacement due to relatively poor fabrication quality and low stiffness of the hulls. However, as fabrication technique and mechanical properties were improved, FRP composite materials can be applied to larger patrol boat, minecountermeasure vessels, and corvettes.

Because the usage of FRP composite materials in underwater environments enable various applications to have good benefits in terms of stealth function on SONAR and RADAR detection, this research is focused on immersion effects of the underwater applications: submarine, underwater robot fish and unmanned underwater vehicle. These applications in immersion can cause the severe degradation compared to exposure to the moisture and moisture-immersion mixed environments. Polymer such as an epoxy can be modified into superhydrophobic surface by nano pattern. The sueprhydrophobic surface of the polymer can bring out the lower water uptake than any other matrices. Carbon mentioned above has good stealth function by superior electric conductivity. For this reason, carbon/epoxy composite materials were investigated to evaluate the mechanical property, water uptake, and diffusion coefficient after immersing in seawater and deionzed water for one and a half years.

1.2. Seawater Effect

There are a lot of researches related to carbon- and glass-fiber composites immersed in saline conditions. In this section, the effects exposed to various environments will be introduced from summarizing of many studies toward tensile strength and modulus.

T.S Grant et al.[1] investigated the effect of immersion in seawater on transverse tensile properties of three graphite/epoxy composite materials. The transverse tensile strength was found to be reduced by 17% in one of the systems with essentially no change in the other two systems studied. The 17% decrease in transverse tensile strength was associated with degradation of the interfacial strength. Also they found that little difference was found in the behavior of composite immersed in distilled water and in seawater at ambient pressure or seawater at 20.7 MPa pressure. Leif A. Carlsson et al.[2] accomplished tests using carbon/bismaleimide-epoxy, E-glass/epoxy, E-glass/polyphenylsulfide and carbon-epoxy for immersion more than 4000 hrs in distilled water and natural seawater at room temperature and 35°C. Transverse modulus, E_2 was not significantly changed after water absorption except for E-glass/polyphenylsulfide that lost about 60% of its dry modulus, despite of its low water absorption. The substantial reduction of E_2 was attributed to extensive fiber/matrix debonding induced by water. Also, all composites experienced large reduction in transverse tensile strength due to water absorption. The maximum reduction took placed in E-glass/polyphenylsulfide with 85% decrease of its dry strength. An E-glass/carbon/epoxy interlayer hybrid composite has been aged by immersion in simulated seawater for varying lengths of time and then tested in transverse tension by C.A Wood et al.[3].

The flexural test measures the force required to bend a beam under three point or four point loading conditions. The data is often used to select materials for parts that will support loads without flexing. Flexural modulus is used as an indication of a material's stiffness when flexed. E.P. Gellert and D.M. Turley[4] performed flexural test using the polyester, phenolic and vinylester glass-fiber reinforced polymer materials immersed in 30°C seawater either unloaded or loaded at 20% of the maximum strain at flexural failure ($0.2\varepsilon_f$) to examine the effects of loading on flexural property. Flexural strength continued to degrade for the unloaded polyester and vinylester GPRs as water uptake continued toward saturation, where strength losses between 15% and 21% occurred. The unloaded phenolic lost 25% of initial strength at saturation. For ageing, loading affected the strength of only phenolic GRP with strength loss advancing from 25 to 36% loss from the initial strength. Wayne C. Tucker and Richard Brown[5] generated all data on the vinlyester/graphite composite material immersed in natural filtered seawater in tubs in the laboratory at 1 atmosphere of pressure and in natural filtered seawater pressurized to a depth of 2000feet of seawater. The flexural strength and stiffness of the composite material were decreased by the high pressure exposure. In contrast, atmospheric seawater exposure did not produce any strength decrease. Both the enhanced moisture uptake and the strength decrease at high pressure exposure were thought to be due to mechanical damage induced by the increased pressure. In 5% and 10% salt solution, the flexural properties (strength and modulus) of pultruded glass-fiber reinforced vinylester matrix composite coupons were measured for the 90° specimens as-received and after ageing by K.Liao et al.[6].

2. Experimental Setup

2.1. Material specification

The carbon/epoxy composites were comprised of Tyfo S Epoxy and Tyfo SCH-41 reinforcing fabric supplied by FYFE Co. Composite panels comprising of two layers as shown in Figure 1 were fabricated using a mannual wet layup process with cure under ambient conditions. Tests were conducted after a minimum of 7 days cure. The fiber content was approximately 60% by weight. The fabric reinforcement was consisted of a primarily unidirectional fabric of 644 g/m^2 areal weight.

The fibers had a nominal tensile strength, modulus and density of 3.79 GPa, 230 GPa, 1.74g/cm^3, respectively. The resin system was a two-component epoxy with a viscosity of 600-700 cps at 25°C. After curing and then after 72 hours of post cure at 60°C, the glass transition temperature, tensile strength, modulus and elongation are specified as 82°C, 72.4 MPa, 3.18 GPa, and 5.0%, respectively.

All tests were performed using unidirectional 2 layer panels. Because composite panels were fabricated in the field, they were uniformly preconditioned at 23°C and 30% relative humidity (RH), in a humidity chamber to set a uniform baseline.

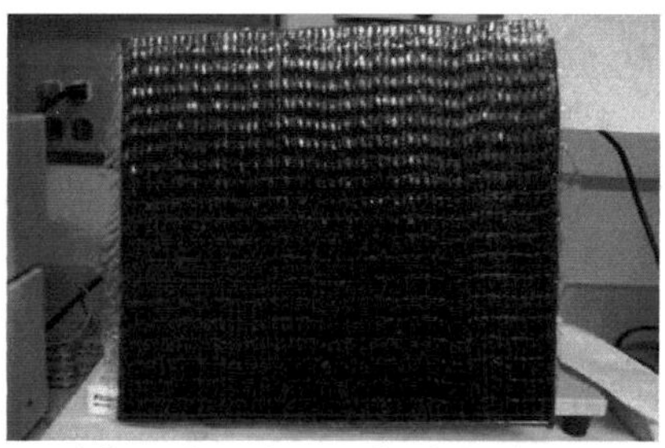

Figure. 1 Unidirectional Panel of Carbon/Epoxy Composites

2.2. Test Conditions

After cutting composites panels with a tile saw, all specimens for mechanical properties and immersion testings were stored in a humidity chamber (Figure 2 (a)) to ensure an initial condition for 2 weeks. Carbon/epoxy composite specimens were exposed thermally to ambient, 66, 93, 121, 149, 177, 204, 232, and 260°C under ageing times of 1, 2, 4, 8, 16, 24, 48, and 72 hrs. Test specimens were kept in an oven (Figure 2(b)) until they reached a set time and temperature, and then they were removed to ambient conditions for testing.

After the simulation of exposure to high temperatures, accurate evaluations are necessary to assess whether composite materials degraded by high temperatures can operate at the required level of functionality in seawater. Therefore, the specimens aged thermally at each temperatures were immersed in seawater under ambient temperature for up to 72 weeks. Simultaneously, test specimens were immersed in deionized (DI) water in the same conditions to provide a base-line comparison.

IOP Conf. Series: Materials Science and Engineering 1117 (2021) 012018 doi:10.1088/1757-899X/1117/1/012018

Seawater from La Jolla shores was used and was periodically changed in the water bath. Figure 3 shows the specimens for water uptake tests and SBS tests.

Moisture uptake tests were performed using gravimetric analysis. Test specimens were taken out from water bath to weigh the mass, periodically. In the case of short beam shear tests, tests were conducted after immersion in seawater and DI water under specified conditions (exposure to Ambient, 66, 93, 121, 149, 177, 204, 260°C for 8 hr and exposure to 232°C for 1, 2, 4, 8, 16, 24, 48 and 72hr). Water uptake and SBS test are summarized in Table 1.

(a) (b)

Figure. 2 (a) Humidity chamber for initial condition and (b) Oven for environmen

(a) (b)

Figure. 3 (a) Test specimens for water uptake tests and (b) Test specimens for SBS tests immersed in seawater and DI water.

Table 1. Test specimens for immersion tests

Test	Specimens No	Size (mm)
Water Uptake	Seawater (73) + DI water (73) = 143	25.4×25.4×3
SBS	16(temp)×8(ageing)×2(case)×5(Set) = 1,280	18×12.7×3

2.3. Test Methodology

Test methods available for the determination of interlaminar shear include short beam shear, four-point shear, double notch shear and v-notch beam shear. The short beam shear test was applied to determine the interlaminar shear strength in this research.

The short beam shear tests were accomplished following ASTM D2344[7] using specimens which were cut to dimension of 6 mm in width and 18 mm in length, using test fixture, and an Instron 5583 equipment (Figure 4). According to ASTM, the following geometries are recommended:

- Specimen length = thickness ×6
- Specimen width = thickness ×2

The test span in this case was 14 mm. A cross head speed of 1 mm/min was applied for all tests and the load was applied until failure of the specimens was attained.

For measuring the moisture uptake of the specimens immersed in sea water and de-ionized (DI) water, all samples exposed to elevated temperatures were removed from each immersion environment and kept at room temperature until measuring. All unidirectional specimens were cut to dimensions of 25.4 mm by 25.4 mm for gravimetric measurement. Wet samples from immersion environments were wiped for dryness with a paper towel prior to weighing. Weight measurements were undertaken using a Sartorius Analytical Balance with a resolution of 10^{-5} grams. Weights were recorded when the LCD display of the balance kept a stable value for 5 seconds to ensure the consistency in test method. After measurement of weight, all samples were returned to the original environment for further exposure.

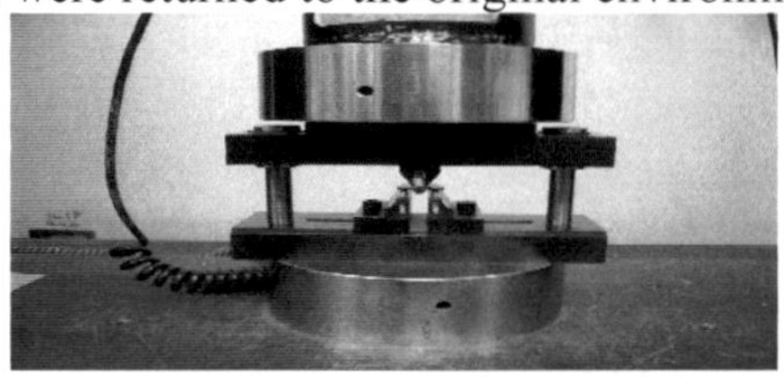

Figure. 4 Short Beam Shear Test Fixture and Instron 5583

3. Results

3.1. Diffusion

In general definition, diffusion is the movement of molecules from a region of high concentration to a region of low concentration by means of random molecular motion. Fick's laws provide a theoretical basis for the diffusion of a fluid into a distinct sorbing medium from a higher concentration to a lower concentration. Also, Fick's second law provides a theory for non-steady-state diffusion. Fick's law refer to that the mass of absorbed water increases linearly with the square root of time and then gradually slows until an equilibrium plateau or saturation is reached. The rate of diffusion and the attainment of an equilibrium content can be affected by materials characteristics, processing factor, environmental condition, and geometry. Since Fickian diffusion assumes no chemical reaction between the diffusion solution and composite materials, composites technically do not follow Fick's law. However, in a number of researches, the diffusion of moisture in fiber reinforced composites and crosslinking resin has been shown Fickian behavior[8,9]. Fickian diffusion has following features.

- Linear in the initial stage and the linear region until at least $M_t/M_m=0.6$

where:

M_t = the moisture absorbed by the composites at time t

M_m= the maximum moisture content absorbed by the composite

- The decrease of the rate of diffusion until an equilibrium of moisture content
- Diffusion coefficient as a function of temperature

$$D = D_0 \exp\left(\frac{-E_a}{RT}\right)$$

where:

D = diffusion coefficient

D_0= a constant

E_a= activation energy

R = the universal gas constant

3.2. Water Uptake

Glass fiber laminates have low Young's modulus, which makes it difficult to build ultralight marine structures including naval vessels with adequate stiffness. Therefore, marine composite structures requiring high stiffness are often made of carbon fiber composite. However, a little of published papers are available on the effect of long-term seawater immersion on carbon fiber composite[10]. In addition, atmospheric ageing at high humidity has been reported to cause water uptake similar to that from immersion for epoxy laminates[11], while Gutierrez reported that ageing in sea air was as severe as in seawater for a range of marine composite. Especially, this study is focused on the effect of long-term seawater immersion on carbon fiber composites exposed to elevated temperatures for up to 72 hrs of ageing time using comparison with the effect of immersion in deionized water.

Figure 5 shows the maximum weight gain (%) on specimens immersed in deionized water and seawater at atmospheric temperature for 72 weeks. There is a little difference in the maximum weight gain. In particular, the maximum weight gain of all specimens which ageing time is less than 8 hrs in entire exposure temperatures was almost identical. All data of the maximum weight gain in these conditions existed between 1.2% and 2.8%. In both immersion conditions, the maximum weight gains were slightly increased in proportion to ageing time and exposure temperature. The difference among the maximum weight gains in lower (ambient, 66 and 93°C) and higher (121, 149, 177, 204, 232°C) exposure temperatures existed on specimens immersed in deionized water. However, this gap was not shown on specimens immersed in seawater. This means that the unreacted chemical species in partially cured composite were released more rapidly into deionized water. In addition, it should be pointed out that the maximum weight gains in seawater were slightly lower than those in deionized water in overall environmental conditions. Apparently, the sorption of salts into the epoxy by diffusion and along fiber-matrix interface debonds and bulk material cracks by wicking can be resulted in the higher mass retention in seawater while the organic species leached from the specimens were separated from the salty residue by solvent extraction. The organic species were even found visually in seawater container. Consequently, mass loss by leaching of organic species than mass uptake by sorption of salts largely contributed to lower maximum weight gain in seawater compared to the values of deionized water.

Figure 6 shows comparison of diffusion coefficients on specimens immersed in deionized water and seawater for 72 weeks. Overall diffusion coefficients calculated for deionized water immersion were higher than those for seawater immersion in all environmental conditions. Diffusion coefficients in deionized water were widely distributed with increase of ageing time and exposure temperatures. In other words, the degree of cure on specimens was strongly dependent on the diffusion coefficient. On the other hand, the variation of diffusion coefficients in immersion of seawater was less than that indeionized water. Lower diffusion coefficient in seawater seems to be attributed to mass loss by leaching out of organic species.

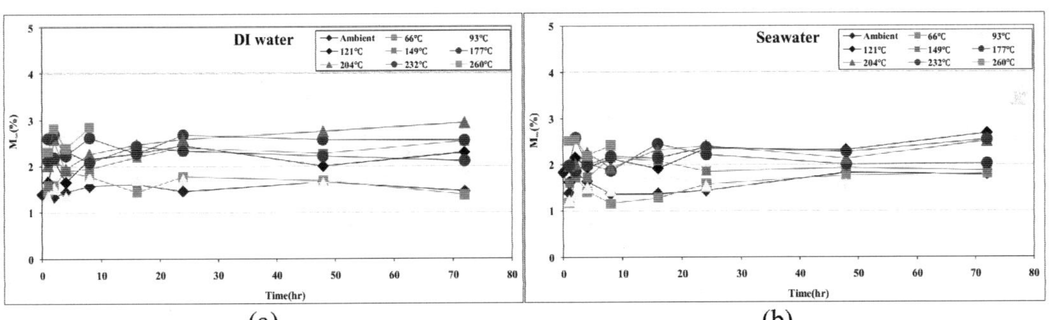

(a) (b)

Figure. 5 The maximum weight gain (%) on specimens immersed in (a) DI water and (b) seawater for 72 weeks

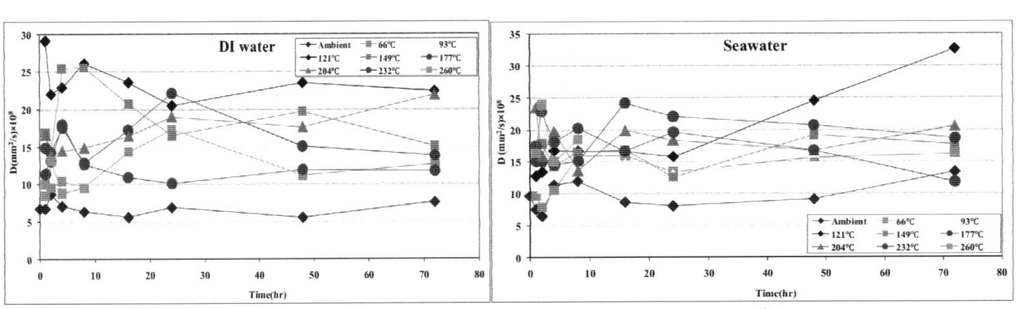

(a) (b)

Figure. 6 Diffusion coefficients on specimens immersed in (a) DI water and (b) seawater for 72 weeks

3.3. Short Beam Shear Testing

Comparison of short beam shear strength on specimens immersed in deionized water for 72 weeks after exposure to elevated temperatures in 8 hrs of ageing time are presented in Figure 7. Except for the specimens immersed in deionized water after exposure to 260°C for 8 hrs, all specimens showed similar characterizations which had initially decrease (~ 16 weeks), asymptotic trend or slightly increase (16 ~ 48 weeks) and rapidly decrease after 48 weeks in terms of short beam shear strengths. Initially rapid drop and asymptotic trend were corresponded to Fick's law which the mass of absorbed water increases linearly with the square root of time and then gradually slows until equilibrium. The rate of decrease in short beam shear strengths after immersion from original state to 72 weeks in deionized water was 23.0, 28.1, 35.7, 38.7, 36.0, 33.7, 35.2, 35.9 and 30.7% on specimens exposed to ambient, 66, 93, 121, 149, 177, 204, 232 and 260°C, respectively. Post-cured specimens in intermediate exposure temperatures (121, 149, 177, 204, 232°C) had higher rate of decrease in short beam shear strengths compared to other conditions since specimens post-cured from the increase of ageing time and exposure temperature showed the rapid saturation and the higher maximum weight gain compared to un-cured specimens. Contrary to specimens in other conditions, specimens exposed to 260°C for 8 hrs showed initially a little of increase in short beam shear strengths. However, after 48 weeks in immersion time, rapid decrease of short beam shear strengths occurred by delamination between 2 layers due to moisture uptake.

From many researches, it is well known that epoxy has the superior durability and the lowest diffusion coefficient compared to vinylester, polyester and phenolic resins in water environmental conditions. In addition, it is known that glass fibers chemically react with water while carbon fibers do not absorb moisture and are resistant to any corrosive effects of water. Therefore, intrinsic properties of epoxy and carbon fiber against water resulted in a slight decrease or asymptotic trend in terms of short beam shear strengths until 48 weeks in immersion period. However, it appears that catastrophic drop of short beam shear strengths after 48 weeks of immersion was derived from irreversible degradations such as hydrolysis, microcracking, microvoids and epoxy relaxation.

To compare of results of short beam shear test immersed in deionized water, short beam shear tests were conducted using specimens immersed in seawater under same environmental conditions. Comparison of short beam shear strength on specimens immersed in seawater for 72 weeks after exposure to elevated temperatures in 8 hrs of ageing time are shown in Figure 8. Although a little difference between specimens immersed in deionized water and seawater existed with regard to diffusion coefficient and the maximum water uptake, the results of short beam shear strengths on specimens immersed in seawater were in good agreement with the data immersed in deionized water as shown in Figure 7and Figure 8. In addition, it seems that salt or salinity involving in seawater did not influence deterioration and degradation in short beam shear property.

The rate of decrease in short beam shear strengths after immersion from original state to 72 weeks in seawater was 22.7, 25.2, 32.8, 36.3, 40.7, 33.4, 31.0, 35.3 and 22.7% on specimens exposed to ambient, 66, 93, 121, 149, 177, 204, 232 and 260°C, respectively. These rates of decrease in short beam shear strengths were slightly lower than the values after immersion in deionized water. However, the difference of rate of decrease was negligible.

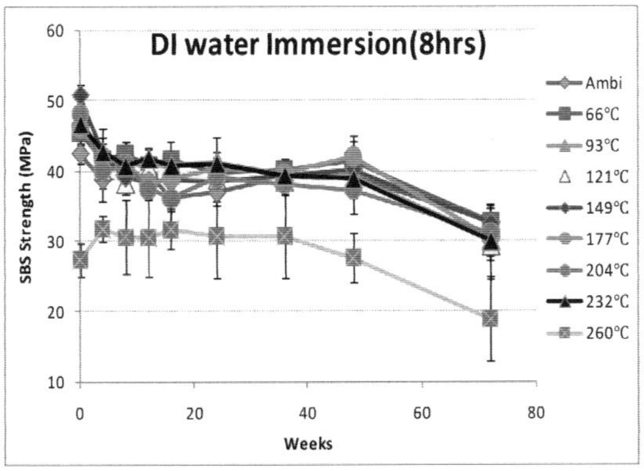

Figure 7. Short Beam Shear Strengths on specimens immersed in DI water for 72 weeks after exposure to elevated temperatures in 8 hrs of ageing time.

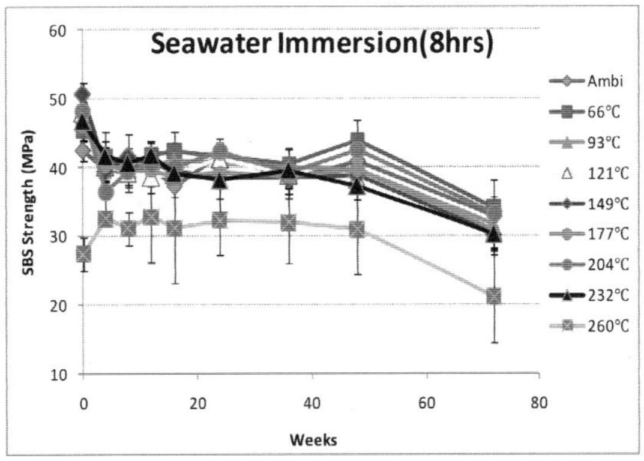

Figure 8. Short Beam Shear Strengths on specimens immersed in sea- water for 72 weeks after exposure to elevated temperatures in 8 hrs of ageing time

4. Conclusion

Experimental results can be summarized as follows,

- The results of the gravimetric measurements on specimens immersed in deionized water and seawater showed a Fickian response in all conditions.
- Specimens post-cured from the increase of ageing time and exposure temperature showed the rapid saturation and the higher maximum weight gain compared to un-cured specimens.
- The partially cured composite could be expected to have a greater concentration of unreacted chemical species with the epoxy resin and it appears that these species were released more rapidly into water resulting in a slower net mass gain.
- Mass loss by leaching of organic species than mass uptake by sorption of salts largely contributed to lower maximum weight gain in seawater compared to the values of deionized water.
- Overall diffusion coefficients calculated for deionized water immersion were higher than those for seawater immersion in all environmental conditions. Diffusion coefficients in deionized water were widely distributed with increase of ageing time and exposure temperatures.

5. Acknowledgments

This research is part of the result carried out by support of LIG nex1.

References

[1] Grant, T.S. and W.L. Bradley, In-Situ Observations in SEM of Degradation of Graphite/Epoxy Composite Materials due to Seawater Immersion. Journal of Composite Materials, 1995. 29(7): p. 852-867.

[2] Leif A. Carlsson, F.P., *Influence of sea water on transverse tensile properties of PMC, NIST Special Publication 887*. National Institute of Standards and Technology, 1995: p. 203-221.

[3] Wood, C.A. and W.L. Bradley, *Determination of the effect of seawater on the interfacial strength of an interlayer E-glass/graphite/epoxy composite by in situ observation of transverse cracking in an environmental SEM*. Composites Science and Technology, 1997. 57(8): p. 1033-1043.

[4] Gellert, E.P. and D.M. Turley, *Seawater immersion ageing of glass-fibre reinforced polymer laminates for marine applications*. Composites Part A: Applied Science and Manufacturing, 1999. 30(11): p. 1259-1265.

[5] Tucker, W.C. and R. Brown, *Moisture Absorption of Graphite/Polymer Composites Under 2000 Feet of Seawater*. Journal of Composite Materials, 1989. 23(8): p. 787-797.

[6] Liao, K., C.R. Schultheisz, and D.L. Hunston, Effects of environmental aging on the properties of pultruded GFRP. Composites Part B: Engineering, 1999. 30(5): p. 485-493.

IOP Conf. Series: Materials Science and Engineering 1117 (2021) 012018 doi:10.1088/1757-899X/1117/1/012018

[7] ASTM D2344/D2344M-00, Standard Test Method for Short-Beam Strength of Polymer Matrix Composite Materials and Their Laminates. Amercian Society of Testing Materials, 2006. Vol.15.03.
[8] Tsotsis, T.K. and S.M. Lee, Long-Term Durability of Carbon and Glass-Epoxy Composite Materials in Wet Environments. Journal of Reinforced Plastics and Composites, 1997. 16(17): p. 1609-1621.
[9] Liao, K., Schultesiz, C. R., Hunston, D. L., Brinson, L. C, Long-Term Durability of Fiber-Reinforced Polymer-Matrix Composite Materials for Infrastructure Applications: A Review. Journal of Advanced materials, 1998. 30, No.4: p. 3-40.
[10] Mouritz, A.P., A. Kootsookos, and G. Mathys, Stability of polyester- and vinyl ester-based composites in seawater. Journal of Materials Science, 2004. 39(19): p. 6073-6077.
[11] Bonniau, P. and A.R. Bunsell, A Comparative Study of Water Absorption Theories Applied to Glass Epoxy Composites. Journal of Composite Materials, 1981. 15(3): p. 272-293.

Manufacturing lateral flow tests for tuberculosis diagnosis: choosing a reactants completion and sensing regime

N A Byzova[1], D V Sotnikov[1], A V Zherdev[1], B B Dzantiev[1]

[1]A.N. Bach Institute of Biochemistry, Research Centre of Biotechnology, Russian Academy of Sciences, Leninsky prospect 33, 119071Moscow, Russia

sotnikov-d-i@mail.ru

Abstract. To develop technologies for rapid out-of-laboratory diagnoses of tuberculosis based on the level of specific antibodies, the requirements for the antibody preparations and their combinations used in the analysis were studied, as well as procedures for the manufacture of a multimembrane composite. A comparative characterization of various recombinant antigens in terms of diagnostic sensitivity and specificity was carried out. The possibility of increasing the diagnostic sensitivity to 15%–20% when using multi-antigen mixtures (cocktails) of antigens was shown. The requirements for the manufacture of multi-membrane composites—test strips for diagnostics—were determined. The correspondence of the minimum detectable level of specific antibodies to the diagnostic threshold of 0.2 mg/ml was shown. Testing of test systems via the most effective cocktails of antigens on blood samples from healthy and sick subjects was carried out.

1. Introduction

Each year globally, approximately 3 million people die from tuberculosis, and 6 million new cases are detected. Tuberculosis ranks first among infectious diseases in terms of mortality among the adult population [1]. The complexity of its diagnosis is because of the different etiology of the disease, the polymorphism of pathogens, and the frequency of these pathogens' presence in a healthy body. The formation of a population-wide testing system for early diagnosis of tuberculosis is a priority task in public health. Correct and prompt diagnosis allows the identification of patients during the early stages of the pathological process, and starting therapy on time prevents the development of progressive forms and reduces the disease's spread [2-3]. At the same time, carrying out mass testing without the involvement of specialized laboratories requires the production of autonomous analytical systems that function without additional reagents, equipment, and multistage manipulations.

Immunochromatography appears to be a promising analysis format that ensures rapid testing, methodological simplicity, and the possibility of registering results outside specialized stationary laboratories [4-5]. The use of colored particles as markers can lead to direct visual registration of results. The duration of the immunochromatographic analysis does not exceed 15 min, and its labor intensity is extremely minimal—screeners apply a sample to the strip and register the result without resorting to any additional manipulations and reagents. Currently, different types of immunochromatography are used extensively for early detection of pregnancy, markers for cardiological and oncological diseases, and narcotic compounds [6-7]. In addition, several commercial test systems are known for their ability to diagnose tuberculosis; these systems register the presence of specific antibodies to the causative agent of the disease in the blood (serodiagnostics) [8]. However, their effectiveness remains controversial because of (i) the lack of a unique biomarker for diagnosis and (ii) significant variability in antibody titers and affinity in the subjects [9-10]. Therefore, technological solutions are needed for the manufacture of immunochromatographic test systems that make it possible to effectively combine various antigenic preparations applied to the membrane components of the test, as well as provide interaction conditions during the movement of fluid along the multimembrane composite for effective detection of both low- and high-affinity antibodies. The solution to this problem will provide the possibility of scalable technologies to produce test kits for reproducible, sensitive, and reliable serodiagnosis of the disease. In this study, the possibilities of using various antigen drugs and multi-antigen composites as antibody-binding components used in the manufacture of immunochromatographic test systems for the diagnosis of tuberculosis are considered.

2. Materials and Methods

Recombinant proteins MTB-12, MTC-28, P-30, P-34, MPT-64, Ag-85, and MTSA (provided by the State Research Center of Applied Microbiology [Obolensk, Russia]) were used, as well as Mycobacterium tuberculosis growth factors of the Rpf family (Rpf, Rpf-100, Rpft1, Rpft2) obtained from the Laboratory of Microorganisms Stress Research Centre of Biotechnology of the Russian Academy of Sciences (Moscow, Russia) [11-12]. In addition to specific antigens, the complete set of test systems being developed includes antibodies against human immunoglobulins labeled with colloidal gold and secondary anti-species antibodies to control the reactivity of anti-human antibodies (the formation of a control zone on the membrane).

Colloidal gold was obtained by the Frens method [13]. One ml of a 1% aqueous solution of $HAuCl_4$ was added to 100 ml of water. The mixture was heated to boiling, and then 1.5 ml of a 1% aqueous solution of sodium citrate was added with vigorous stirring. The solution was boiled for 30 min, then cooled and stored at +4ºC.

Immobilization of antibodies on colloidal gold was carried out by means of adsorption. Antibodies were transferred into a 10 mM Tris-buffer solution, pH 9.0, after which they were introduced into the nanoparticle solution with vigorous stirring. The mixture was incubated for 30 min, then a 10% bovine serum albumin (BSA) solution (vol conjugate: vol BSA = 40:1) was added to the resulting conjugate and incubated for another 15 min with vigorous stirring. Gold nanoparticles with antibodies immobilized on them were precipitated by centrifugation at 13000 g and 4ºC for 15 min. The supernatant was removed, and the precipitate was transferred to a 50 mM Tris-buffer solution, pH 7.5, with 1% BSA and 1% sucrose. Sodium azide was added to a final concentration of 0.05% and stored at 4°C.

Membranes FF 60/100, FF 85/100, AE 99 (Schleicher & Schuell, Germany), HF07504, HF13504, HF24004 (Millipore, USA), and CNPC-SS12 (Advanced Microdevices, India) were characterized to form a multimembrane composite. The complete set of the test system included a) a nitrocellulose working membrane, b) a glass fiber conjugate support, c) a cellulose acetate separating membrane for applying a sample, d) a cellulose acetate final adsorbent membrane, and e) an adhesive laminating protective film.

Solutions of antigens and their mixtures were applied to a nitrocellulose working membrane using an IsoFlow dispenser (Imagene Technology, USA). Marker–anti-species antibody conjugates were applied to the glass fiber membrane. The membranes were then dried for 1 day at room temperature. Next, the working glass fiber for applying the sample and the final adsorbing membrane were collected in a multimembrane composite, and test strips 3.5 mm wide were prepared using an Index Cutter-1 automatic guillotine (A-Point Technologies, USA). Together with a desiccant (silica gel), the test strips were hermetically sealed in bags of a combined material based on aluminum foil using an FR-900 mini-conveyor sealer (Wenzhou Dingli Packing Machinery, China). The test strips were prepared and packaged in a special room with a relative humidity of no more than 30% at 20–22°C.

Experimental samples of test systems were characterized using a panel of positive (n = 25) and negative (n = 12) samples. The collecting and testing of human serum samples was performed following the International Ethical Guidelines for Biomedical Research Involving. Human Subjects.

Digital imaging was used to document and quantify binding. The test strip was scanned on a Lide 90 scanner (Canon, Japan) with a resolution of 600 dpi and without automatic contrast or color correction. The intensity of staining in the analytical zone was calculated using TotalLab v2.01 software (Nonlinear Dynamics, UK).

3. Results and Discussion

To study the diagnostic capabilities of various antigens and their composites, the classical format of serodiagnostics was implemented, in which antigenic preparations are immobilized on the working membrane in the analytical (test) zone, and a conjugate of anti-species antibodies with gold nanoparticles is applied to the glass fiber membrane. Contact of the test strip with the serum sample causes the immunoglobulins in the sample to interact with the anti-species antibodies. The front of the liquid moves along the test strip under the action of capillary forces. Upon reaching the analytical zone, antibodies specific to tubercular antigens bind to immobilized antigenic preparations, leading to the development of color (see figure 1).

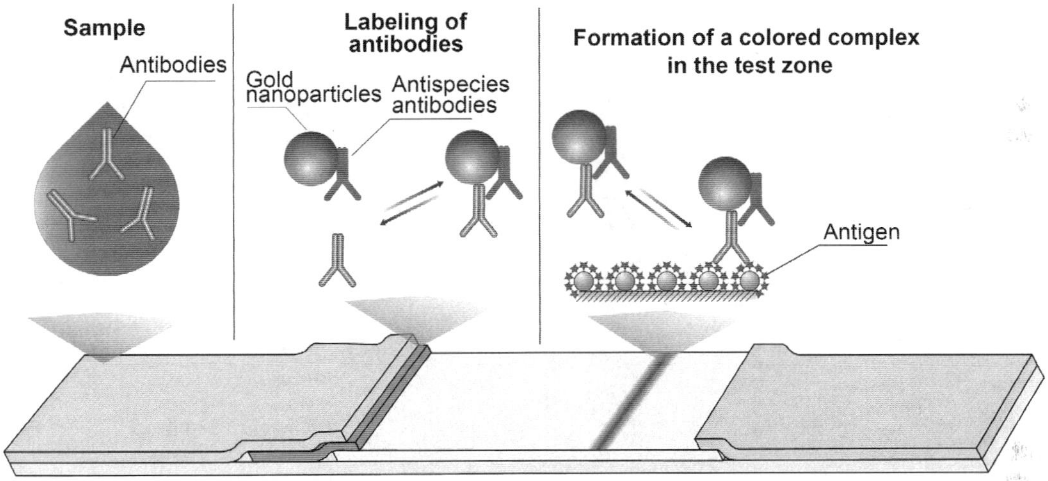

Figure. 1 The principle of serodiagnostic immunochromatography.

In immunochromatographic systems, colloidal gold (the most commonly used marker for visualization of binding zones) was used. For colloidal preparations, adsorption isotherms of antibodies were obtained, and conjugates with antibodies against human immunoglobulins (anti-IgA, anti-IgG, anti-IgM, anti-[IgA + IgG + IgM]) were synthesized for the reagent ratios accorded to plateou levels of the obtained adsorption isotherms. Analysis of the antibody/colloid molar load and the efficiency of binding between the conjugates and immobilized antibodies during lateral diffusion showed that, for serodiagnostic purposes, it was optimal to use sufficiently large particles that provided the possibility of monolayer adsorption of at least 15-20 immunoglobulin molecules.

Using the reagents obtained—antigenic preparations, standard preparations of antibodies, and conjugates with colloidal markers—experimental samples of immunoanalytical systems for tuberculosis serodiagnostics were made.

To ensure comparability of data on the sensitivity and specificity of testing using various antigenic drugs and their combinations, the results of screening a panel of blood samples from patients and healthy people were considered in qualitative ("yes-no") terms. At the same time, we were guided by the threshold level of specific antibodies of 0.2 mg/ml. This level is generally accepted in diagnostics of tuberculosis and reflects the severity of the infection process [14]. Figure 2 summarizes the test results.

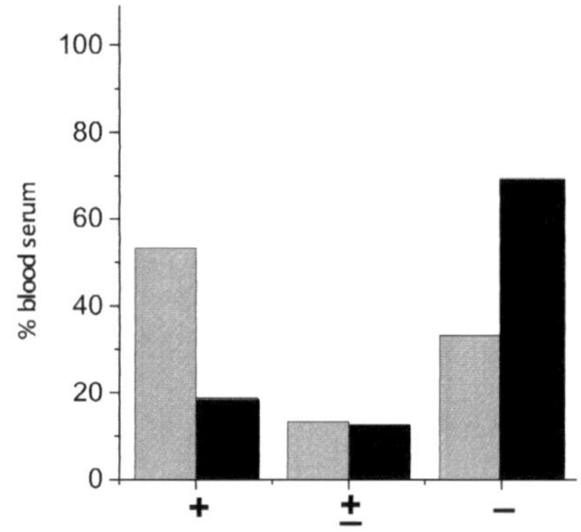

(a) Ag 85 antigen

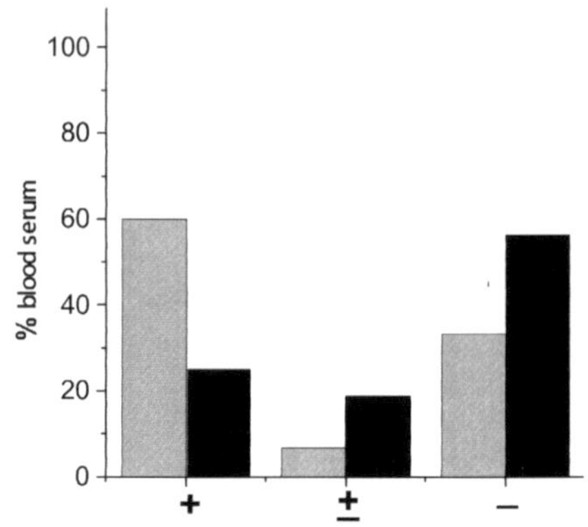

(b) MPT 64 antigen

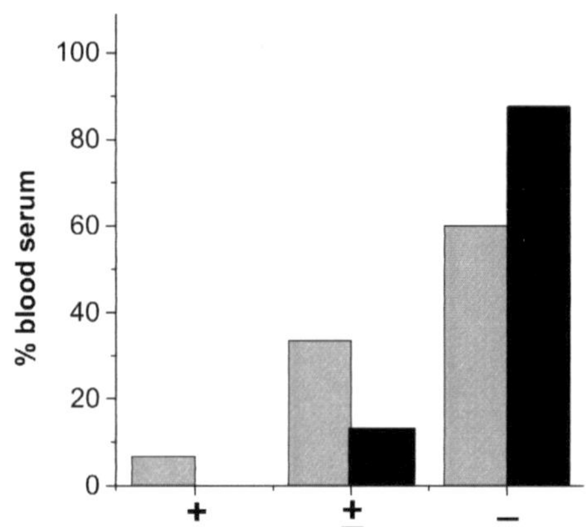

(c) MTSA

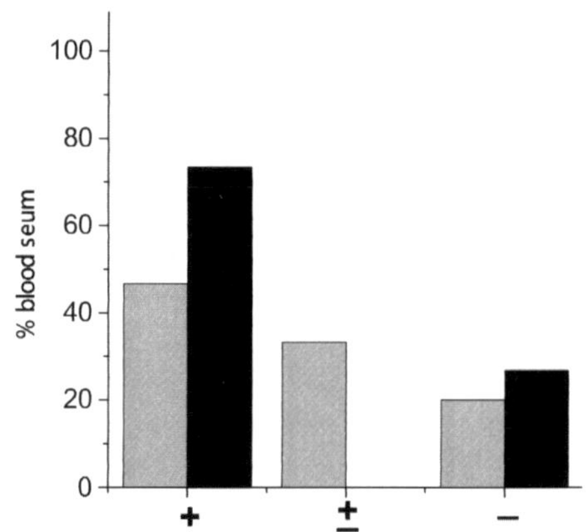

(d) MTB 12 antigen

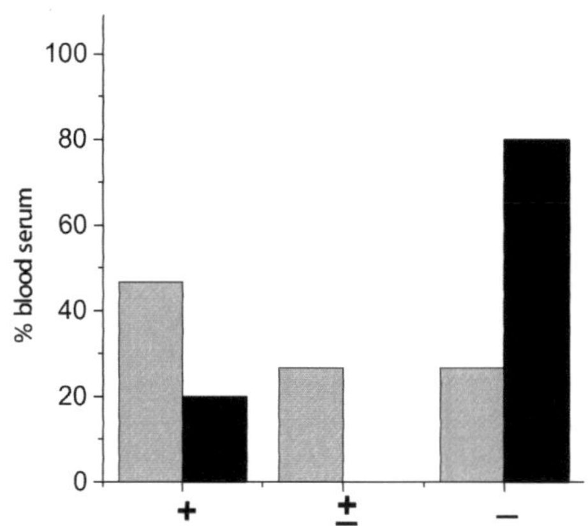

(e) MTC 28 antigen

Figure. 2 Binding of specific human immunoglobulins to proteins of the tuberculosis pathogen M. tuberculosis in blood serum of healthy people (black bars) and patients with tuberculosis (grey bars).

IOP Conf. Series: Materials Science and Engineering 1117 (2021) 012019 doi:10.1088/1757-899X/1117/1/012019

In characterizing the information content of the analysis using various recombinant antigens, it should be noted the diagnostic efficiency is low when using the MTC-28 and MTSA proteins. The rest of the tested antigens provided the assay sensitivity in the range of 50% to 70% (the maximum for MPT-64). Thus, the test is not sufficiently reliable if only one of the M. tuberculosis antigens is used.

To increase the sensitivity and specificity, the use of multicomponent combinations (cocktails [15]) of antigenic preparations immobilized on membrane carriers was proposed. This approach increased the percentage of correct diagnoses by 10%-20% (figure 3). Thus, the sensitivity and specificity of the analysis reached the level of 85%-90%, which is not inferior to known foreign analogues and meets the requirements of diagnostic practice. Varying the cocktail ratio prevented the potential threat that the switch to multi-antigen mixtures would increase the rate of false positives. As followed from the data obtained, it was advisable to use the M. tuberculosis P-34, MPT-64, and Ag-85 antigens as cocktail components for the scalable production of test systems.

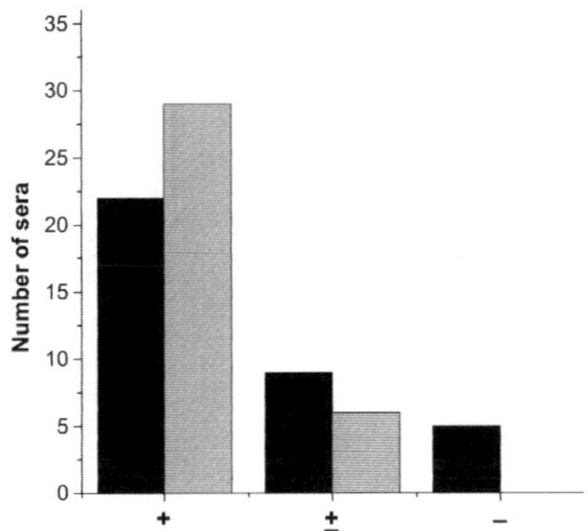

(a) Cocktail 1 (Rpf: Rpf-100: Ag 85: Rpft1: Rpft2 = 1: 1: 1: 1: 1)

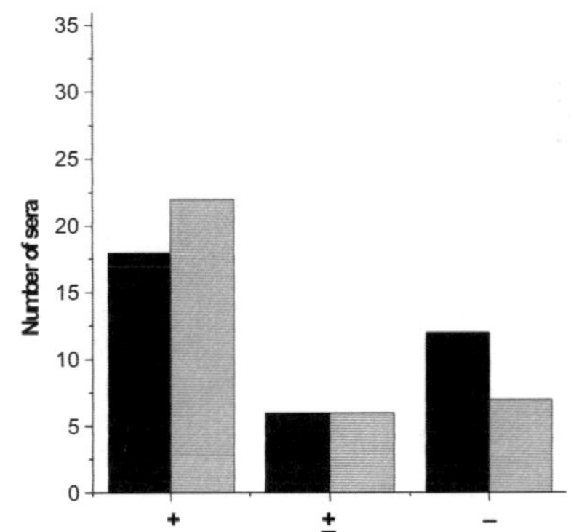

(b) Cocktail 2 (Rpf: Rpf-100: MPT 64 = 1: 1: 1)

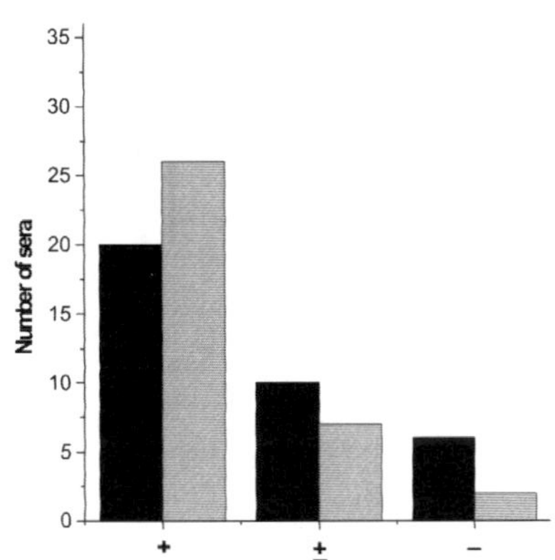

(c) Cocktail 3 (MPT 64: Rpft1: Rpft2 = 1: 1: 1)

Figure. 3 Testing of interaction with cocktails of recombinant antigens of blood serum of patients with tuberculosis for testing the presence of IgG (black columns) and sum of immunoglobulins (grey columns).

Optimizing the conditions for immunochromatographic detection was carried out using multivariate gradient analysis approaches by varying the concentration of soluble reagents and the surface density of the immunochemical system's immobilized components. The optimization criteria were the signal amplitude (the brightness of the detected colored band in the analytical zone of the test strip), the limit of reliable detection of antigen in the sample, and the threshold for the qualitative difference between positive and negative samples. Using these criteria, the effect of changes in the ratio of specific reagents on the efficiency of antigen and antibody detection was monitored. The concentrations of the components of the analytical system, specific antibodies, and conjugates varied in a range corresponding to a 10–20-fold change in the absolute amounts of reagents and were determined by the membranes' sorption properties.

The choice of the colloidal conjugate's concentration was carried out, guided by the requirement of clear non-device visualization of colloid binding in both the control and the analytical zones of the immunochromatographic test strip. Experimental data made it possible to recommend the use of colloid dilutions corresponding to an optical density of 580 nm in the range of 1.0-2.0 optical units. Preparations of the obtained colloidal conjugates at these concentrations exhibited almost no nonspecific binding to the carrier outside the membrane zones containing immobilized reagents. This allowed a clear qualitative analysis with an instrument-free (visual) registration of results.

The choice of the immobilized antigen's concentration was determined by two conditions: (i) the possibility of correct visual detection of colloid binding in the analytical zone and (ii) the compliance of the threshold level of difference between positive and negative samples with the requirements of medical diagnostics. Based on the obtained concentration dependences of the number of bound immune complexes and the sensitivity of specific antibodies' determination in the serum, the optimal conditions for the immunochromatographic analysis were established. Under these conditions, it was possible to work with sera in dilutions of 1:5-1:15 and obtain visually or instrumentally (reflectometrically) controlled results within 10-15 min.

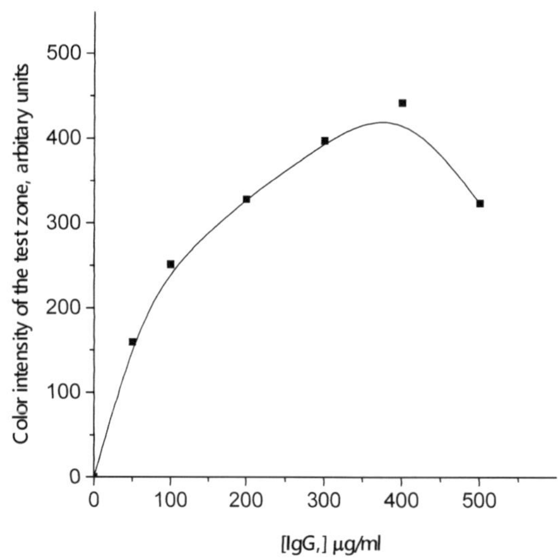

Figure. 4 Dependence of the color intensity of the analytical zone on the concentration of specific IgG in the serum sample.

Preliminary tests of test systems' experimental samples included testing samples with different amounts of determined specific antibodies. The characterization of the level of binding in the analytical zone of a standard preparation of affinity-purified antibodies at different concentrations (figure 4), as well as a serial dilution series of positive sera, was carried out. As can be seen from the presented data, the test system allowed researchers to reliably determine the level of antibodies in the diagnostically significant range of 100-400 μg/ml and to work with dilutions of blood samples up to 1:10. Although they allowed correct high-quality registration of results, large dilutions such as 1:20-1:40 led to a significant decrease in the amplitude of the detected signal, which negatively affected its accuracy.

Experimental samples were characterized using a panel of positive (n = 25) and negative (n = 12) samples. The analysis time was 10 min. The maximum brightness of the bands (associated colloidal label) in the control and analytical zones of the test strip was observed during a time interval of 10-15 min after the start of testing, which ensured the maximum reliability of the analysis. Under the selected conditions, the samples' matrix did not interfere with the formation of immune complexes in the control and analytical zones and did not lead to the development of nonspecific membrane staining. Nonspecific stain prevention was also ensured by introducing an additional membrane into the multimembrane composite, which cut off the cellular elements of the blood.

4. Conclusion

The study demonstrated efficacy of the use of multiantigenic mixtures as applied preparations in manufacturing immunochromatographic testes for serodiagnosis, The effect of diagnostic sensitivity increasing was be reached by simple mixing od absorbed proteins in equal ratio. The chosen conditions of immune interactions at multimembrane test strip provided distinguishing positive and negative samples in the accordance with common cut-off level of specific antibodies content for tuberculosis.

Acknowledgements

The study was financially supported by the Russian Science Foundation (Grant 16-15-00245).

We are grateful to L.V. Barshevskaya from Laboratory of Immunobiochemistry (A.N. Bach Institute of Biochemistry, Research Centre of Biotechnology, Russian Academy of Sciences) for the help in preparing this manuscript.

References

[1] Floyd K, Glaziou P, Zumla A and Raviglione M 2018 *Lancet* Respir Med. **6(4)** 299-314.
[2] Gupta S and Kakkar V 2018 *Biosens. Bioelectron.* **115** 14-29.
[3] Garcia-Basteiro A L 2018 *Pulmonology* **24(2)** 73-85.
[4] Urusov A E, Zherdev A V and Dzantiev B B 2019 *Biosensors* **9(3)** 89.
[5] Mak W C, Beni V and Turner A P F 2016 *Trends Anal. Chem.* **79** 297-305.
[6] Jiang N, Ahmed R, Damayantharan M, Unal B, Butt H and Yetisen A K 2019 *Adv. Healthc. Mater.* **9(14)** 1900244.
[7] Kozel T R and Burnham-Marusich A R 2017 *J. Clin. Microbiol.* **55(8)** 2313-2320.
[8] Long B, Liang S Y, Koyfman A and Gottlieb M 2020 *Am. J. Emerg. Med* **38(5)** 1014-1022.
[9] Barshevskaya L V, Sotnikov D V, Zherdev A V, Khassenov B B., Baltin K K, Eskendirova S Z, Mukanov K K; Mukantayev K K and Dzantiev B B 2019 *Biosensors* **9**, 115.
[10] Sotnikov D V, Zherdev A V and Dzantiev B B 2017 *Anal. Chem.* **89(8)** 4419-4427.
[11] Nikitushkin V D, Demina G R and Kaprelyants A S 2016 *Biochemistry (Moscow)* **81(13)** 1719-1734.
[12] Korolyova-Ushakova A G, Baranova E V, Ignatov S G, Soloviev P V, Kondakov N N, Mel'nikova T M, Abronina I, Podval'nyi N M, Kononov L O and Biketov S F 2019 *Appl. Biochem. Microbiol.* **55** 696-703.
[13] Frens G 1973 *Nat. Phys. Sci.* **241** 20-22.
[14] Awoniyi D O, Baumann R, Chegou N N, Kriel B, Jacobs R, Kidd M., Loxton A G, Kaempfer S, Singh M and Walzl G 2017 *Oncotarget* **8** 37525.
[15] Lyashchenko K P, Gortázar C, Miller M A and Waters W R 2018 *Vet. Microbiol.* **214** 89-92.

IOP Conf. Series: Materials Science and Engineering 1117 (2021) 012020 doi:10.1088/1757-899X/1117/1/012020

Development and Characterization of Bioplastic Film from *Salacca zalacca* Seed Starch

S Ma'mun, D N Alamsyah, and A S Pribadi

Department of Chemical Engineering, Universitas Islam Indonesia
Jl. Kaliurang Km. 14.5, Yogyakarta 55501, Indonesia

E-mail: sholeh.mamun@uii.ac.id

Abstract. The world community has given a particular concern to environmental issues such as pollution of the marine environment due to petroleum-based plastic materials. Since petro-based plastics cannot easily be degraded in nature, bioplastics' development is a must. Bioplastics can be made from various renewable feedstocks such as polysaccharides, protein, and fat. Salacca zalacca is one of the starch sources that can be used to manufacture bioplastics. The purpose of this study was to develop and characterize the bioplastic by use of Salacca zalacca starch. The bioplastic films were made by mixing all the raw materials at 70 oC for 40 minutes. The sample was then put into a mold and heating it in the oven at 50 °C for 24 hours. The results show that an optimum condition with 4 wt% of PVA occurs at the starch concentration of 1.8 wt% with the tensile strength and the elongation of 20 N and 53%. Moreover, the optimum condition with 15 wt% of PVAc gives the tensile strength of 20 N with lower elongation (15%).

1. Introduction

One of the recent environmental problems facing the world community is the marine environment's pollution due to petroleum-based plastic materials in which about 70 – 80% of plastic waste is found in the oceans that originates from land [1]. The petroleum-based plastics cannot be degraded both in the landfills and aquatic environments. Since petroleum-based polymers are very durable materials, they will remain intact as polymers for a long time [2].

Policies to reduce plastic use have been implemented in many countries worldwide to minimize plastic waste. However, these policies have proven insufficient to solve the problem; therefore, the substitution of petroleum-based polymers with biodegradable ones is necessary. Various renewable feedstocks, which are environmentally friendly, such as polysaccharides (starch, cellulose, chitin), protein, and fat, can be used as raw materials for biodegradable plastics manufacture [3-12]. Moreover, biodegradable synthetic materials such as polyvinyl alcohol (PVA), polyhydroxyalkanoate (PHA), polylactic acid (PLA) have also been used as materials for biodegradable plastics. PVA is a synthetic polymer that is soluble in water and easily degraded. PVA has widely been used as a promising alternative packaging material because of its excellent properties in packaging formation, resistance to oil and grease, high tensile strength as well as high flexibility [12]. Besides, plasticizers such as glycerol will improve the material performance because it reduces the inter-and intra-molecular strength but increases the film's mobility and flexibility.

Salacca zalacca, known as snake fruit, is a native Indonesian plant. The fruit can be made into several derivative products such as chips, syrup, sweets, and *pudding*. *Salacca zalacca* derivative products' production has resulted in a large amount of seed waste that needs to be utilized. Since *Salacca zalacca* seed contains about 81 wt% of starch [13], the *Salacca zalacca* seed waste can then be used as a raw material in manufacturing bioplastics. The purpose of this study is to develop and characterize the bioplastic film by use of *Salacca zalacca* seed starch as a matrix in addition to PVA and polyvinyl acetate (PVAc).

2. Materials and Method

2.1. Materials

The materials used were *Salacca zalacca* seeds obtained from a domestic industry in Sleman Yogyakarta, Indonesia, polyvinyl alcohol, polyvinyl acetate, chitosan, glycerol, acetic acid, and deionized water.

2.2. Apparatus and Procedure

Salacca zalacca seed starch was prepared by cutting the seeds into small pieces and dried them in the sun. The dried samples were then ground and sieved using a 100-mesh screening. A certain weight of seed flour was mixed with deionized water and left the mixture for about 24 hours until the starch was formed; the water was then removed while the sediment was taken and dried.

The bioplastic film was made by mixing 100 mL of deionized water, 10 mL of 5 wt% chitosan, and 2 mL of glycerol with a certain amount of the *Salacca zalacca* seed starch and a certain amount of either PVA or PVAc for about 40 mins at 200 rpm and 70 °C. The mixture was then poured into a mold and dried in an oven at 50 °C for 24 hrs. The bioplastic film was then taken from the mold to be further characterized.

3. Results and Discussion

The bioplastic films were made from *Salacca zalacca* seed starch (0.7 – 8.3 wt%) in the mix together with either PVAc (4 – 15 wt%) or PVA (4 wt%) by use of glycerol (2 wt%) as plasticizer and chitosan (0.4 wt%) as filler. The *Salacca zalacca* seed starch was used as the matrix to provide the desired plastic properties. Glycerol was used as the plasticizer because of its hydrophilic properties, which are readily soluble in water and hygroscopic, absorbing water in the soil to affect the film's degradation process. The bioplastic films produced were then characterized by measuring the tensile strengths σ, elongation ε, and degradability factors.

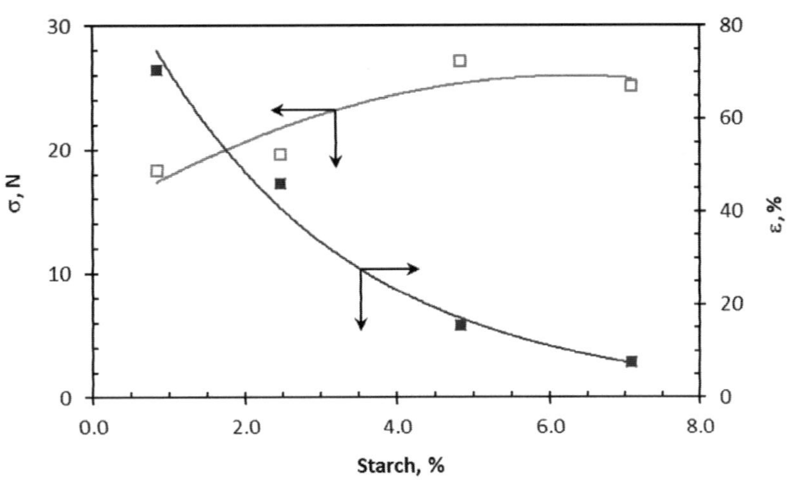

Figure 1. Effect of starch variation on the tensile strength σ dan elongation ε with 4 wt% of PVA

Figure 1 shows *Salacca zalacca* seed starch's effect on the tensile strength σ and elongation ε with 4 wt% of PVA. It can be seen that starch concentration positively affects the tensile strength in which the matrix becomes more compact and more robust. It requires more force to make it break up. Besides, as the starch concentration increases, the film thickness slightly increases; this leads to a shorten elongation factor. An optimum condition was observed at the starch concentration of 1.8 wt% with the tensile strength and the elongation of 20 N and 53%, respectively.

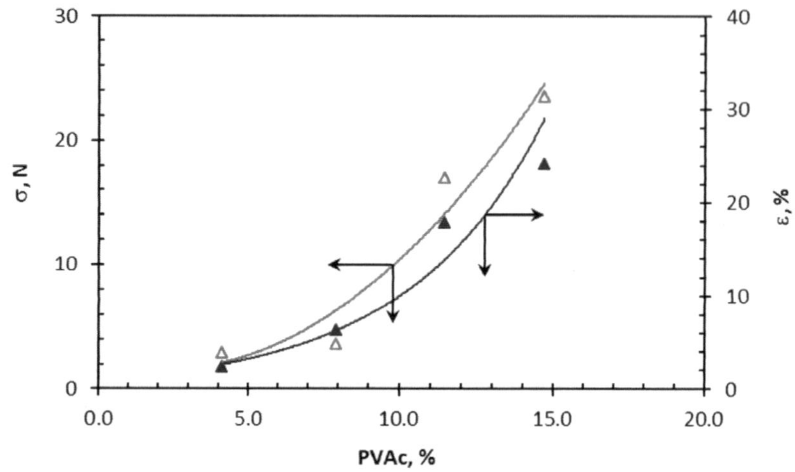

Figure 2. Effect of PVAc variation on the tensile strength σ dan elongation ε with 2 wt% of starch

In addition to PVA, PVAc was also used as another matrix with concentrations ranging from 4 – 15 wt%. Figure 2 shows PVAc concentration's effect on the tensile strength and the elongation with 2 wt% of starch. It is shown that as the PVAc concentration increases, both the tensile strength and the elongation factor also increase. This is due to PVAc's nature as white glue that could have similar properties as those of cassava starch. Using PVAc 15 wt%, the effect of the *Salacca zalacca* seed starch concentration on the tensile strength and the elongation was also evaluated. As seen in Figure 3, the higher the starch's concentration, the higher the tensile strength. Similar to the PVA matrix, PVAc also provides a compact matrix structure leading to higher force to break it up. This condition also affects shortening the elongation factor. The optimum condition is slightly better than the one of the PVA matrix, which gives the tensile strength of 22 N and the elongation factor of 15% with the starch concentration of 2.7 wt%.

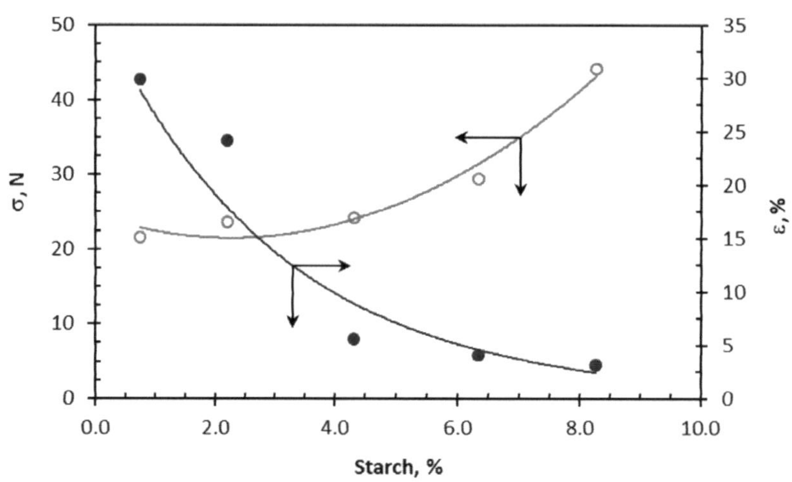

Figure 3. Effect of starch variation on the tensile strength σ dan elongation ε with 15 wt% of PVAc

Anaerobic degradation test was conducted to evaluate a change in the biofilm's morphological structure consisting of 12 wt% of PVAc and 2.2 wt% of the *Salacca zalacca* seed starch. Starch is mainly degraded by glycoside hydrolases, a large group of enzymes that catalyze glycosidic bonds' hydrolysis [14]. The biofilm was buried in the soil for 20 days. Figure 4 shows the morphological structure change before and after the degradation test. It can be seen from Figures 4a and 4b, the biofilm was physically changing in which the biofilm structure was significantly damaged. Moreover, based on the Scanning Electron Microscope (SEM) results, as shown in figures 4c and 4d, the biofilm was damaged due to starch's hydrophilic property.

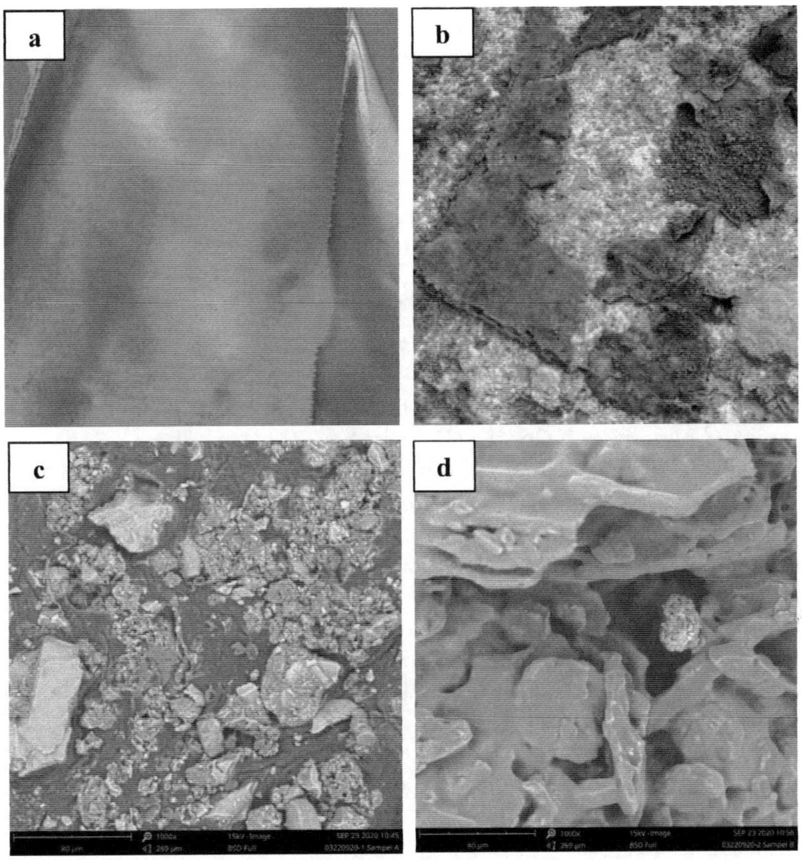

Figure 4. Effect of starch variation on the tensile strength σ dan elongation ε with 15 wt% of PVAc: 4a and 4b, sample before and after degradation test; 4c and 4d, SEM results before and after degradation test.

4. Conclusion

Bioplastic films from *Salacca zalacca* seed starch were developed with PVA and PVAc as the matrices and chitosan as filler. The biofilms made were then characterized toward the tensile strength, the elongation, and degradation process in the soil. The experimental results show that as the starch concentration increases, the tensile strength also increases while the elongation decreases. This occurs due to the compactness of the matrix, which makes it stronger. A degradation test was also conducted to evaluate the degradability factor when the biofilm was buried in the soil for a particular time. The result shows that the biofilm was damaged due to the hydrophilic property of starch.

Acknowledgment
This work was financially supported by the Department of Chemical Engineering, Universitas Islam Indonesia through Research Grant No. 086/Kaprodi/70/TK/VII/2020.

References

[1] Calabrò PS and Grosso M 2018 Bioplastics and waste management *Waste Management* **78** 800-01.

[2] Honhenblum P, Bettina L and Marcel L 2015 Plastic and microplastic in the environment Report REF-0551 Vienna, Austria: Umweltbundesamt.

[3] Luckachan GE and Pillai CKS 2011 Biodegradable polymers - a review on recent trends and emerging perspectives *Journal of Polymers and the Environment* ***19*** *637–76.*

[4] Amin MR, Chowdhury MA and Kowser MA 2019 Characterization and performance analysis of composite bioplastics synthesized using titanium dioxide nanoparticles with corn starch *Heliyon* **5** e02009.

[5] Jimenez-Rosado M, Zarate-Ramírez LS, Romero A, Bengoechea C, Partal P and Guerrero A 2019 Bioplastics based on wheat gluten processed by extrusion *Journal of Cleaner Production* **239** 117994.

[6] Azmin SNHM, Hayat NAbM and Nor MSM 2020 Development and characterization of food packaging bioplastic film from cocoa pod husk cellulose incorporated with sugarcane bagasse fibre *Journal of Bioresources and Bioproducts* **5** 248–55.

[7] Rocha CJL, Alvarez-Castillo E, Yanez MRE, Bengoechea C, Guerrero A and Ledesma MTO 2020 Development of bioplastics from a microalgae consortium from wastewater *Journal of Environmental Management* **263** 110353.

[8] Thiruchelvi R, Das A and Sikdar E 2020 Bioplastics as better alternative to petro plastic, *Materials Today: Proceedings* (In press: https://doi.org/10.1016/j.matpr.2020.07.176).

[9] Yamada M, Morimitsu S, Hosono E and Yamada T 2020 Preparation of bioplastic using soy protein International *Journal of Biological Macromolecules* **149** 1077–83.

[10] Emadian SM, Onay TT and Demirel B 2017 Biodegradation of bioplastics in natural environments *Waste Management* **59** 526–36.

[11] Zimmermann L, Dombrowski A, Volker C and Wagner M 2020 Are bioplastics and plant-based materials safer than conventional plastics? In vitro toxicity and chemical composition *Environment International* **145** 106066.

[12] Maryam, Rahmad D and Yunizurwan 2019 Micro synthesis of bacterial cellulose as a reinforcement in bioplastic composite with PVA matrix (polyvinyl alcohol) *Journal of Chemical and Packaging* **41** 110–18.

[13] Karta IW, Susila LANKE, Mastra IN and Dikta PGA 2015 Nutrition content in snake fruit (*Salacca zalacca*) seed coffee produced by Abian Salak farmers of Sibetan village which has the potential as a competitive local food product containing antioxidant *Virgin Journal* **1** 123–33.

[14] Warren RAJ 1996 Microbial hydrolysis of polysaccharides *Annual Review of Microbiology* **50** 183–212.

IOP Conf. Series: Materials Science and Engineering 1117 (2021) 012021 doi:10.1088/1757-899X/1117/1/012021

Study of engineering properties of lightweight geomaterial (LWGM) stabilized by Waste of Buton Asphalt

N Marfu'ah[1,2], T Harianto [1], R Irmawaty[1], A B Muhiddin[1]

[1] Civil Engineering of Engineering Faculty, Hasanuddin University, Gowa 92171, South Sulawesi, Indonesia

nurulmarfuah.nm@gmail.com

Abstract. Material with high strength but has the lightest density that can be tolerated is the main purpose of doing a research about soil improvement. This research aims to develop a lightweight geomaterial (LWGM) by substituting of the soft soil with Expanded Polysterene (EPS) which is stabilized with Waste of Buton Asphalt (WBA). The variation of the WBA used was 3%, 5%, 7% and 9%, while the variation of the EPS used was 0.15% and 0.30%. Curing of the specimens was carried out for 7 and 28 days to see the effect of curing on the pozzolanic reaction to the specimens strength. The test results shows that soil substitution with EPS reduce the density by 10% - 25%. Even though it decreased comparing to the strength of the sample without additional of EPS, the CBR and UCS LWGM values are still classified as good performance and can be tolerated. Based on the results of curing samples, during the curing period of 7 days, the CBR value increased up to 5 times at 0.15% EPS and 4 times at 0.30% EPS. Meanwhile, during the 28-day curing period, the CBR value increased up to 10 times at 0.15% EPS and 6 times at 0.30% EPS. Meanwhile, the UCS value after curing 7 days is 12 times - 48 times and after curing 28 days is 15 times - 63 times greater than untreated soils.

1. Introduction

Soil has an important role to support construction system such as dams, buildings, roads, backfill on retaining wall systems, bridges even landfill cover. That makes soil as supporting system must have high bearing capacity. Soft soil is very abundant in Indonesia, however they tend to have low shear strengths and also easy to lose shear strength. In order to increase its strength, soil improvement is needed. The interest in research about soil improvement with natural materials has grown rapidly nowadays. The most popular soil improvement methods known is a chemical stabilization using binder materials such as: cement [1], lime [2], fibres [3], fly ash [4] [5] and for this paper we used waste of Buton Asphalt (WBA) as another new binder materials.

Material with high strength but has the lightest density that can be tolerated is the main purpose of doing a research about soil improvement. In road constructions, high-density materials will easily cause deformation to the layer beneath them due to the effects of loading [5] while in the backfill of retaining wall systems it will increase lateral pressure to its retaining wall. To decrease its density, reducing its self weight with lightweight filling can be a solution. Lightweight material as an alternative for road pavement layer has been started since 1980 in Japan by using Expanded Polysterene (EPS) in the form of geoblocks. However, its utilization is still limited because it's expensive,has high buoyancy, and it's not flexible geometrically [6]. By considering the difficulties of utilization of geoblocks in some situation, the effort to mix soil with EPS beads by chemical stabilization methods is become a

IOP Conf. Series: Materials Science and Engineering 1117 (2021) 012021 doi:10.1088/1757-899X/1117/1/012021

consideration now. Moreover, soil mixed with EPS beads are easier to be made with the density we desired

2. Materials and Method

2.1. Materials

2.1.1. Soil

Soil sampling in this study was taken around Hasanuddin University campus in Gowa district, South Sulawesi, Indonesia. This soil then put into a container to maintain its natural water content then transported to Hasanuddin University's Geoenvironmental Laboratory. The physical and mechanical characteristics of the soil was tested referring to ASTM testing standard which will be explained further in the results and discussions section.

2.1.2. Waste of Buton Asphalts (WBA)

In the stabilization process, as the alternative binder materials, we used waste of Buton Asphalt. Waste of Buton Asphalt as it's named, it's originally from Buton Island and processed on Waste of Buton Asphalt treatment plant on Baubau Island (geographically lies between latitude 5.21 ° S – 5.33 ° S and longitude 122.30 ° E – 122.47 ° E).

Figure. 1 Location of WBA's processing plant

(a) (b)

Figure. 2 WBA (a) boulder, (b) passed #100 WBA

In the process of extracting Buton Asphalt, besides producing bitumen it also produces a residual called Mineral of Buton Asphalt/ Waste of Buton Asphalt (WBA). WBA is one of natural material which is originally a byproduct of Buton Asphalt. At first, the WBA is in the sludge form which is dried at room temperature (28°C) and it turns into boulder as shown in Figure 2. then crushed into powder passes No. 100 sieves before used. This is based on the research of Janz [7] which states that the level of particle fineness greatly affects the reactivation of the binder material. The higher is the specific surface of the binder, the faster the reaction between the soil and the binder material is. Furthermore, mineral contents of WBA from X-Ray Diffraction test results is shown at Table 1 below.

Table 1. Chemical Composition of WBA [9] (Rauf,2019)

Compound	Unit	Value
Gypsum (CaSO4)	%	63,63
Kalsit (CaCO3)	%	15,54
Quartz SiO2	%	16,46
Oldhamite (Ca,Mg,Fe,S)	%	2,78
Magnesite (MgCO3)	%	1,60

Based on Table 1, it can be seen that the mineral contents of WBA are dominated by gypsum, so we expect that WBA has a similar behaviour with lime in general. Therefore, we consider it as a potential materials to be used as binder agents in future that need to be researched more. From this XRD results, we determined the amount of WBA based on the reference which will be explained in Mix Design part.

2.1.3. Expanded Polystyrene
Expanded Polystyrene or EPS which also known as Expanded Polystyrol is another form of geofoam materials which is very popular since early 90s. EPS is material that used as a lightweight fill in the constructions. It has so many advantages compared to the other foam because it's eco-friendly, more economical and also available worldwide [9]. It's even easier to be used in field because it's already in beads form so it's more flexible and we can avoid EPS blocks cutting to fit particular application. The density of EPS that we used is around 0,17 kN/m^3.

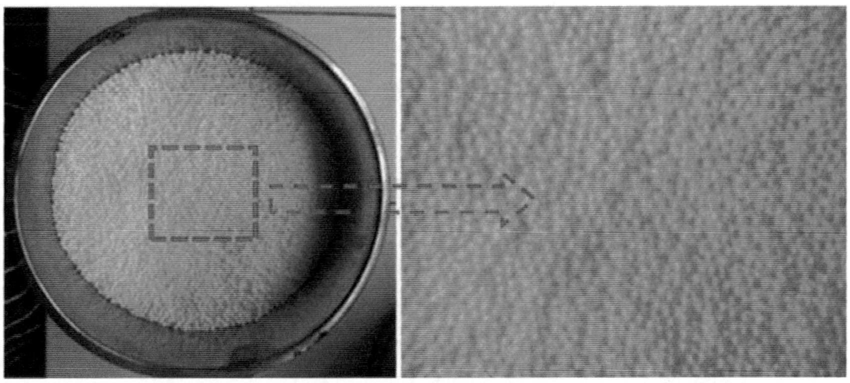

Figure. 3 EPS Bead

2.2. Methods

2.2.1. Mix Design
The specimens making and testing for these samples were using static compaction method as it's more accurate [10] [11], easy to control and approach the density that we desired. The following in Table 2. shows the mix design of the specimens.

Table 2. Mix Design

Composition	EPS %	WBA %	Curing time (days)
Untreated Soil	0	0	0
WBA- treated soil	0	3,5,7,9	7, 28
Lightweight Geomaterial(LWGM) 0,15%	0.15	3,5,7,9	7,28
Lightweight Geomaterial(LWGM) 0,30%	0,30	3,5,7,9	7,28

Based on Table 2, there are 4 types of samples tested for UCS and CBR. UCS and CBR specimens were made based on the OMC and MDD values obtained from standard proctor test results. CBR specimens were made with dimensions of 15.2 cm x 17.8 cm meanwhile UCS specimens were made with d= 5,5 cm and h= 11 cm. The amount of WBA were determined based on 3 references which are INDOT, Ingles and Metcalf, and Alaska Department of Transportation and Public Facilities Research and Technology Transfer [11].
The UCS test is carried out to determine the compressive strength while CBR test is carried out to determine the bearing capacity of the specimens, especially those that aims to be used as road pavement layer. The UCS test is the standard test that usually carried out for cement-treated soil [12], since WBA is also pozzolanic materials which similar to lime and cement, the specimens will be cured for 7 days, 14 days, until 28 days to optimize the pozzolanic reaction.

3. Result and Discussions

3.1. Soil Properties

The following in table 3. shows the data on the characteristics of the original soil used in this study.

Table 3. Soil Properties

No.	Name	Standard Method (ASTM)	Value	Unit
	A. Physical Properties			
1.	Specific Gravity	D-162	2,71	-
2.	Natural Water Content	D2216-98	12,17	%
3.	Classification System			
	a. USCS	C-136-06	CH	-
	b. AASHTO		A-7-6	-
4.	Atterberg Limit			
	a. Liquid Limit (LL)	D-423-66	58,37	%
	b. Plastic Limit (PL)	D-424-74	29,08	%
	c. Plasticity Index (PI)	D-4318	29,29	%
	B.Mechanical Properties			
5.	Standard Proctor Test	D-698		
	a. ɤdry max (MDD)		14,12	kN/m3
	b. Optimum Moisture Content (OMC)		25,28	%
6.	Unconfined Compression Strength (UCS)	D-633-94	57	kN/m2
7.	California Bearing Ratio Test (CBR)	D-1833	6,84	%

Table 3 shows the data on the characteristics of untreated soil, both its physical and mechanical characteristics. Based on the results of the sieve analysis and hydrometer tests and the Atterberg limit test, the soil used in this study are classified as CH according to the USCS and A-7-6 according to AASHTO classification, which means that the soil is very soft clay with high plasticity.

3.2. Unconfined Compression Strength (UCS)

The following is the effect of additional of WBA as a binder material shown in Figure 4. Based on the data obtained by Rauf [12], the additional of WBA increased the UCS value with the maximum UCS value obtained at the additional of 7% WBA and decreased along with the additional of 9% WBA. This is happened both in specimens with or without EPS. It can be seen in Figure 4. Adding EPS to the specimens decreased the UCS value, but this value is still above untreated soil which is only 57 kN /m^2.

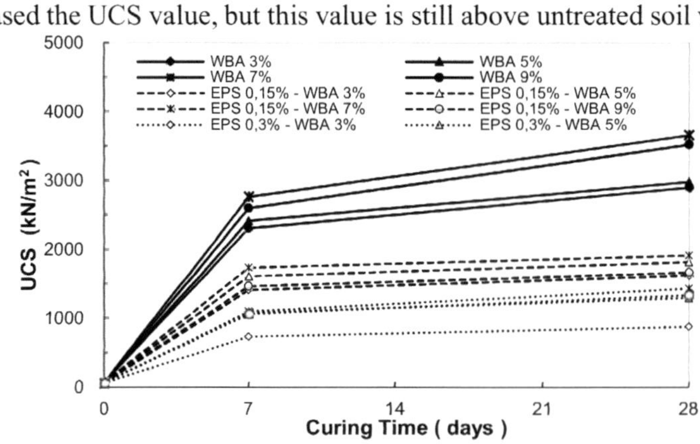

Figure. 4 Compressive Strength of Various Sample

The additional of WBA increased the UCS value 12 times - 48 times in the 7-day curing period and increased 15 times - 63 times in the 28-day curing period. Based on Figure 4. we can conclude that the UCS value tends to increase along with increasing curing period, the same behavior is also shown by cement-treated soil.

3.3. California Bearing Ratio

The following Figure 5. is the result of WBA-treated soil specimens which we compared to LWGM 0.15% and LWGM 0.30% at the 7-day curing period. [13]

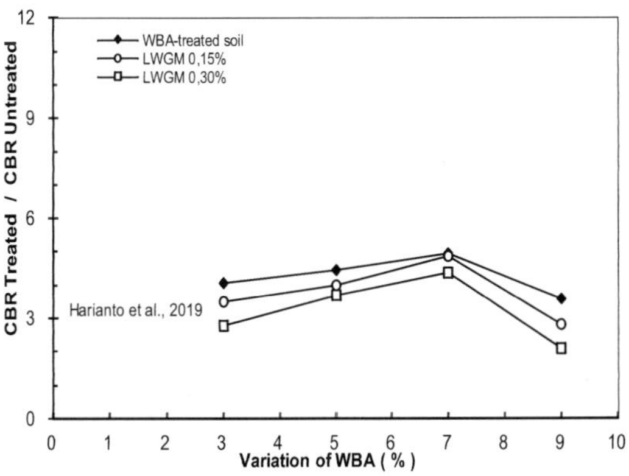

Figure. 5 CBR Ratio sample cured for 7 days

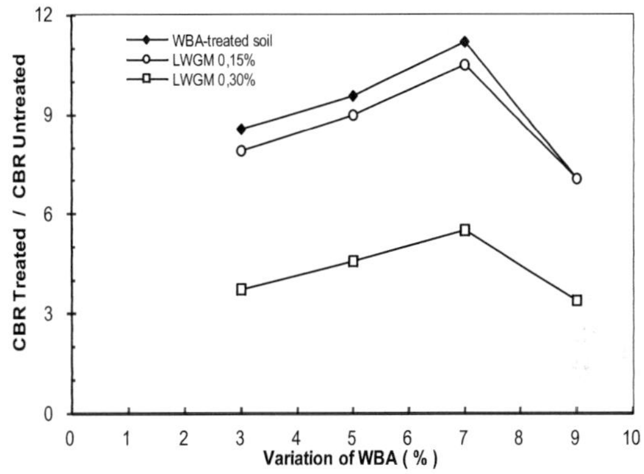

Figure 6. CBR Ratio sample cured for 28 days.

Based on the data shown in Figure. 5, both WBA-treated soil and LWGM increased the CBR value with maximum value obtained at 7% WBA addition and it decreased after additional of 9%. However, the CBR value of WBA-treated soil was higher than the value of LWGM specimens due to additional of EPS. WBA-treated soil had the highest CBR value of 4.9 times that of untreated soil, while LWGM 0.15% and LWGM 0.30% were 4.8 times and 4.3 times, respectively. Meanwhile, in Figure 6. is shown specimens testing results after curing 28 days. For WBA-treated soil specimens, the CBR value increased up to 11 times, while for LWGM 0.15% was 10.5 times and LWGM 0.30% was 5.5 times compared to untreated soil. These results proved that the treated samples especially samples that treated with pozzolanic materials need time to react optimally. That's why curing period are needed for specimens treated with chemical stabilizations.

3.4. Correlation of Mechanical Properties Parameters

The relationship between all parameters (UCS, CBR and E) is shown in Figure 7. The figure below is a comparison of 3 cases which are 1.) WBA-treated soil (without additional of EPS), 2.) LWGM which stabilized by WBA and 3.) BM+C treated soil which are taken from previous research of Harianto [1].

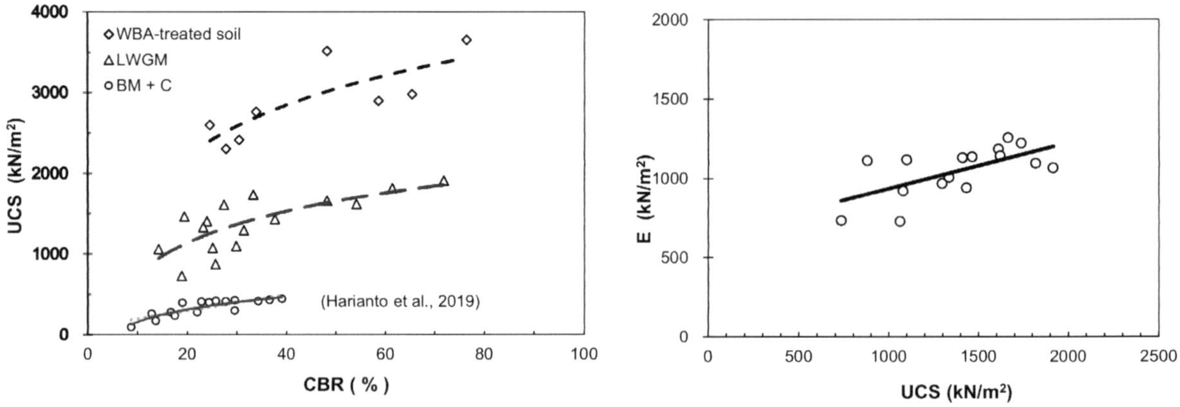

Figure 7. Relationships beetween parameters: (a) UCS and CBR (b) E and UCS LWGM

In those figures above, we can see the performance of each sample with different treatments. WBA-treated soil has the best performance when compared to LWGM and BM + C treated soil with the highest CBR value of 76.38% and UCS value of 3653.9 kN / m^2. LWGM performed below the WBA-treated soil sample with the highest CBR value of 71.820% and UCS value of 1913.9 kN / m^2, but the results were still above the performance of the BM + C treated soil sample which had the highest CBR value of 39% and UCS value of 446,7 kN / m2. From Figure 8 (b) we can see that the modulus elasticity tend to increase along with increasement of UCS value. This is related to calcium hydrate silicate (CHS) formation which are sides products of hydration and hydrolitics reactions on cement/pozzolanic-treated soil. These compounds are crystallized and get hardening over time [16]

3.5. The Impact to Density Reductions

The following in Figure 8 shows the density ratio of all samples compared to untreated soil.

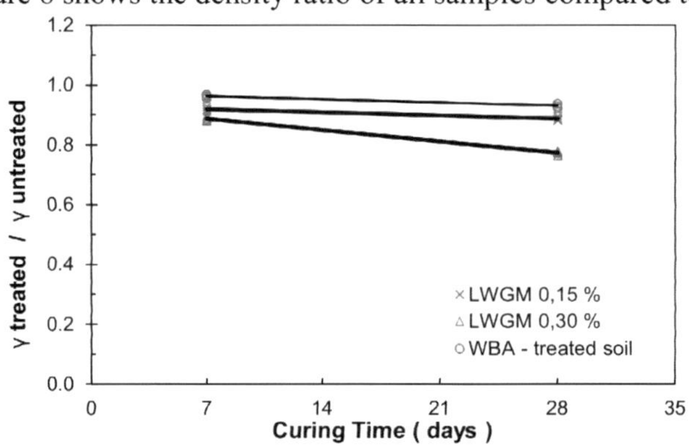

Figure 8. Ratio γ sample

As we can see, the WBA-treated soil is the heaviest and the closest to the density of untreated soil with a ratio of 0.922 - 0.970. While for LWGM 0.15% and LWGM 0.30%, respectively 0.881-0.928 and 0.764-0.839. EPS is added to a sample to obtain a lightweight sample. The EPS used in this study has a density 0.17 kN / m^3 while the soil's density is 14.12 kN / m^3. EPS is 83 times lighter than soil within the same volume unit. This is the reason why EPS is used as a substitute material to obtain lightweight geomaterials. However, the additional of 0,15% and 0,30% EPS for this research still have good performances by referring to its UCS and CBR results.

Conclusion

- Additional of WBA as the binder materials affect the UCS and CBR Value more than the impact of BM+C on treated soil because WBA's pozzolanic content is more than BM+C.
- Additional of EPS affect the UCS and CBR value. The less is the amount of EPS, the greater UCS and CBR Value were obtained. These amount of EPS (0,15% and 0,30%) still can be tolerated by referring its UCS and CBR Value. However, it's still needs further research for the other characteristics and also another amount of EPS.

References

[1] T. Harianto, F. Sitepu and Jasruddin, "Strength improvement of cement stabilized soil by binder mineral additive," *Lowland Technology International,* vol. 21, no. 2, pp. 90-97, 2019.

[2] Z. Saing, L. Samang, T. Harianto and S. Patanduk, "Study on characteristics of laterite soil with lime stabilization as a Road Foundation," *International Journal of Applied Engineering Research,* vol. 2, no. 14, pp. 4687-4693, 2017.

[3] T. Harianto, Y. Du, S. Hayashi, D. Suetsugu and Y. Nanri, "Geotechnical properties of soil-fibre mixture as a landfill cover barrier material," *J. Southeast Asian Geo. Society,* vol. 39, no. 3, pp. 137-143, 2008.

[4] R. M. Brooks, "Soil Stabilization with Fly Ash and Rice Husk," *International Journal of Research and Reviews in Applied Sciences,* vol. 1, no. 3, pp. 210-217, 2009.

[5] A. Kumar, B. J. Walia and A. Bajaj, "Influence of Fly Ash, Lime, and Polyester Fibers on Compaction and Strength Properties of Expansive Soil," *J.Mater.Civ.Eng. ASCE,* vol. 19, pp. 242-248, 2007.

[6] J. S. Horvath, "The Compressible Inclusion Function of EPS Geofoam," *Geotextiles and Geomembranes,* pp. 77-120, 1997.

[7] D. Negussey and M. Jahanandish, "Comparison of Some Engineering Properties of Expanded Polysterene with Those of Soils," 1993.

[8] M. Janz and S. Johansson, "The Function of Different Binding Agents in Deep Stabilization," Swedia, 2002.

[9] J. S. Horvath, "Expanded Polystyrene (EPS) Geofoam: An Introduction to Material Behavior," *Geotextiles and Geomembranes,* pp. 263-280, 1994.

[10] B. Sharma and A. Deka, "Static Compaction Test and Determination of Equivalent Static Pressure," India, 2016.

[11] B. Venkatarama Reddy and K. Jagadish, "The Static Compaction of Soils," 1993.

[12] H. C. Hardiyatmo, Stabilisasi Tanah untuk Perkerasan Jalan, Yogyakarta: Gadjah Mada University Press, 2010.

[13] R. S. Millard, "Cement and lime stabilization," *Road building in the tropics,* vol. 9, pp. 183-185, 1993.

[14] I. Rauf, L. Samang, T. Harianto and A. Arsyad, "Pengaruh Pemeraman terhadap Nilai Kuat Tekan Bebas Tanah Lempung Stabilisasi Limbah Aspal Buton," Surakarta, 2019.

[15] T. Harianto, S. H. Nur, I. Rauf and N. Marfu'ah, "Studi Nilai CBR Geokomposit Ringan (Tanah-EPS) Stabilisasi Limbah Aspal Buton," in *Semnas IX,* Yogyakarta, 2019.

[16] S. H. Chew, A. H. M. Kamruzzaman and F. H. Lee, "Physicochemical and Engineering Behavior of Cement Treated Clays," *J. Geotech. Geoenviron. Eng. ASCE,* pp. 696-706, 2004.

The Effect of Modified LTO Synthesis using Salt Assisted Method with Various of Salt Metal Materials

Soraya Ulfa Muzayanha[1], Agus Purwanto[1], Muhamad Iqbal Putra[1], Harry Kusuma Aliwarga[2]

[1]Chemical engineering department, Advance Material Laboratory, Surakarta, Central Java, Indonesia

[2]UMG Idea Lab Indonesia, Jl. Tangkas Baru Blok E No.2, Jakarta, Indonesia.

Corresponding authors e-mail: iqtraaa@student.uns.ac.id

Abstract. This study aims to determine the effect of modified LTO synthesis using salt assisted method with various related materials. Salt assisted $Li_4Ti_5O_{12}$ for lithium-ion batteries with stable rate-capacity were prepared by simple solid state-mixing, followed by high temperature calcination. This process effected the performance of Li-ion kinetics and stability. Therefore, the salt assisted LTO, which was doped with 1wt% $CaCl_2$, $MgCl_2$, and NaCl presented various discharge capacity, and the highest was shown by LTO-Ca (82.41 mAh g^{-1}, at 0.1C). While the stable rate-capability at 0.5C and 1C were 59.28 mAh g^{-1} and 58.72 mAh g^{-1}, respectively.

1. Introduction

$Li_4Ti_5O_{12}$ (LTO) is an anode material with an outstanding behaviour, because of its specifications in using high-current electronic devices. At 1.5V vs. Li + / Li, LTO work stably by avoiding electrolyte decomposition, and exhibit safety concerns by the use of carbon that produces a film of surface passivation [1], [2]. The surface passivation cause capacity loss and decreased battery life cycle [3], [4]. Also, LTO has superior structural stability where there is no change in its structure or volume during the lithiation / de-lithiation process [5]–[7].

There are many methods of synthesizing LTO materials, such as sol gel [8], [9], spray pyrolysis [10], hydrothermal [11], microwave-assisted synthesis [12], [13], and solid state reactions [5]–[7]. The studies of Yin et al. (2015) and Bai et al. (2008) stated that the application of molten salt method improved the crystallinity of the LTO material [14], [15]. In several reports, the addition of other metals, such as Ca, Mg, and Na were carried out. The research by Sattar et al (2020) found that, the addition of Mg through high temperature solid state technique, resulted in superior discharge ability and improved electrochemical efficiency [16]. In the Zhang et al (2020) report, the addition of Ca through the co-precipitation method showed that, Ca increased electrochemical stability, capacity, and reversibility [17]. Furthermore the addition of NaCl increased the crystal formation of ZnO, $ZnFe_2O_4$, and strontium ferrite [18]–[20]. Meanwhile, the increased material crystallinity boosted the electrochemical performance [21].

In this study, the salt-assisted method was adopted with 1wt% $CaCl_2$, $MgCl_2$, and NaCl, and it was known that this technique has not been used in the development of LTO materials. Using a CR2032 coin cell and $LiFePO_4$ as the cathode, electrochemical performance was evaluated.

2. Material and Methods

2.1 Materials $Li_4Ti_5O_{12}$

The material used for the synthesis of $Li_4Ti_5O_{12}$ (LTO) were LiOH (Merck, Darmstadt, Germany) and TiO_2 (Merck, Darmstadt, Germany) used as sources of Li and Ti in the $Li_4Ti_5O_{12}$ material. Analytical grades of $MgCl_2$, NaCl, and $CaCl_2$ (Merck, Darmstadt, Germany) were used as salt addition, while methanol or CH_3OH (Merck, Darmstadt, Germany) were used as dispersants in the solid-state mixing process.

2.2 $Li_4Ti_5O_{12}$ Preparation

LTO powders were synthesized via solid-state reaction. TiO_2 and LiOH were mixed with a Li:Ti molar ratio of 4:5. The investigation carried out in this study was to compare the effects of adding LTO-Na, LTO-Ca, and LTO-Mg, to determinethe electrochemical performanc e, as well as the addition of NaCl, $CaCl_2$, and $MgCl_2$ (1% wt). LiOH, TiO_2, $CaCl_2$/$MgCl_2$/NaCl, and methanol were mixed in a ball mill for 4 hours. The purpose of using methanol as a material solvent. After the mixing process was complete, the precursor was dried in the oven to evaporate the methanol and sintered at 800°C for 12 hours with O_2 flow (Samator, Jakarta, Indonesia). After this, the material was pounded and sieved using 100 mesh.

2.3 $Li_4Ti_5O_{12}$ Characterization

The crystalline structure of the LTO powder was analyzed using X-Ray Diffraction/XRD (D2 Phaser Bruker, Germany) and Cu-K-alpha radiation (= 1.5418 Å) with two theta in the range 10° - 80°. The morphological identification of LTO powder was carried out using scanning electron microscopy/SEM (Jeoul JSM-6510LA, Tokyo, Japan).

2.4 $Li_4Ti_5O_{12}$/$LiFePO_4$ Cylindrical Cell Assembly

The analysis of electrochemical performance characteristics was carried out using a CR2032 coin cell, and the cathode used was Nickel Mangan Cobalt/NMC (MTI, Richmond, CA, USA). LTO powder was mixed with carbon (acetylene black / AB, MTI, Richmond, CA, USA), and binders (carboxylmethyl cellulose (CMC), as well as styrene butadiene rubber (SBR) with a weight ratio of 80: 10 :2:8 in water. This was carried out to make slurry coated onto copper foil using doctor's blade technique, followed by drying to evaporate the water. Then, the coin cell was fabricated in the glove box with Ar. $LiPF_6$ (1:1 v/v of EC / DMC) and cellgard (MTI, Richmond, CA, USA) were used as electrolytes and separators, respectively. The charge and discharge test was performed using the MTI BST-CH8-3A battery analyzer (1.0 V to 3.5 V potential range).

3. Results and Discussion

3.1 XRD Characterization

Table 1. The structure properties of LTO materials.

Sample	Lattice Constant (Å)	FWHM (°)			Crystallite Size (nm)		
		(111)	(311)	(440)	(111)	(311)	(440)
LTO-Ca	8.36179	0.103	0.116	0.071	81.266	84.218	244.9
LTO-Na	8.35483	0.120	0.132	0.075	68.783	74.076	232.2
LTO-Mg	8.35896	0.113	0.149	0.076	74.089	65.600	229
LTO	8.36698	0.188	0.192	0.306	44.846	50.718	56.67

The characteristics of LTO material structure was carried out using XRD. Figure 1. showed the XRD patterns of the LTO material at various salt additions at 2θ in the 10°-80° range. All samples showed the overall formation of LTO peaks according to JCPDS Card No. 42-0207, with the cubic spinel structure and the space group of F3dm [43-45]. The XRD results of adding $CaCl_2$ salt showed the formation of $CaCl_2$ peaks at 2θ at 25°, and 33° (JCPDS No. 49-1092 and 12-0056). The addition of $CaCl_2$ and $MgCl_2$ indicated the formation of $MgCl_2$ peaks at 2θ and 33° respectively (JCPDS No. 25-1156). Table 1 showed that the lattice constant value was approximately 8.36179-8.35483 Å, and the value was close to that of LTO constant lattice from other literature samples (8.395-8.3538 Å) [22]–[24].

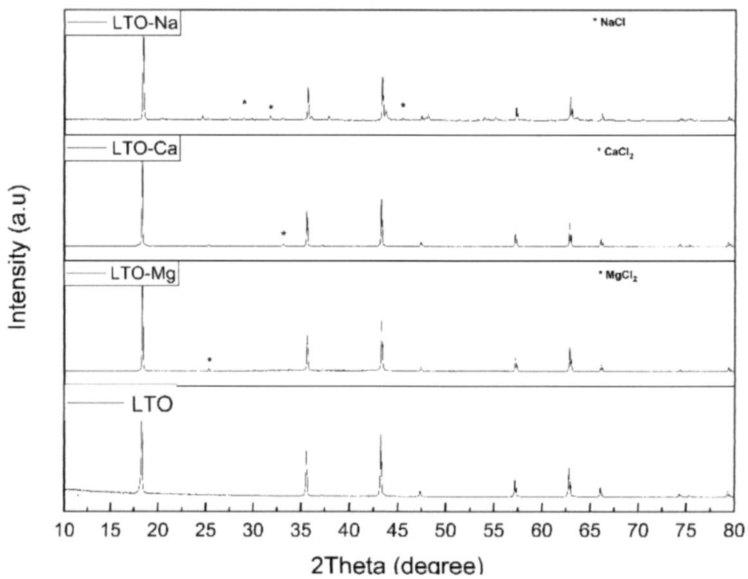

Figure. 1 XRD Pattern of LTO-Salt Assisted

SEM Characterization

To analyze the material morphology, the material was characterized using SEM. Figure 2 shows SEM images for various LTO materials with variations in the addition of salt metal through a solid state reaction. The size range of all particle sizes of the LTO material was 1.049 μm – 2.377 μm. Figure 2c shows the morphology of the LTO-Ca material with little agglomeration and the smallest crystal size was 1.049 μm.

IOP Conf. Series: Materials Science and Engineering 1117 (2021) 012022 doi:10.1088/1757-899X/1117/1/012022

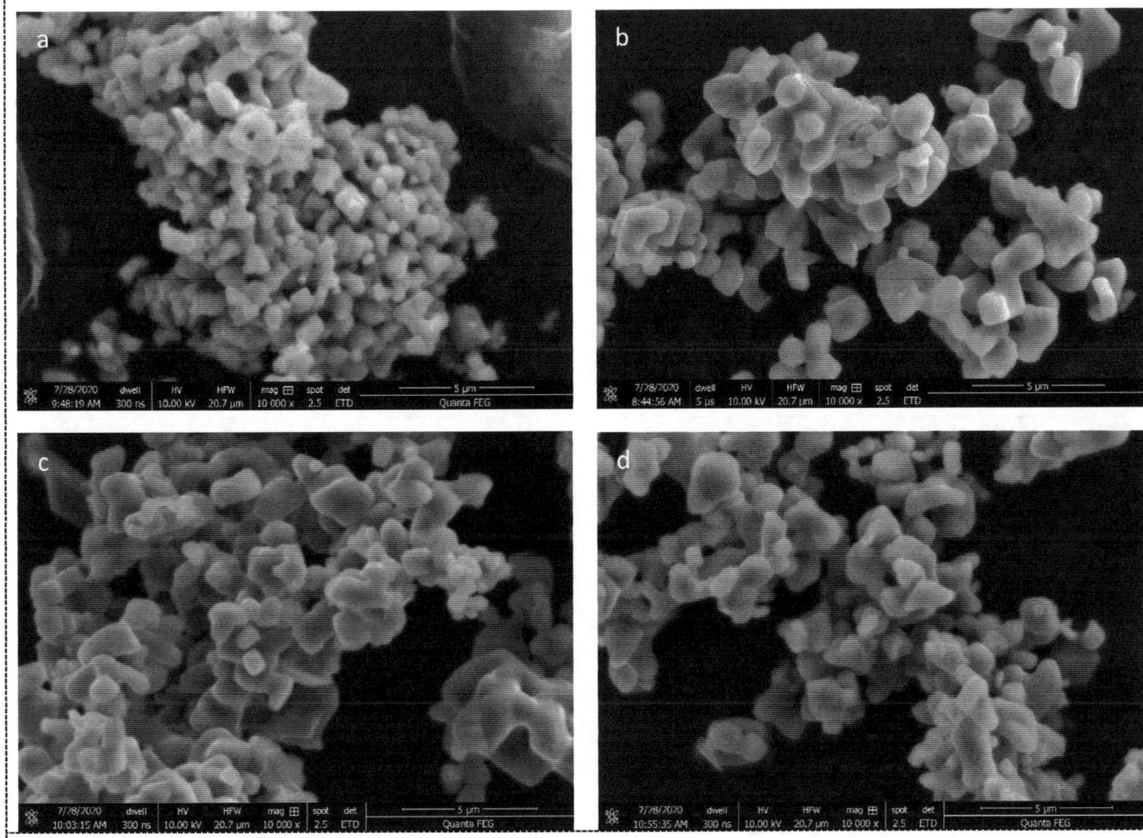

Figure. 2 SEM Image LTO (a) LTO-Na (b) LTO-Mg (c) LTO-Ca (d)

3.2 Electrochemical Characterization

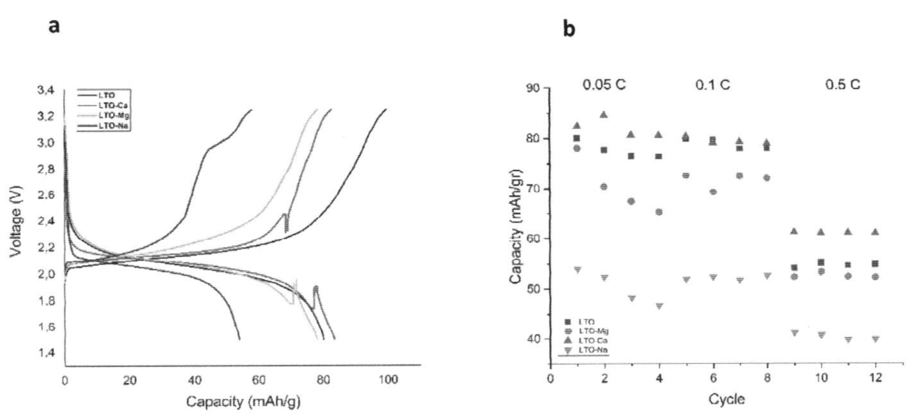

Figure. 3 Charge-Discharge Specific Capacity at 0.1C (a) Rateability of Each Sample (b)

To determine the effect of adding salt ($MgCl_2$, NaCl, $CaCl_2$), electrochemical testing was carried out using a BTS Analyzer. Figure 3 showed the electrochemical performance of 0.1C formation, where $CaCl_2$ indicated the best performance with a capacity of 82.18 mAh / gr. While the lowest was obtained in the LTO with the addition of NaCl with a capacity of 47.66 mAh / gr. The analysis was continued by rating the three samples at 0.5C, to ascertain the retention capacity of each sample. The results showed that the obtained retention capacity for each sample was the highest shown in LTO-Ca of 71.4%, while the lowest was shown in LTO-Na of 70.67%. These results indicated that the addition of $CaCl_2$ has a greater specific capacity, and produced a material with a better rateability compared to LTO-Na and LTO-Mg. To measure the cycleability resistance of the LTO-Ca material, cycle tests were carried out 100 times. The results showed that after 100 cycles the capacity was reduced to 81%. To determine the performance of LTO-Ca, a rate of 0.1C, 0.5C, 1C, 5C, and 10C was carried out, as in Table 2 which was presented as follows.

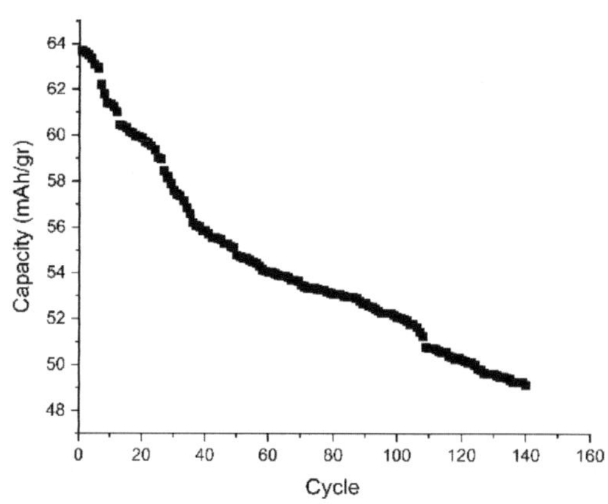

Figure. 4 Cycle Test LTO-Ca

Table 2. Table of Charge and Discharge Specific Capacity of LTO-Ca at Various Rates

Sample	Charge (mAh/gr)					Discharge (mAh/gr)				
	0.1C	0.5C	1C	5C	10C	0.1C	0.5C	1C	5C	10C
LTO-Ca	81.47	57.62	59.72	51.6	27.48	82.18	59.28	58.72	53.3	27.54

4. Conclusion

The addition of salt has a different effect on the electrochemical performance of modified LTO synthesis. LTO-Ca has the highest retention capacity as well as specific discharge and charge capacities with 82.18 mAh / gr at 0.1C rate. Therefore, the addition of $CaCl_2$ salt was better than NaCl, and $MgCl_2$ salts.

5. References

[1] M. Ganesan, 2008, *Li4Ti2.5Cr2.5O12 as anode material for lithium battery, Ionics (Kiel).*, **14**, pp. 395–401.

[2] L. Gao, R. Liu, H. Hu, G. Li, and Y. Yu, 2014, *Carbon-decorated Li4Ti5O12/rutile TiO 2 mesoporous microspheres with nanostructures as high-performance anode materials in lithium-ion batteries, Nanotechnology*, **25**.

[3] H. G. Jung *et al.*, 2011, *Microscale spherical carbon-coated Li4Ti5O 12 as ultra high power anode material for lithium batteries, Energy Environ. Sci.*, **4**, pp. 1345–1351.

[4] Y. Qi, Y. Huang, D. Jia, S. J. Bao, and Z. P. Guo, 2009, *Preparation and characterization of novel spinel Li4Ti5O12-xBrx anode materials, Electrochim. Acta*, **54**, pp. 4772–4776.

[5] T. Yuan, R. Cai, R. Ran, Y. Zhou, and Z. Shao, 2010, *A mechanism study of synthesis of Li4Ti5O 12 from TiO2 anatase, J. Alloys Compd.*, **505**, pp. 367–373.

[6] A. Guerfi, P. Charest, K. Kinoshita, M. Perrier, and K. Zaghib, 2004, *Nano electronically conductive titanium-spinel as lithium ion storage negative electrode, J. Power Sources*, **126**, pp. 163–168.

[7] J. W. Shin, C. H. Hong, and D. H. Yoon, 2012, *Effects of TiO 2 starting materials on the solid-state formation of Li 4 Ti 5 O 12, J. Am. Ceram. Soc.*, **95**, pp. 1894–1900.

[8] Y. H. Rho and K. Kanamura, 2004, *Li+ ion diffusion in Li4Ti5O12 thin film electrode prepared by PVP sol-gel method, J. Solid State Chem.*, **177**, pp. 2094–2100.

[9] C. M. Shen, X. G. Zhang, Y. K. Zhou, and H. L. Li, 2003, *Preparation and characterization of nanocrystalline Li4Ti5O12 by sol-gel method, Mater. Chem. Phys.*, **78**, pp. 437–441.

[10] S. H. Ju, J. H. Kim, and Y. C. Kang, 2010, *Electrochemical properties of LiNi0.8Co0.2-xAl xO2 (0≤x≤0.1) cathode particles prepared by spray pyrolysis from the spray solutions with and without organic additives, Met. Mater. Int.*, **16**, pp. 299–303.

[11] A. Nugroho, S. J. Kim, K. Y. Chung, and J. Kim, 2012, *Synthesis of Li 4Ti 5O 12 in supercritical water for Li-ion batteries: Reaction mechanism and high-rate performance, Electrochim. Acta*, **78**, pp. 623–632.

[12] L. V. Nowack, O. Waser, O. Yarema, and V. Wood, 2013, *Rapid, microwave-assisted synthesis of battery-grade lithium titanate (LTO), RSC Adv.*, **3**, pp. 15618–15621.

[13] J. Li, Y. L. Jin, X. G. Zhang, and H. Yang, 2007, *Microwave solid-state synthesis of spinel Li4Ti5O12 nanocrystallites as anode material for lithium-ion batteries, Solid State Ionics*, **178**, pp. 1590–1594.

[14] Y. Bai, F. Wang, F. Wu, C. Wu, and L. ying Bao, 2008, *Influence of composite LiCl-KCl molten salt on microstructure and electrochemical performance of spinel Li4Ti5O12, Electrochim. Acta*, **54**, pp. 322–327.

[15] S. Y. Yin *et al.*, 2015, *ChemInform Abstract: Molten Salt Synthesis of Sodium Lithium Titanium Oxide Anode Material for Lithium Ion Batteries., ChemInform*, **46**, p. no-no.

[16] T. Sattar, S. H. Lee, S. J. Sim, B. S. Jin, and H. S. Kim, 2020, *Effect of Mg-doping on the electrochemical performance of LiNi0.84Co0.11Mn0.05O2 cathode for lithium ion batteries, Int. J. Hydrogen Energy*, **45**, pp. 19567–19576.

[17] M. Zhang *et al.*, 2020, *A high-energy-density aqueous zinc-manganese battery with a La-Ca co-doped ϵ-MnO2cathode, J. Mater. Chem. A*, **8**, pp. 11642–11648.

[18] L. Shen, L. Guo, N. Bao, and K. Yanagisawa, 2003, *Salt-assisted Solid-state Chemical Reaction. Synthesis of ZnO Nanocrystals, Chem. Lett.*, **32**, pp. 826–827.

[19] P. Rodionovs, J. Grabis, and A. Krumina, 2018, *Microwave-assisted hydrothermal synthesis of znfe2o4/tio2 composite and photocatalytic properties, Key Eng. Mater.*, **788 KEM**, pp. 102–107.

[20] T. Y. Hwang, G. H. An, J. H. Cho, J. Kim, and Y. H. Choa, 2015, *Effects of different salts on salt-assisted ultrasonic spray pyrolysis (SA-USP) calcination for the synthesis of strontium ferrite, J. Nanosci. Nanotechnol.*, **15**, pp. 8062–8069.

[21] L. Battery, A. Purwanto, S. U. Muzayanha, and C. S. Yudha, *applied sciences High Performance*

of Salt-Modified – LTO Anode in, **c**.

[22] Z. Liu, N. Zhang, and K. Sun, 2012, *A novel grain restraint strategy to synthesize highly crystallized Li 4Ti 5O 12 (~20 nm) for lithium ion batteries with superior high-rate performance*, *J. Mater. Chem.*, **22**, pp. 11688–11693.

[23] K. Ariyoshi, R. Yamato, and T. Ohzuku, 2005, *Zero-strain insertion mechanism of Li[Li1/3Ti 5/3]O4 for advanced lithium-ion (shuttlecock) batteries*, *Electrochim. Acta*, **51**, pp. 1125–1129.

[24] T. F. Yi, S. Y. Yang, and Y. Xie, 2015, *Recent advances of Li4Ti5O12 as a promising next generation anode material for high power lithium-ion batteries*, *J. Mater. Chem. A*, **3**, pp. 5750–5777.

Acknowldgements

This research was funded by UMG Idea Lab Indonesia, through Investment of Start Up Funding 2019-2021.

Incident Clustering in the Warehouse Workspaces by Using Text Mining

F A Syaani[1], Irhamah[2*], A Mukarromah[3], K Fithriasari[4]

[1,2,3] Department of Statistics, Institut Teknologi Sepuluh Nopember Surabaya
Jl. Raya ITS, Kampus ITS Sukolilo, Surabaya, Indonesia 60111

Corresponding Author Email: irhamah@statistika.its.ac.id[2*]

Abstract. Incidents can occur anytime and anywhere whether intentional or unintentional, especially in the industrial area. To prevent the incident, an analysis is needed to find out the patterns that formed based on the information from incident reports. The goals of company are to be able to identify and take action to deal quickly if some incident happens. Text Mining is done by applying the bigram technique to form the incident patterns. The analytical method used is K-Means Clustering and Hierarchical Clustering. In addition, a feature selection is also used by applying the Genetic Algorithm method to obtain optimal features. The results obtained state that the feature selection process is very influential on the formation of incident clusters. When compared, the K-Means Clustering and Hierarchical Clustering methods have different effects on each warehouse sector. The best results in the Sector of Life Style and Sector of Technology are formed by using the K-Means Clustering method while in the Sector of Consumer, Retail and SPL the best cluster results are obtained based on the Hierarchical Clustering method.

1. Introduction

Incidents can happen anytime and anywhere. As it is true with traffic incidents, industrial incidents are also relatively easy to happen. Although some incidents may arise from unexpected things, however, most incidents have a similar pattern to incidents that were happened before. Therefore, studying past incidents is very important to determine any preventive measures to be taken in the future. Based on this principle, many people, particularly those engaged in industrial sector, have started to collect data on incidents reports that have occurred in the past [1]. Individuals or companies that have studied the causes of previous occurrence of incidents are generally able to identify and take immediate measure to deal with any occurrence of incidents at any time of such occurrence [2]. In the process of identifying an accident in the past, any related information is needed. Such information can be in the form of legitimate company reports pertaining accident investigations in the work environment [3]. Usually, such reports are provided in the form of text data which are relatively large in size so that if done manually it will take a long time. One of the techniques that can be used to collect information based on text data is Text Mining [4]. Text Mining facilitates the extraction of information hidden in a set of data quickly and accurately [5]. In the process of information extraction, several techniques are needed to obtain interesting patterns by applying the Bigram technique to perform word combination.

Bigram can provide information more effectively because it is formed from words that are related to one another [6].

The analytical method applied in this study is cluster analysis by comparing between Ward's Clustering with K-Means Clustering. K-Means method can work quickly on relatively large datasets based on the Euclidean distance [7]. Meanwhile, Ward's Clustering method can effectively minimize the variance between group members based on the sum of square values [8]. In determining the optimal number of clusters, the value of the Silhouette Coefficient is used for each clustering process. Both methods then shall be subject to an evaluation of clustering by comparing the value of icd-rates in each cluster formed. In improving the accuracy of the cluster formed thereby, a feature selection method is also carried out using a Genetic Algorithm. This algorithm is used to find features that can be used in determining cluster type labels. Such process is performed by minimizing any features that are considered as uninformative features in order to obtain the main features in describing a class. Genetic Algorithm method is able to improve the quality of the results of a text summarization where the objective function (fitness) in the Genetic Algorithm is able to find words with more optimal weight when compared to ordinary method [9].

The final results of the research describe incident clusters based on their respective characteristics so that it can find out what problems are often reported based on each the acquired mapping. It is hoped that this research can provide information that can be applied in minimizing the problem of accidents in the workplace. Thus, the company can provide any appropriate preventive measures and response for the safety, health and security of their employees.

2. Literature Review

2.1. Text Mining

Text Mining is a process extracting information from data during the certain period and processed using the analytical method that correspond. Text Mining is used to extract the information from unstructured text data sources through identification and exploration of interesting patterns [10]. There are several stages before data processing in order to make the data can be processed more easily. This stage is better known as text pre-processing. Text pre-processing is an initial process before further analysis to change unstructured data to be more structured [11].

2.1.1. Data Cleaning. Data Cleaning is a process to clean the data without any noise anymore. In the text pre-processing, this process shall include changing of capital letters to lowercase (case folding), removing punctuation, removing numbers, removing irrelevant words (stop-words removal), removing double spaces (whitespace), translating of report sentences into a certain language desired, and breaking the sentences into words (tokenizing) [12].

2.1.2. Stemming. Stemming is a process of mapping and decomposition from various forms of words into their root words [12]. A set of rules in stemming in Indonesian shall be applied to remove affixes contained in a words, for example, "membetulkan", shall be changed into "betul" and "membaca" shall be changed into "baca". The Library which is used in the Python in the stemming process for Indonesian documents is Sastrawi (literal). Before carrying out a process of weighting onto each word in the document, it is necessary to calculate the frequency of each word therein or sentence and form a matrix.

2.1.3. n-Grams. *n*-Grams is a technique which is used to combine closely related *n* words together [13]. *n*-Grams with $n = 2$ is called bigram. Examples of the use of Bigram are such as "atap bocor", “lantai lubang”, “sampah berserakan", etc. In general, *n*-Grams ($n > 1$) will provide information more effectively because they are made up of words that are closely related to one another.

2.1.4. Term Weightening. The term weightening is a method used to obtain the frequency value of each word in a document so that we can get a comparison of one certain word's frequency to that of another due to difference of level of importance of the words respectively [14]. One method which is frequently used to calculate the weight of a term or word in a document is the Term Frequency-Inverse Document Frequency (TF-IDF). Term Frequency-Inverse Document Frequency (TF-IDF) is a measurement used to measure how important a word is to a document. Term Frequency (TF) is used to summarize the appearance of a word in a document. Meanwhile, the Inverse Document Frequency (IDF) is used to calculate the frequency of appearance of a word in a set of documents entirely [11]. There are 7 stages in weighting process with TF-IDF method that shall include Document Indexing, Term Frequency Weighting, Log Term Frequency Weighting, Document Frequency, Inverse Document Frequency, Term Frequency-Inverse Document Frequency (TF-IDF) and the final value of TF-IDF.

2.1.5. Feature Selection. Feature selection is one of the stages in data pre-processing which purpose to select influential features and remove features that do not have any role in data modelling. One of the methods that can be used is Genetic Algorithm. Genetic Algorithm is a meta-heuristic method in the field of Evolutionary Algorithm which is inspired by Darwin's theory of evolution. The concept in the Genetic Algorithm method is to find a way to perform search techniques to get an approximate solution based on the process of natural evolution [15]. The process of determining the best feature using the Genetic Algorithm method is as follows [16]:

- Population initialization, which is a stage of randomizing chromosome strings for a large number of predetermined populations;
- There are two main genetic operators used to maintain the diversity of the population in the Genetic Algorithm, include:
  - Crossover, the technique used in one-cut-point in which it is selected two parent chromosomes in a population and then the value of the contents of the chromosome subset is exchanged starting from a certain index point. This process is repeated until it is obtained the offspring number that meets the required amount from the crossover rate;
  - Mutation, which is a process carried out to prevent the algorithm from being trapped in the local optimum solution by adding a very small random value with a relatively low chance for hereditary variable to emerge.
- Selection, which is a process of evaluating fitness of reproduction results. This process purpose to determine the individual who will be selected. A fitness value is needed in the evaluation process to see the quality of the chromosome.

2.2. Text Clustering

Text clustering is an unsupervised process used to group several datasets in the form of text by applying different clustering algorithms.

2.2.1. Ward's Clustering. Ward's Clustering is a clustering method that works based on the principle of the Agglomerative Hierarchical Clustering algorithm. Ward's method is the only method out of other several methods in Agglomerative Clustering that has a calculation of the number of clusters based on the sum of square criteria [8].

2.2.2. K-Means Clustering. K-Means clustering is a clustering technique that works based on Partitioned Clustering. The principle of Partitioned Clustering is to group documents randomly because they are influenced by centroids. Furthermore, in each iteration of the grouping based on Partitioned Clustering allows combined selection of documents to happen more than once [17]. The K-Means method has the ability to classify data with a fairly large amount. The time required to perform this clustering technique is also relatively fast and efficient [10].

IOP Conf. Series: Materials Science and Engineering 1117 (2021) 012023 doi:10.1088/1757-899X/1117/1/012023

2.3. Evaluation of Clusters
Evaluation of clustering results is carried out using two approaches, namely internal evaluation which is based on the Silhouette Coefficient value and external evaluation with the ic-drate value. The Silhouette Coefficient is used to measure the equations that occur in an object by comparing certain clusters with other clusters. Matters that need to be known when determining the silhouette value is the result of partitioning or clustering results and grouping all proximity between objects [11]. The formula used is as follows:

$$s(g) = \frac{b(g) - a(g)}{\max\left\{a(g), b(g)\right\}}, \text{if} \left|C_g\right| > 1 \quad (1)$$

With:

- $a(g)$ = the smallest value of the average distance between documents g and all objects in one cluster;
- $b(g)$ = the smallest value of the average distance between documents g and all objects in another cluster.
- C_g = cluster of one data object g

The best cluster performance results in the clustering method can use the icd-rate value. The icd-rate (internal cluster dispersion rate) is an indicator that can be used to measure the level of dispersion between clusters formed thereby [18]. The icd-rate value calculation can use the following formula:

$$\text{icd-rate} = 1 - \frac{\text{SSB}}{\text{SST}} = 1 - R^2 \quad (2)$$

with SSB is the total sum of square between clusters while SST is total sum of squares from partition [19].

2.4. Word Cloud
Word Cloud is a simple visualization technique that purpose to provide information on a word that often appears in a document. Technically, the word cloud describes the words that have the most frequent occurrences in a document. The word is displayed in size and color that will attract the reader's attention at the very first time they see the word cloud presented [20].

2.5. PT. X
PT. X is a multinational company engaged in global logistics. In 2018, PT. X already has 39 warehouses that are fully operational and actively running a goods logistics system. Like other industrial companies, PT. X has several reports that need to be analyzed, one of which is an incident report which is reported regularly in a monthly basis. The report is differentiated based on its level of treatment provided thereupon which is described in brief writing explaining the condition at the time. There are three categories of reports according to its level of treatment, namely low severity, mid severity and high severity. Each of these categories is subdivided into several types of accidents, for example unsafe act and unsafe condition in the low severity category. Unsafe act is a type of action which is dangerous for workers while unsafe act is a condition which is considered dangerous for workers [21].

3. Methodology

3.1. Data Sources
In this research, the data sources that will be used are primary data obtained from the monthly reports of PT. X regarding the conditions of damage that occurred in its work environment at 39 warehouses owned by PT. X. The description of incident is reported directly by employees in the company's reporting system at each warehouse who has been granted the access such system.

3.2. Data Structure
The data taken came from the Hazard Report owned by PT. X from January 1st, 2018 to September 29th, 2018 with total of 6,598 data. The sector categories contained in the incident report are described in table 1.

Table 1. Category of Warehouse Sector

Sector	Description	Number of Warehouse	Number of Report
1	Consumer	16	2,591
2	Life Style	3	321
3	Retail	6	640
4	Spare Part Logistic (SPL)	8	2,740
5	Technology	5	281

3.3. Steps of Analysis
The analysis steps that will be carried out in this research are as follows:

- Collect incident description report data in each warehouse which is presented in the form of text data;
- Perform text pre-processing of incident report data, which includes data cleaning, stemming, bigrams, weighting (weighting), and feature selection;
- Obtain the optimum number of clusters based on the overall data as well as data from the feature selection using the Silhouette Coefficient value;
- Perform clustering analysis using Ward's Clustering and K-Means Clustering method;
- Evaluating the best cluster results;
- Make conclusions on the results of research analysis.

4. Result and Discussion

4.1. Characteristics of Incident Report Data
The characteristics of the incident report data can be described using a descriptive statistical analysis. The analysis is carried out based on the number of reports submitted as well as the classification of the description of the incident that occurred.

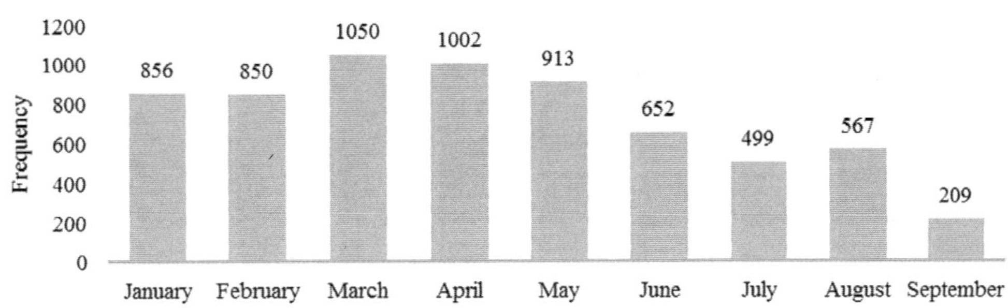

Figure 1. Frequency of Incident Reports for January 1st, 2018 – September 29th, 2018

The number of incident reports recorded on January 1st, 2018 – September 29th, 2018 has a pattern that tend to decrease every month. It appears that most of reported incidents occurred in March by 1,050 incidents or about 16% of the total reports. Meanwhile, the lowest number of reports was in September with only 209 incidents reports. This condition can be expected as a result of any corrections and preventive measures that the company regularly takes to minimize the number of incidents. In addition, there is a classification on the description data incident. The classification can be divided into two categories, namely unsafe act and unsafe condition.

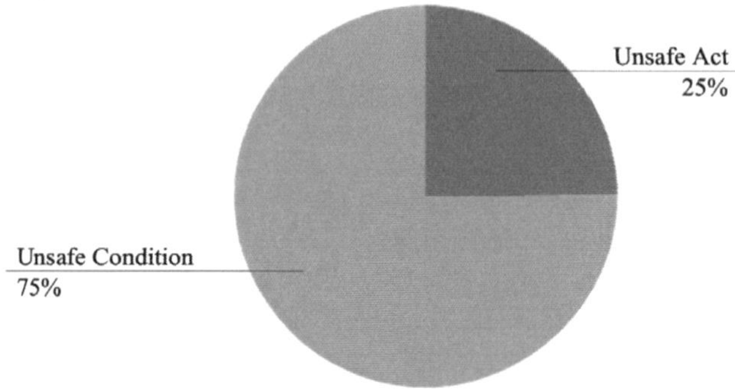

Figure 2. Proportion of Reported Incident Category

When compared, this two proportions shown in the figure 2 have an unbalanced proportion. It is known that the category with the highest proportion based on the total reports received by the system is incidents with the category of unsafe condition, which is 75% or 4,958 reports, while for the category of unsafe act it is only around 25% of the total number of reports.

4.2. Data Pre-processing

Data pre-processing purpose to collect data which is ready to be analyzed. Data pre-processing includes data cleaning, stemming, and term-weighting. Data cleaning is a process of removing the unimportant characters so that the data is free of unnecessary symbols, numbers and words. The methods used to perform this stage includes changing capital letters to lowercase letters (case folding), removing symbols (punctuation), deleting numbers, deleting irrelevant words (stop words removal), and removing double spaces (whitespace). Then the words resulted from the data cleansing process shall be changed to their base word form. After being converted into their base word form, then these words are combined with each other using the n-Grams technique as many as $n = 2$ (bigrams). The next step is to calculate the frequency of appearance in the document. The frequency of each word is formed in a matrix called Document-Term Matrix. The set of frequencies obtained in the Document-Term Matrix is then weighted using the Term Frequency-Inverse Document Frequency (TF-IDF).

4.3. Data Exploration

The incident characteristics observed are the number of words formed into a feature based on the description of reported data in each warehouse sector. It is known that the number of features that are formed in each warehouse sector are as in Table 2.

Table 2. Number of Features in Each Sector

No	Warehouse Sector	Number of Features
1	Consumer	523
2	Life Style	923
3	Retail	1,253
4	SPL	1,483
5	Technology	561

The words (features) with the highest occurrence frequency can have a significant effect on cluster formation. The frequency of these words was previously obtained based on calculations in the Document Term-Matrix. The appearance of the words is presented with the word cloud visualization as figure 3 to 7.

Figure 3. Wordcloud visualization for Consumer Sector

Figure 4. Wordcloud visualization for Life Style Sector

Based on figure 3 and figure 4, in the Consumer Sector, event that often results in incidents is pallet damage due to careless placement or overuse. Pallets and other facilities shall be regularly maintained and controlled in order to maintain their quality. In the Life Style Sector, there are many incidents that occur in the pedestrian area, such as employees who do not walk on the determined existing lanes, as well as loading and unloading activities that are not carried in appropriate places so that the activities disturb other road users who want to pass. Furthermore, in the warehouse entrance area, there were many officers who did not place the goods according to the determined standards and prevented other officers from entering the warehouse.

Figure 5. Wordcloud visualization for Retail Sector

Figure 6. Wordcloud visualization for SPL Sector

Figure 7. Wordcloud visualization for Technology Sector

In the Retail Sector, there are many employees who do not comply with safety requirement standards, such as wearing shoes improperly by stepped on their folded end part of their shoes. On the other hand, there are also several damaged pallets that actually needs for replacement and or repaired/maintained. In the SPL sector, many incidents are caused by broken lights and scattered objects interfering the workers' daily activities. The employees shall improve their awareness on the importance of keeping the warehouse clean and tidy. In the Technology Sector, a lot of damage has occurred in the warehouse or warehouse area, such as damaged floors and facilities therein and many workers who smoke while at works which is endangering the surrounding environment.

4.4. Data Clustering

Data clustering purpose to classify incidents that occurred in the warehouse environment as of January 1st, 2018 to September 29th, 2018 according to the similarities of the events. The clustered data is the incident description data in each warehouse sector. Clustering is carried out using two approaches, Hierarchical Clustering and Partitioned Clustering. In the Hierarchical Clustering approach, by using the Ward's Clustering method, while for the Partitioned Clustering approach, by using the K-Means Clustering method. The results of these two methods will obtain the optimal number of clusters in each method based on the Silhouette Coefficient value. In addition, optimal cluster results are also determined based on the comparison between the clusters with Genetic Algorithm as feature selection and without the same. The purpose of feature selection is to get features that match the conditions in the warehouse environment. The following is the number of features obtained using the Genetic Algorithm method.

Table 3. Number of Featured in Each Sector after Feature Selection

No	Warehouse Sector	Method	Number of Features
1	Consumer	K-Means	52
		Hierarchical	21
2	Life Style	K-Means	452
		Hierarchical	369
3	Retail	K-Means	557
		Hierarchical	464
4	SPL	K-Means	305
		Hierarchical	194
5	Technology	K-Means	235
		Hierarchical	159

4.5. *Evaluation of Cluster*

Evaluation of cluster is carried out by comparing the Silhouette and icd-rate values obtained from each cluster method in each sector. The following is a comparison of the Silhouette value and icd-rate resulted for each sector after the implementation of cluster analysis.

Table 4. Comparison of Silhouette Coefficient Values

Sector	Method	Number of Cluster	Feature Selection	Silhouette Coefficient	icd-rate
Consumer	K-Means	17	No	0.2	0.885
			Yes	0.907	0.556
	Hierarchical	20	No	0.198	0.854
			Yes	0.999	0.0005
Life Style	K-Means	20	No	0.063	0.854
			Yes	0.248	0.739
	Hierarchical	20	No	0.066	0.920
			Yes	0.337	0.898
Retail	K-Means	17	No	0.090	0.876
			Yes	0.337	0.838
	Hierarchical	20	No	0.056	0.924
			Yes	0.761	0.790
SPL	K-Means	18	No	0.106	0.911
			Yes	0.736	0.735
	Hierarchical	20	No	0.054	0.813
			Yes	0.919	0.600
Technology	K-Means	14	No	0.082	0.880
			Yes	0.365	0.756
	Hierarchical	11	No	0.117	0.866
			Yes	0.690	0.771

In table 4, it is known that the results of feature selection using the Genetic Algorithm method are able to provide the highest Silhouette value. In the Consumer, Fashion, Life Style, Retail, SPL, and

Technology sectors each has the greatest Silhouette value for K-Means and Hierarchical Clustering methods. This proves that feature selection using the Genetic Algorithm can significantly improve the cluster performance. Also, the presence of feature selection is able to provide a relatively low icd-rate value compared to that which uses overall feature data. In term of the goodness of the method, in the Consumer Sector it is known that the lowest icd-rate value is generated by cluster analysis using the Hierarchical Clustering with icd-rate value of 0.0005. It's similar to the results for the Retail Sector and the SPL Sector. In both sectors, the best cluster results were obtained by using the Hierarchical Clustering method with icd-rate values of 0.790 and 0.600, respectively. Unlike the previous three sectors, the best cluster results in the Life Style Sector and Technology Sector were obtained using the K-Means Clustering method with icd-rate values of 0.739 and 0.756.

5. Conclusions and Suggestions

5.1. Conclusions

The characteristics of the incident data obtained indicate that the company has played an active role in minimizing incidents in its environment. The clusters resulted by using feature selection that have been selected by the Genetic Algorithm method tend to have better results, as evidenced by the value of the Silhouette Coefficient in each sector which tends to be higher than that resulted by using the overall feature data. Based on the goodness of the resulted clusters by using the icd-rate value, the best results in the Life Style and Technology Sector were formed with the K-Means Clustering method, while in the Consumer, Retail and SPL Sectors the best clusters resulted by using the Hierarchical Clustering method.

In the Consumer Sector, the cluster with the most members is related to the damaged warehouse facilities, the pallets, in particular. Similar to the Consumer Sector, in the Life Style Sector there is also a lot of damaged pallets and there are officers who are negligent during the course of their duties such as placing object disorderly. In the Retail Sector, incidents that often occur according to the number of clusters formed are many employees who do not use safety shoes properly while working and there are facilities that need to be repaired immediately because they can interfere with the company performance itself. In the SPL sector, the incidents that often occur based on the resulted clusters are that related to broken lights and other facilities requiring repair immediately. In addition, there are scattered objects around the warehouse area. In the Technology Sector, there are many incidents related to dangerous activities carried out by employees while working resulting in potential accidents.

5.2. Suggestions

Suggestions can be given regarding the analysis result is that we need to understand problem accurately and deeply during the course of data pre-processing because the results obtained from this process will greatly affect the results of the cluster analysis formed. While for PT. X, it is advised to consider the incident reporting process based on the cluster formed so that it can make it easier to handle and prevent any recurring occurrence.

Acknowledgement

This work was supported by Ministry of Research, Technology, and Higher Education, Republic of Indonesia and Institut Teknologi Sepuluh Nopember Surabaya Indonesia under "Penelitian Dasar". The authors greatly appreciate the support.

References

[1] Veltman L M 2008 *Incident Data Analysis using Data Mining Techniques* (Texas: Texas A&M University)

[2] Nakata T 2017 *Text-Mining on Incident Reports to Find Knowledge on Industrial Safety* 2017 Annual Reliability and Maintainability Symposium (RAMS) (Orlando) pp 1-5

[3] Safety C f C P 2016 *Guidelines for Implementing Process Safety Management* (New York: WILEY)
[4] Allahyari M, Pouriyeh S, Assefi M, Safaei S, Trippe E D, Gutierrez J B and Kochut K 2017 *A Brief Survey of Text Mining: Classification, Clustering and Extraction Techniques* (Canada: KDD Blgdas)
[5] Talib R, Hanif M K, Ayesha S and Fatima F 2016 *Text Mining: Techniques, Applications and Issues* International Journal of Advanced Computer Science and Applications (IJACSA)
[6] Jurafsky D and Martin J H 2018 *N-gram Language Models* In Speech and Language Processing 3rd Edition (USA: Stanford University)
[7] Zhang C and Fang Z 2013 *An Improved K-Means Clustering Agorithm* Journal of Information & Computational Science pp 193-199
[8] Murtagh F 2014 *Ward's Hierarchical Agglomerative Clustering Method: Which Algorithms Implement Ward's Criterion?* Journal of Classification pp 274-295
[9] Meena K Y and Gopalani D 2015 *Evolutionary Algorithm for Extractive Automatic Text Summarization* Intelligent Computing, Communication & Convergence pp 244-249
[10] Feldman R and Sanger J 2007 *The Text Mining Handbook : Advanced Approaches in Analyzing Unsrtructed Data* (New York: Cambridge University Press)
[11] Jannah S Z, Fithriasari K and Usagawa T 2018 *Clustering and Visualizing Surabaya Citizen Aspirations by Using Text Mining* (Surabaya: Institut Teknologi Sepuluh Nopember)
[12] Tala F Z 2003 *A Study of Stemming Effects on Information Retrieval in Bahasa Indonesia* M.Sc. Thesis (Amsterdam: Universiteti van Amsterdam)
[13] Nguyen V H, Nguyen H T, Duong H N and Snasel V 2016 *n-Gram-Based Text Compression* Hindawi Publishing Corporation Computational Intelligence and Neuroscience
[14] Abualigah L M, Khader A T and Betar M A 2016 *Unsupervised Feature Selection Technique Based on Genetic Algorithm for Improving the Text Clustering* International Conference on Computer Science and Information Technology (CSIT) IEEE
[15] Tsai C F, Eberle W and Chu C Y 2013 *Genetic Algorithms in Feature and Instance Selection* Knowledge-Based System 39 pp 240-247
[16] Ghareb A S, Bakar A A and Hamdan R A 2015 *Hybrid Feature Selection based on Enhanced Genetic Algorithm for Text Categorization* Expert Systems With Applications pp 31-47
[17] Handoyo R, Rumani M R and Nasution S M 2014 *Perbandingan Metode Clustering menggunakan Metode Single Linkage dan K-Means pada Pengelompokan Dokumen* Jurnal SIFO Mikroskil (JSM) STMIK Mikroskil
[18] Mingoti S A and Lima J O 2006 *Comparing SOM Neural Network with Fuzzy c-Means, K-means and Traditional Hierarchical Clustering Algorithms* European Journal of Operational Research pp 1742-1759
[19] Everitt B S 2001 *Cluster Analysis* (New York: John Wiley & Sons)
[20] Castella Q and Sutton C 2013 *Word Storm : Multiples of Word Clouds for Visual Comparison of Documents* Computer Research Repository (CoRR)
[21] X P 2017 *Welcome to My Site PT. X Indonesia Induction 2017* (Jakarta: PT. X)

IOP Conf. Series: Materials Science and Engineering 1117 (2021) 012024 doi:10.1088/1757-899X/1117/1/012024

Transparent for microwave electrodes for applications in nonlinear devices based on ferroelectric ceramic

A G Alltynnikov[1], R A Platonov[1,*], A E Komlev[1], A V Tumarkin[1], A M Sosunov[1], P.M.Trofimov[1] and A A Tcymbaliuk[1]

[1]Saint Petersburg Electrotechnical University "LETI", Saint Petersburg 197022, Russia

*Corresponding author e-mail: raplatonov@etu.ru

Abstract. This article provides the results of an experimental investigation of characteristics of thin-films transparent electrodes deposited on ferroelectric ceramic plates. The experimental investigation of parameters of ZnO and resistive alloy of Si-Ti-Ce deposited on ferroelectric BSTO plates was performed. The influence of transparent electrode deposition on properties of ferroelectric ceramic based on a solid solution of barium-strontium titanate was investigated. The measured low insertion loss of the Si-Ti-Ce thin-film deposited on BSTO ceramic substrate allows considering this material as the perspective for the transparent for microwaves electrodes realization.

1. Introduction

The term "transparent electrodes" is commonly used in optical electronics [1,2]. Such electrodes are produced based on a thin-film or thick-film material that meets two main requirements: relatively high electrical conductivity and high transparency in the optical wavelength range. The ITO (indium tin oxide) is the typical material to manufacture such electrodes [3]. However, in addition to optical electronics, the term "transparent electrodes" is applicable in microwave techniques. As a rule, such electrodes are used to construct lens devices and tunable nonlinear elements [4,5]. These electrodes must be transparent for radiation in the microwave frequency range, and therefore, in contrast to electrodes used in optoelectronics, they should demonstrate a high resistance. ZnO material is suitable for transparent (in the microwave and optical ranges) electrode applications. Another perspective class of materials for microwave transparent electrode applications is resistive alloys. These materials are usually used for thin-film resistors manufacturing. Notice that these materials are not optically transparent but produce thin-film coating with high sheet resistance values up to 10 MOhm. Such high resistance values allow suggesting small insertion loss induced by electrode layers made of resistive alloys. Of particular interest is choosing the material of transparent electrodes and the technology of their deposition for use in devices based on nonlinear materials. The transparent electrode deposition procedure directly to the surface of a nonlinear material can change its nonlinear properties, which is a negative effect. This work is devoted to the investigation of the influence of transparent electrode deposition on properties of ferroelectric ceramic based on a solid solution of barium-strontium titanate [6,7]. The ZnO and resistive alloy based on Si-Ti-Ce were chosen as a material for transparent electrodes [8]. Despite a great number of articles devoted to preparing thin-films on different substrates, the compatibles of ZnO and resistive alloys with BSTO ceramic plates are unclear from the point of view of transparency in the microwave and millimeter-wave ranges. This article provides the results of an experimental investigation of characteristics of thin-films transparent electrodes deposited on ferroelectric ceramic plates. The thin

IOP Conf. Series: Materials Science and Engineering 1117 (2021) 012024 doi:10.1088/1757-899X/1117/1/012024

films of ZnO and resistive alloy of Si-Ti-Ce were deposited on ferroelectric BSTO plates for experimental investigations.

2. Samples preparation

The resistive alloys such as silicides of chromium, nickel, iron as well as binary and ternary systems on their base are traditionally used for deposition of resistive thin films. A wide range of thin-film surface resistivity is achieved by the variation of Si concentration from 15% to 95% in the resistive alloys. Such variation allows for providing control over permissible dissipation power and temperature coefficient of resistance of deposited resistive thin films.

The methods of thermal vacuum resistive evaporation and ion-plasma sputtering can be used for deposition of resistive thin films. The use of magnetron sputtering has the several advantages over resistive evaporation. The significantly higher adhesion of films to the substrate, associated with the energy of the condensing particles and additional activation of the substrate surface caused by the plasma. The higher accuracy in controlling the thin film's thickness (including films thinner than 100 nm). It is quite important as far as the film's thickness determines its resistance.

In this work, the properties of two transparent electrode materials deposited on a ferroelectric substrate were investigated. The first of them is the resistive alloy containing Si-76%, Ti-20%, and Ce-4% (Si-Ti-Ce). The deposition of thin-film Si-Ti-Ce transparent electrodes was performed by the vacuum magnetron sputtering method. The residual pressure in the chamber did not exceed 6×10^{-4} Pa. High-purity argon at a pressure of 0.1 Pa was used as a plasma-forming gas. The sputtering of the target was carried out at a constant current at a power of 0.25 kW. The deposition temperature was kept constant at 300 C. The growth rate was 50 nm/min. As a result of the experiments, the samples of high-resistive films with a thickness of 40-100 nm were deposited on BSTO ceramic plates.

The second material for transparent electrodes was ZnO. ZnO films were deposited by the method of high-frequency ion-plasma magnetron sputtering of ceramic ZnO target. The residual pressure in the chamber did not exceed 10^{-3} Pa. High-purity oxygen at a pressure of 1 Pa was used as a plasma-forming gas. The target-to-substrate distance was 25 mm. The substrate holder is heated from below by a system of heaters made of nichrome. Plates of polycrystalline aluminum oxide and bulk ceramics $BaSrTiO_3$ were used as substrates. The substrate temperature was varied in the range of 600-800°C.

3. Experimental results

The first stage of measurements was devoted to observing the transparent electrode's sheet resistance and its variation with time. Measurements show that Si-Ti-Ce thin-films deposited on BSTO ceramic plates demonstrate higher sheet resistance (~1.5 MOhm/square) than the ZnO films (~1.1 MOhm/square). Each sample was tested to observe the thin-film coating quality, which was observed that sheet resistance variation per sample does not exceed 10% along the sample surface. This result shows a high homogeneity of the obtained thin films on BSTO substrates. During the experimental investigations, the change of sheet resistance value in time was observed in the case of the ZnO thin-films deposited on the BSTO ceramic plate. At the same time, Si-Ti-Ce did not demonstrate such a negative tendency. The obtained experimental results of changing sheet resistance in time are presented in Figure 1.

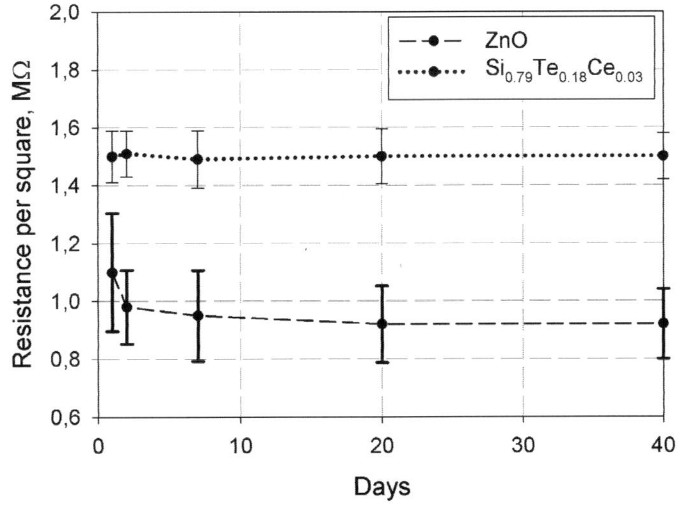

Figure. 1 The comparison of thin films' sheet resistance in time for ZnO and resistive alloy of Si-Ti-Ce.

One can see that the ZnO thin-film electrode demonstrates a significant decrease in sheet resistance value for 10 days after manufacturing. The decrease down to 85% of the value measured in the first day was observed for ZnO thin films. In comparison, Si-based resistive alloy thin film is more stable in time and demonstrates a slightly lower deviation of sheet resistance per ceramic sample (see error bars in Figure 1). Note that observed sheet resistance degradation with time is not typical for ZnO thin-films widely used in electronic devices. In our case, the nature of this effect can be explained by oxygen-containing compounds in the substrate. Films of ZnO were deposited on BSTO ceramic at relatively high temperatures, and this could lead to formation of an oxygen-defective interface BSTO\ZnO. This defective surface can be the reason of sheet resistance variation with time. It should be noted that the presented explanation is hypothetic, and the observed effect requires additional investigation.

Since the substrate is a BSTO ceramic, it is important to investigate the influence of transparent electrode deposition procedure on ceramic nonlinear properties. The characterization of BSTO ceramic nonlinearity is provided by capacitance-voltage (CV) dependency measurements of BSTO capacitors under dc control voltages. Because of the demonstrated degradation of the sheet resistance of ZnO thin-films, further measurements were focused on Si-Ti-Ce thin-film electrodes. The ceramic parallel plate capacitors were manufactured on the base of BSTO with deposited Si-Ti-Ce film and without it. Estimation of the Si-Ti-Ce film deposition procedure influence on ceramic properties was performed by comparison of CV characteristics of manufactured capacitors. The silver ink was used to form the electrodes of parallel plate capacitors. Experimental results of measured CV characteristics at 1 MHz are presented in Figure 2. In accordance with measured CV-characteristics presented in Figure 2, one can conclude that the Si-Ti-Ce deposition procedure influence on nonlinear properties of BSTO ceramic is negligibly low.

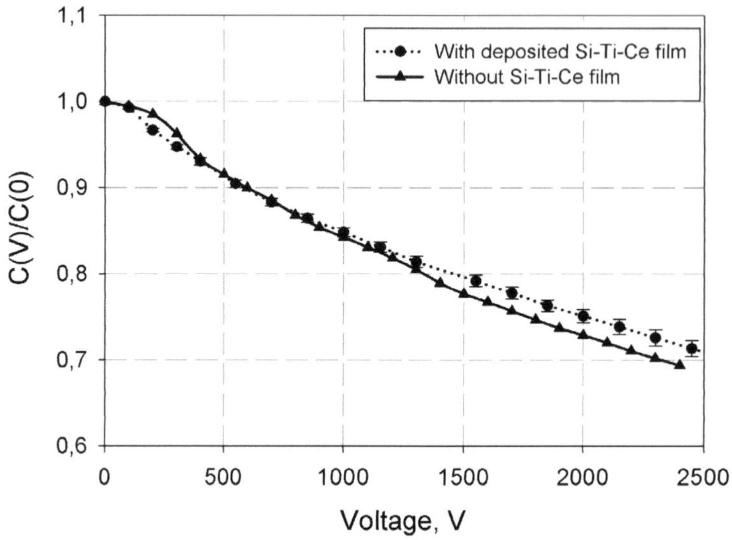

Figure. 2 Measured CV-characteristics of parallel plate capacitor based on BSTO with Si-Ti-Ce film and without it.

One of the main parameters of transparent electrodes from the microwave application point of view is the insertion loss value per electrode layer. This parameter was measured at the frequency of 30 GHz by the method of two-port waveguide measurements. The results show that the insertion loss of deposited Si-Ti-Ce thin-film is about 0.1-0.3 dB.

4. Conclusion

The ZnO and Si-Ti-Ce materials were considered as transparent for electrodes for microwave applications in nonlinear devices based on BSTO ceramic. The samples of both thin-film materials were prepared on the BSTO ceramic plates to measure sheet resistance and insertion loss values. The decreasing of sheet resistance in time of ZnO electrodes deposited on the BSTO ceramic was observed in contrast to Si-Ti-Ce material. The experimental investigations showed that there is no degradation of nonlinear properties of BSTO ceramic after deposition of Si-Ti-Ce thin-film. The measured low insertion loss of the Si-Ti-Ce thin-film deposited on BSTO ceramic substrate allows considering this material as the perspective for the transparent for microwaves electrodes realization.

Acknowledgments
The authors gratefully acknowledge the financial support provided by the Russian Science Foundation under grant 18-79-10156.

References

[1] Ellmer K 2012 *Nature Photonics* **6** 12 p 809
[2] Cao W et al. 2014 *Journal of Photonics for energy* **4** 1 p 040990
[3] Hofmann A I, Cloutet E, Hadziioannou G 2018 *Advanced Electronic Materials* **4** 10 p 1700412
[4] Platonov R, Altynnikov A, Kozyrev A 2020 *Coatings* **10** 2 p 180
[5] Scheele P et al. 2007 *IEEE transactions on microwave theory and techniques* **55** 2 p 383
[6] Nadaud K, Borderon C, Gillard R, Fourn E, Renoud R, Gundel H W 2015 *Thin Solid Films* **591** p 90
[7] Altynnikov A G et al. 2019 *Technical Physics Letters* **45** 6 p 540
[8] Kolesnikova I G, Korotkov V G, Kuz'mich Y V 2016 *Russian Metallurgy (Metally)* **2016** 5 p 472

An Empirical Study on Smart Steel Stockyard Management System through Character Recognition on Steel Materials using CNN

GiYeong Cho[1] and NamHyun Yoo[2]

[1]ForceTEC Co., Ltd., 6 Jungang-daero 228 Beon-gil, Uichang-gu, Changwon-si, Gyeongsangnam-do 51439, Korea
[2]School of Computer Science & Engineering, 7 Kyungnamdaehak-ro, Masanhappo-gu, Changwon-si, Gyeongsangnam-do 51767, Korea

hyun43@kyungnam.ac.kr

Abstract. In shipyards wherein large ships and plants are constructed, the timely supply of steel is a critical factor that can decrease the production costs and shorten the construction period. However, the standard size of the steel materials utilized in shipyards exceeds at least 10m, and the weight of the steel is heavy; therefore, it is difficult to readjust the steel once it is deployed. Moreover, if any part of the production plan is changed, the steel that has been deployed according to the existing plan must be relocated. However, when cumbersome, heavy steel is vertically stacked in several layers, it is difficult for a shipyard worker to directly identify the desired steel's location. Therefore, it is not easy to spot the location of the steel to be found. In cases such as this, one attempts to recognize characters on the steel's surface using OCR, but it is difficult to apply this to an actual system as the recognition rate is only about 50 to 70% on average. Therefore, through this study, we intend to develop a model that recognizes characters printed on the steel's surface using CNN and to check whether it can be employed in a smart steel stockyard management system.

1. Introduction

Timely supply of steel in the shipyard production system is a key factor in reducing production period and cost [1]. In medium to large ships, most steels are large in size, so it is most efficient to arrange them according to the ship manufacturing process in order to manage them. However, the shipyard production process is frequently changed due to the request of the ship-owner, delay in delivery from the supplier, and occurrence of safety accidents, which in turn causes difficulties in managing the steel arrangement. In addition, if the steel material is large and heavy, and the height of the steel material stacked or overlapped increases, it becomes difficult for the manager to directly check it with the naked eye. This leads to difficulties in managing the steel arrangement. When steel is shipped from a steel company, production information is printed on the side and top in text format. If this data is used to link with the steel material management and production process management system, it is possible to immediately check in which process the steel material can be used. If it is possible to check the location of the steel material in real time, efficient arrangement of the steel material working together with the production process is possible. Therefore, in this paper, we derive a model and related algorithm that can recognize more than 90% of characters printed on the steel surface in real time using CNN so that it can be applied to the smart steel stockyard management system designed and developed by STX

Offshore & Shipbuilding [2,3]. The reason for applying CNN technology in this paper is that the result of recognition using the existing OCR technology shows 50%~70% level of performance [4]. It is estimated that the recognition rate is low due to corrosion of steel materials, damage of the printed characters due to outdoor environment such as rainwater, high temperature and humidity. Therefore, we intend to derive an optimal model and algorithm for the recognition of characters printed on the steel surface by applying CNN, which is widely used in the image recognition field.

2. Related work

OCR technology is a technology that traditionally recognizes characters within an image. In the case of printed characters in English, it has a recognition rate of almost 100%. And in the case of Korean, it has a recognition rate of 95% or more, although the recognition rate is partially low. However, in the case of character recognition printed on steel, it is difficult to apply it in the field as it shows recognition rate of 50 to 70% in its own test using an existing commercial tool conducted by STX Offshore & Shipbuilding.

2.1. Convolutional Neural Network (CNN)

The neural network is a proposed algorithm that mimics the behavior of neurons in the human brain. The first proposed neural network was a single layer perceptron with only one input and one output[5]. However, since a single layer perceptron is difficult to solve even a simple XOR problem, a multilayer perceptron was proposed[6]. The multilayer perceptron can solve complex problems by configuring the input and the output part into a multiple structure with two or more layers. In order to solve a complex problem, the number of layers constituting a neural network is increased, which soon requires more parameters and causes an over fitting problem. The proposed method to solve this problem is CNN. CNN is similar to a multilayer perceptron, but it solves the over fitting problem by introducing a convolution (convolution product) layer in part of the layer. The convolutional layer solves the over fitting problem by reducing the number of parameters, by connecting through the kernel rather than connecting all nodes of the previous layer and the current layer to each other.

$$G[i,j] = \sum_{u=-k}^{k} \sum_{v=-k}^{k} I[u,v]F[i-u,j-v] \quad (1)$$

Figure. 1 Product information recognized on steel

3. Implementation of a character recognition system using CNN

As shown in Figure 1, the name of the steel manufacturer, product serial number, and size are commonly displayed on the steel material, and for the remainder, different data are printed for each company. Although similar in each production company, different fonts are used, and each company has its own writing format. Figure 1 shows some corrosion, but in some steel materials, corrosion occurs in the printed text, making it difficult to recognize.

In order to implement a character recognition system using CNN, the creation of an Image Set, which will be provided as character recognition learning data, must be created first, the trained model will be created using the generated Image set. Then, the accuracy of the recognition result is evaluated by using this model to determine whether to use the model. In this research project, it aims to achieve a recognition rate of 90% or more because there is not enough data to be used for learning. To implement the system, the operating system from Microsoft, such as Windows 10, Intel Core i7-9700 3.0GHz, NVIDIA Geforce GT 710, CUDA Release 10.1, CUDNN-10.1, Labeling V1.8.1, Python 3.7.2, TensorFlow 2.1.0, and OpenCV-Python 4.2.0.34, etc. were used[7].

3.1. Generating an ImageSet for learning

Data for learning is required for character recognition using TensorFlow [8]. Accordingly, 79 steel image files with characters printed at the steel stockyard were secured, and only the dotted line rectangle in Figure 1 was extracted and used. Because it needs to be implemented to classify one character for each printed character image, learning is impossible only with the extracted image. For this purpose, Labeling was used to designate a learning area and set classification criteria. Figure 2 is an image preprocessed using Labeling, and the preprocessed information is saved in xml file format.

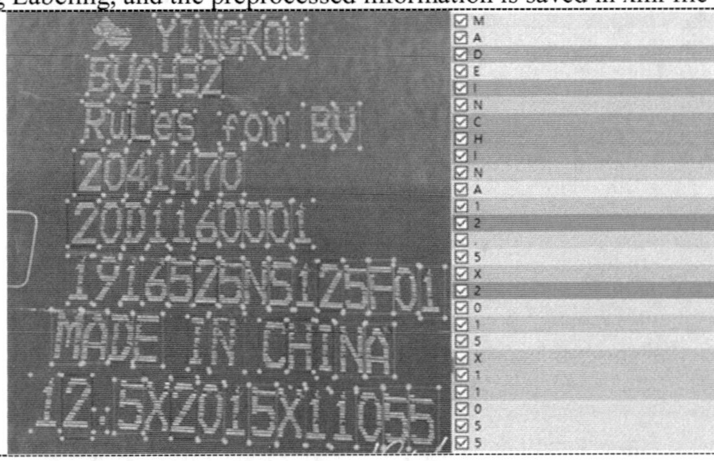

Figure. 2 Text area and label specified by using LabelImg

Table 1. Data list saved in Tfrecord

Key	Value
Image/height	Original image height
Image/width	Original image width
Image/filename	Original image file name
Image/encoded	Original image binary data
Image/format	Original image file type
Image/object/bbox/xmin	Designated character area xmin list (array)
Image/object/bbox/xmax	Designated character area xmax list (array)
Image/object/bbox/ymin	Designated character area ymin list (array)
Image/object/bbox/ymax	Designated character area ymax list (array)
Image/object/class/text	Designated character name (letter name)
Image/object/class/label	Designated character area ID (number, 0-n)

Generating Tfrecord, which is a learning data format to be used in TensorFlow using xml file having original image and preprocessed data. The data list saved in Tfrecord can be found in Table 1. Image data standardization for image recognition was conducted by using this data. The size is adjusted into 28x28, which is suitable size to distinguish characters for image data standardization, and color image was converted into 8 bit black and white image. Furthermore, as image can be obtained in various angles, additional images were created by rotating in angles such as 0°, 90°, 180°, 270°, etc.

3.2. Recognition model generation and recognition training

In this paper, CNN technique was used to recognize characters printed on steel materials. CNN, an algorithm proposed in 1959, is a method to reduce the complexity of the training model by filtering the area with features only. CNN is largely composed of Convolution Layer, Maxpooling Layer, and Fully Connected Layer. The convolution layer is a layer to specify features in data in an image, and designates the size and movement interval of an area to be specified, which is called a filter. The return data is overlapped as many as the number of filters. Learning is performed while adjusting this filter value. The max pooling layer is a layer that reduces the size of image data or a specified area. It designates the size and movement interval of a specified area, and calculate the return values while calculating the largest value in each area. It is a layer to prevent the over fitting problem from the convolution layer. Fully Connected Layer is a layer that returns the actual classification result value, and returns a one-dimensional array with each classification id as an index, and this array has the matching rate for each classification object. The detailed procedure of this learning model is as follows.

- As input data, it is provided in a three-dimensional array of 28 x 28 x 1 (28 px width, 28 px height, black and white (0-1)).
- Set the size of the filter to 3 x 3 x 1 and the moving distance to 1. The number of filters is set to 16. The data to be returned is a 26 x 26 x 16 in a three-dimensional array, and ReLu is used as the activation function. The layer with a result value close to 0 is less contributing to learning.
- It is set to Maxpooling 2 x 2, and the data to be returned accordingly is a 13 x 13 x 16 in a three-dimensional array.
- The size of the filter is set to 3 x 3 x 16 and the moving distance to 1. The number of filters is set to 32. The data to be returned is an 11 x 11 x 32 in a three-dimensional array, and the activation function is designated as ReLu.
- It is set to Maxpooling 2x2, and the data to be returned is a 5x5x32 in a three-dimensional array.
- In order to calculate the actual classification value, a 5 x 5 x 32 in a three-dimensional array is converted into one-dimensional array of size 800.
- Since the total number of recognized characters is 66, a layout that returns a one-dimensional array of size 66 is connected, and converted into activation function softmax.

It reduces over fitting by preventing learning of a certain ratio of layers for each layer while performing the 7-step learning model.

```
Epoch 5/10
256/256 [==============================] - 1s 5ms/step - loss: 0.3041 - accuracy: 0.9064
Epoch 6/10
256/256 [==============================] - 1s 5ms/step - loss: 0.3076 - accuracy: 0.9042
Epoch 7/10
256/256 [==============================] - 1s 5ms/step - loss: 0.3079 - accuracy: 0.9039
Epoch 8/10
256/256 [==============================] - 1s 5ms/step - loss: 0.2980 - accuracy: 0.9084
Epoch 9/10
256/256 [==============================] - 1s 5ms/step - loss: 0.3038 - accuracy: 0.9065
Epoch 10/10
256/256 [==============================] - 1s 5ms/step - loss: 0.3072 - accuracy: 0.9076
1815/1815 [==============================] - 0s 35us/sample - loss: 0.3610 - accuracy: 0.9174
Train for 256 steps
Epoch 1/10
256/256 [==============================] - 1s 5ms/step - loss: 0.3011 - accuracy: 0.9074
Epoch 2/10
256/256 [==============================] - 1s 5ms/step - loss: 0.3005 - accuracy: 0.9094
```

Figure. 3 Learning training progress of the model

While continuously updating the weight value through the loss function, the optimal model was found by searching whether the calculated result coincides with the actual classified result. The various setting values used for learning were Dropout rate 0.3 and learning rate 0.0001. Adam was used for optimization algorithm and categorical cross entropy was used for the loss function. 90% of the previously acquired image sets were used for learning and 10% for testing. The accuracy was checked every 10 trainings, and training was repeated for a total of 500 times. Figure 3 shows the training progress of the model used in this paper.

3.3. Evaluation of the recognition model

The accuracy is monitored every 10 training sessions by using the model proposed in Section 3.2, and if over fitting occurs, the optimal model setting value can be derived by changing the setting value. In the training process, the accuracy was 93%, and in the actual test, a result value of about 92% was derived, confirming that reliable model training was in progress. Figure 4 is an example of misrecognition that occurred while using this learning model. In the case of learning by acquiring more image files, it is considered that the accuracy of the recognition rate can be derived close to 100%. Therefore, STX Offshore & Shipbuilding is collecting text data on steel materials based on the results of this study.

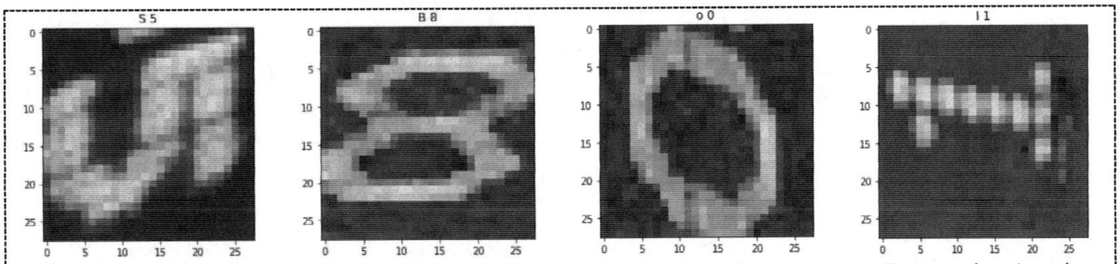

Recognize 5 as S | Recognize 8 as B | Recognize 0 as o | Recognize 1 as i

Figure . 4 Examples of misrecognition of text images

4. Conclusion and future research

STX Offshore & Shipbuilding is designing and developing a smart steel stockyard control system to manage steel stockyards in order to secure production competitiveness through timely supply of steel materials in the shipbuilding and offshore industry where cost competition is fierce. The smart steel stockyard control system is being developed at the same time to check the location of the crane in real time by using a laser sensor, and to check whether the steel is loaded via recognition of the weight of the steel when winding and unwinding the crane by installing a load cell. In the case of using the research results conducted in this paper, the location of each steel material can be checked in real time, and it is judged that the steel materials can be relocated proactively so that they can be shipped in a timely manner by using the production plan from the production management system. Currently, STX Offshore & Shipbuilding continues to collect related data so that the character recognition rate can reach close to 99%, and if relearning is performed after data collection is complete, it is expected to be able to design and develop a character recognition based smart steel stockyard control system.

5. References

[1] Park S, Lee D and Zhu J 2014 *J. Production Economics* **156** 214-22
[2] Albawi S, Mohammed T and Al-Zawi 2017 *Proc. Int. Conf. on Engineering and Technology (Antalya)* vol 1 (IEEE) p 1
[3] Abiodun O, Jantan A, Omolara A, Dada K, Mohamed N and Arshad H 2018 *J. Heliyon* **4** e00938
[4] Nagy G 1992 *Proc. Int. Conf. on the IEEE* vol 80 (NewYork city:U.S. IEEE) p 1093
[5] Hansen L and Salamon P 1990 *IEEE Transactions on Pattern Analysis and Machine Intelligence* **12** 993-1001
[6] Bourlard H. and Kamp Y. 1998 *J. Biological Cybernetics* **59** 291-94

[7] Culjak I., Abram D., Pribanic T., Dzapo H. and Cifrek M 2012 *Proc. Of the 35th Int. Convention MIPRO (Opatija)* vol 1 (IEEE) pp 1725-30
[8] Dillon J., Langmore I., Tran D., Brevdo E., Vasudevan S., Moore D., Patton B., Alemi A., Hoffman M. and Saurous R J. 2017 TensorFlow Distributions arXiv (2017)1711.10604

Acknowledgments
We thank the Software Convergence Cluster 2.0 funded by Ministry of Science and ICT, Gyeongsangnam-do, and National IT Industry Promotion Agency.

IOP Conf. Series: Materials Science and Engineering 1117 (2021) 012026 doi:10.1088/1757-899X/1117/1/012026

Enhancement of self-healing to mechanical properties of concrete

Israa H AbdElzahra and Mohannad H Al-Sherrawi

College of Engineering, University of Baghdad, Iraq

dr.mohannad.al-sherrawi@coeng.uobaghdad.edu.iq

Abstract. Concrete is the main construction material of many structures. Exposing to loads creates cracks in concrete, which reduce the performance and durability. The decrease of concrete cracks becomes a necessity demand to ensure more durability and structural integrity of the concrete structure. Autogenous healing concrete is a kind of new smart concretes, which has the ability to reclose its cracks by means of itself. Concrete self-healing is a type of free repairs processes, which is reduce direct and indirect cost of maintenance and repairing. This work targets to inspect the mechanical properties of concrete after using two combinations of two materials (20 kg/m3 calcium hydroxide Ca(OH)2 and 2 kg/m3 sodium carbonate Na2CO3 and 30 kg/m3 calcium hydroxide and 3 kg/m3 sodium carbonate) as a partial replacement of cement to work as self-healing agents. The paper reports and study the compressive, flexural and splitting tensile strengths and ultrasonic pulse velocity tests' results at 7, 28 and 90 days of concrete age. Use of two combinations of calcium hydroxide and sodium carbonate, as healing agents, enhanced the mechanical properties of concrete and reduced the width of concrete cracks with time. The concrete compressive strength was increased after curing for 90 days by 9.3 and 11.1 %, respectively. The flexural strength was improved by 14.3 and 15.9 %, respectively. Also, the splitting tensile strength was enhanced by 6.9 and 9.3 %, respectively.

1. Introduction

Concrete cracks formulation works as a significant task in the reactions to loads in both compression and tension case. These cracks strongly effect service life of all the structures in contact with water and decrease the capability of concrete cover to shield steel reinforcement against corrosion [1].

Cracks width should not be more than specific limits. Serviceability limits of a concrete structure are largely controlled by spreads of the damages. Concrete cracks might become larger that threaten durability of the structure. So, the periodic checkup, maintenance, and repair of the concrete cracks are necessary. Yet, the repair of inner cracks is complicated [2].

Handling of the concrete cracks with a width of several hundred micrometers figures critical rehabilitation mechanism and performs an influential key role on the materials sustainable design [3]. When a small crack closes without any tangential displacement, it possibly will heal semi-completely in moister condition. The concrete crack resealed is defined as autogenous concrete healing process [4].

Concrete with self-healing has been considered as one of new and smart materials, in which their inner cracks and minor imperfections have the ability to be sealed without external involvement [5].

In 1836, the French Academy of Sciences was the first to notice the autogenous healing process of concrete cracks in structures retaining in water.

Edvardsen (1999) [6] gave details on the crystallization of the calcium carbonate in a cracked fracture surface and supposed that it was the major way for self-heal process of the cracked concrete.

One of the important factors that essentially affect the cracks self-heal process is the amount of water added to the concrete mix. Anhydrate cement particles may react with penetrated moisture through the crack causing to reclose it [7].

The concrete self-healing process can be expanded and accelerated by replacing a small amount of the cement or the sand with additives [5].

The autogenous healing process of young concrete cracks due to some additives and frequently periodic wetting and/or immersion can undergo widths in the range of 0.1 to 0.2 mm through several weeks [8].

Self-heal of concrete may be used to:

1- Concrete durability increases [9].
2- Restraint concrete cracks width and healing them quickly to prevent reinforcements corrosion [10].
3- Maintenance costs reducing.
4- Decrease demand to build new concrete buildings due to increasing service life of concrete structures.

The concrete with self-heal ability, caused by adding healing agents, researches initiated a new type of smart construction material [11].

In the present work, concrete mechanical properties including compressive strength, splitting tensile strength and flexural strength for all mixes (reference and with two types of the healing agents with different dosages) at 7, 28 and 90 days ages) were investigated. Also, ultrasonic pulse velocity (UPV) tests were done to check the recovering of cracked concrete compressive strength.

2. Experimental Work

Four basic concrete materials were used for reference mix, Ordinary Portland Cement, natural sand (Zone 3), natural gravel (Grade 5-20) and potable water. These materials were confirmed to specialized Iraqi Specifications (IQS No. 5/1984 [12] for cement, IQS No. 45/1984 [13] for aggregate and IQS No. 1703/2000 [14] for water). The healing agents, fine white powders supplied from a supplier, that used consisted of two materials (calcium hydroxide $Ca(OH)_2$ and sodium carbonate Na_2CO_3) as a partial replacement of cement.

The weights of the components in the concrete mix, to get a compressive strength of about 35 MPa, is given in Table 1.

Table 1. Weights of the components in the concrete mix according to [15].

Material	Cement	Gravel	Sand	Water
Mix proportion by weight (kg/m^3)	456	1040	555	205
Mix proportion	1	2.281	1.217	0.45

The experimental program contains of casting three mixes. The healing agents' weights in these three mixes are listed in Table 2. M1 represents the reference mix. Several trail mixes had been made to inspect the workability. The slump of all the mixes is within the range of 75-100 mm. The healing agents that used had been chosen based on works of [16], [17] and [18].

Table 2. Weights of healing agents in concrete mixes.

Mix	Calcium Hydroxide ($Ca(OH)_2$) kg/m^3	Crystallization Material (Na_2CO_3) kg/m^3
M1	0	0
M2	20	2
M3	30	3

For each concrete mix, three groups of samples were casted:

Group 1: to find the compressive strength of the concrete mix at 7, 28 and 90 days, 9 cubes with (100×100×100) mm were casted. The average of three cubes has been taken for each concrete mix. Also 30 cubes were casted to determine the self-healing of cracks that introduced due to compressive stresses and cured for 90 days.

Group 2: to find the flexural strength of the concrete mix at 7, 28 and 90 days, 9 prisms with (100×100×400) mm specimens were casted.

Group 3: to find the tensile strength of the concrete mix at 7, 28 and 90 days, 9 cylinders with (150×300) mm specimens were casted.

All tests on concrete specimens were done according to international standards (BS EN 12390-3 [19] for compressive strength, ASTM C293 / C293M-16 [20] for flexural strength and ASTM C496 / C496M-17 [21] for tensile strength) at ambient room temperature.

ACI Committee 224R-09 [22] states that a cracking load (40% of the expected ultimate load) was chosen and the concrete cubes were compressed using (SANS) compression device of (600 kN) maximum capacity in the Department of Civil Engineering laboratories, University of Baghdad.

3. Results and Discussions

Figure 1-a presents the compressive strength tests results for the three concrete mixes at the three ages. From the first week, the increasing in the compressive strength was appeared slightly and continues to final age of test.

The test results indicated that compressive strengths of the self-healing concretes were higher than that of the normal concrete. The greatest increasing appears at age of 90 days that attributed to effect of healing agents. Table 3 presents the increasing percentages in compressive strength, flexural strength and splitting tensile strength due to different content of healing agents compared to reference concrete mix.

Table 3. Percentage of increasing in compressive, flexural and splitting tensile strength.

Mix	Compressive strength Age (days)			Flexural strength Age (days)			Splitting tensile strength Age (days)		
	7	28	90	7	28	90	7	28	90
M2	0.4	0.8	9.3	2.4	5.3	14.3	5.9	6.3	6.9
M3	2.3	2.3	11.1	6.9	9.4	15.9	6.8	7.1	9.3

The increase in cement weight that replaced by healing agents increased the compressive strength of the concrete mix.

Figure 1-b presents flexural strength results for the three concrete mixes at the three ages. The results indicate that flexural strength of self-healing concrete is slightly higher than that of normal concrete at all ages of tests. This result is in agreement with the conclusions of [23].

Figure 1-c presents splitting tensile strength results for cylinders. The results show that the tensile strengths of self-healing concretes were slightly higher than that of the normal concrete at all ages of tests.

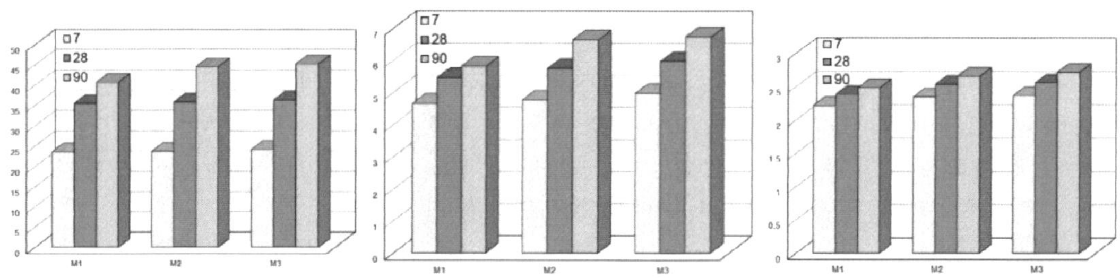

(a) (b) (c)

Figure. 1 Strength results (MPa) with age (a) compressive (b) flexural (c) splitting tensile.

The tests of UPV used to confirm the strength and the quality of concretes and provide data on the concrete properties [24]. UPV has been utilized to distinguish variations between the three concrete mixes that made in this work and determine weakening after cracking process and regaining of the concrete compressive strengths and micro cracks of the concrete cubes after curing. Thirty concrete cubes samples of (100×100×100) mm dimensions have been utilized to evaluate pulse velocities. The variations in UPV readings before and after applying compressive stresses indicate multi inner micro cracks formulations in the concrete. Figure 2 shows the results of ultrasonic pulse velocity for reference concrete (M1). They present that the readings decreased after applying compressive stresses due to inner micro cracks formulations. Later, and after curing for 90 days, the results show very slightly increasing in UPV readings. This was accredited to continuity in the cement hydration progression by hydration of the residue unhydrated cement.

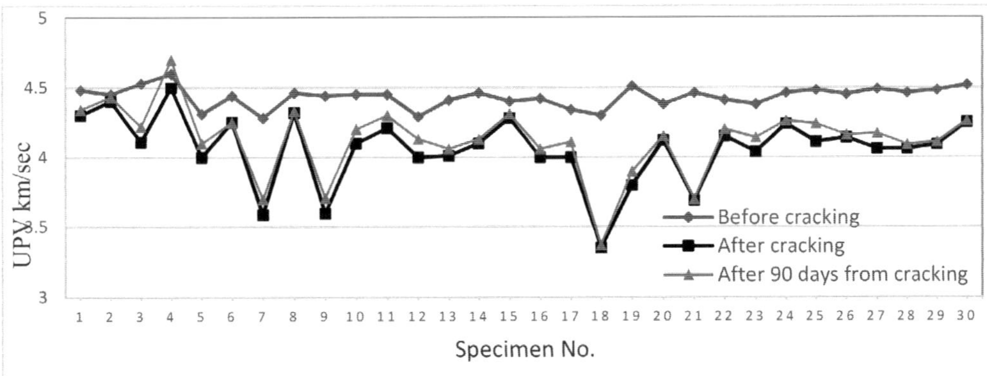

Figure. 2 The values of UPV for concrete mix M1.

Figures 3 and 4 show the results of ultrasonic pulse velocity for concrete mixes M2 and M3 respectively. The results of UPV after cracking are decreased and that indicates to the formation of micro cracks in the concrete cube, meanwhile the continuous curing for 90 days shows perceptible increases in UPV results, and this denotes to the regaining of the concrete compressive strength, and resealing of the concrete micro cracks result from the creations from self-healing reactions, and this in agreement with [25].

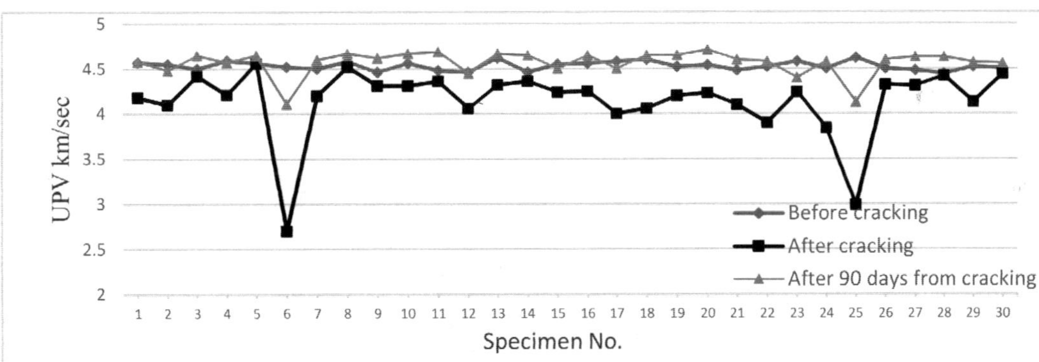

Figure 3. The values of UPV for concrete mix M2.

Figure 5 shows a comparison in the average UPV readings after 90 days from cracking for the three mixes. The results of UPV for concrete mixes M3 and M2 were more than that for concrete mixes M1, this means that amount of increases in the self-healing increases with the adding of the healing agents.

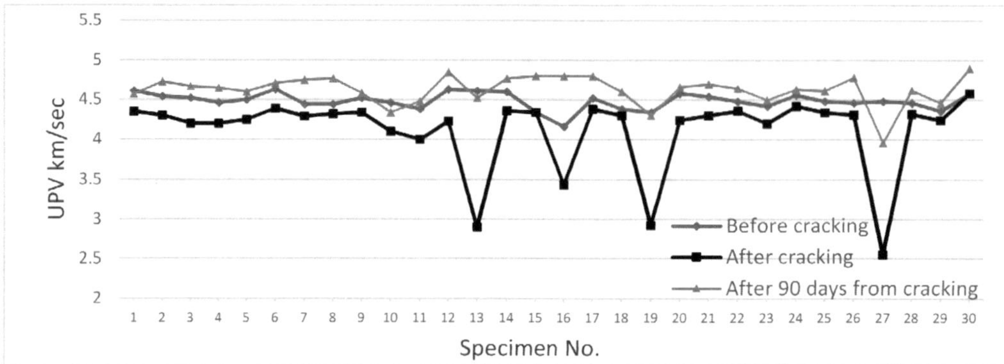

Figure. 4 The values of UPV for concrete mix M3.

The amount of self-healing products increased with increasing period of curing, especially for the first month of curing, and continued to increase but at smaller rate, this is in agreed with [24] and [25]. This attributed for being the self-healing process principles are very similar to the principles of processes that occur during the hydration and hardening of concrete, so the self-healing reactions are relatively rapid at first time but it continues for long time [26].

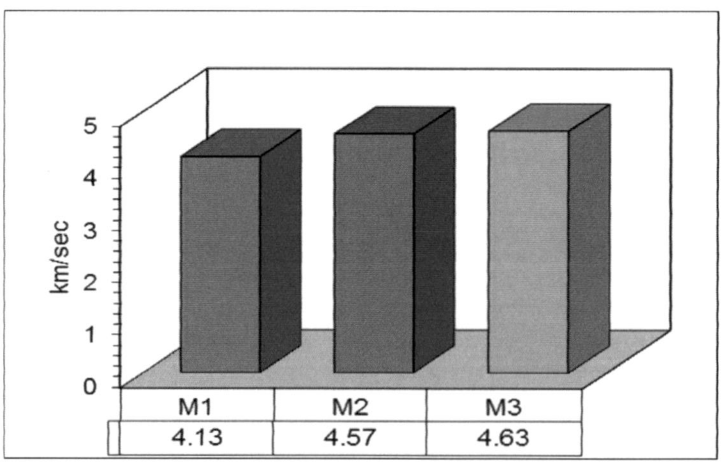

Figure. 5 Average UPV readings after 90 days from cracking.

The compressive strength was tested for the self-healed specimens in order to estimate the self-healing effects on the compressive strength. The results show recovering in compressive strength of self-healed specimens for all concrete mixes, as shown in Figure 6. A recovering was observed in concrete mixes M2 and M3, and that attributed to the effect of two healing agents' materials in sealing the cracks and regain the integrity of cracked specimens. The incorporation of sodium carbonate leads to higher later strength due to relatively low solubility in the water [27]. The produced calcium carbonate gives finer pore structures as well as reduction in the total pore volume and reduce the permeability of the concrete, also Calcium compounds enhanced compressive strength of the concrete samples, so the adding of $Ca(OH)_2$ was useful in terms of the concrete compressive strength [28].

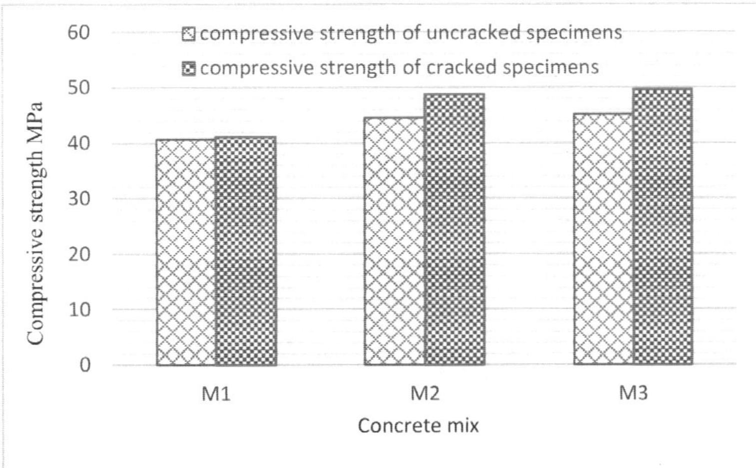

Figure. 6 Recovering in compressive strength.

Figure. 6 shows microscopic image for the self-healing effect (before and after) on micro cracks in a sample from concrete mix M3.

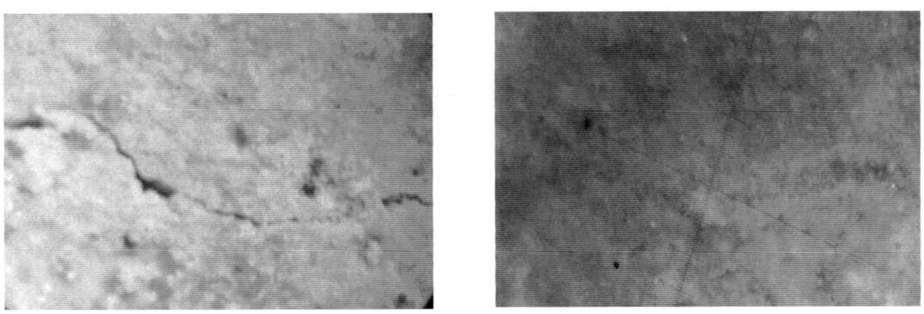

(a) (b)

Figure. 7 Microscopic image for sample from concrete mix M3, (a) before healing (b) after healing.

4. Conclusions

From the experimental results obtained in the experimental work, the following conclusions can be listed:

1- Self-healing of the concrete cracks due to adding two healing agents (calcium hydroxide $Ca(OH)_2$ and sodium carbonate Na_2CO_3) was observed.

2- Test results showed that the compressive strength of the two self-healing concretes was higher than that of normal concrete after curing for 90 days by 9.3 and 11.1%, respectively. The increasing in the concrete compressive strength occurs somewhat in first week of age and continues to final age of tests

3- The increase in the amount of the healing agents at the expense of the cement increased concrete compressive strength.
4- The results indicate that flexural strength and splitting tensile strength of self-healing concretes are slightly more than that of the normal mix at all ages.
5- The results of the UPV for concrete mixes with healing agents appear obvious compressive strength recovering than that for normal concrete.

Acknowledgements

The authors would like to gratefully acknowledge the support of the Department of Civil Engineering at University of Baghdad.

References

[1] Homma D, Mihashi H, and Nishiwaki T 2009 Self-healing capability of fiber reinforced cementitious composites, *Journal of Advanced Concrete Technology*, pp. 217-228.
[2] Tittelboom K, V, Snoeck D, Wang J, and Belie N D 2013 Most recent advances in the field of self-healing cementitious materials, *ICSHM 2013: Proceedings of the 4th International Conference on Self-Healing Materials*, pp. 406-413.
[3] Hameed A A and Al-Sherrawi M H 2018 Torsional strength of steel fiber reinforced concrete beams, *International Journal of Civil Engineering and Technology*, **9**(6), pp. 1388–1396.
[4] Neville A M 2011 Properties of concrete, Fifth and final Edition, 330.
[5] Truong P V, Kish, T, and Kayondo M 2013 Crack self-healing effect on performance of reinforced concrete members subjected to water supply and cyclic loading, *ICSHM 2013: Proceedings of the 4th International Conference on Self-Healing Materials*, pp. 648-651.
[6] Edvardsen C 1999 Water permeability and autogenous healing of cracks in concrete, *ACI Materials Journal*, **96**(4), pp. 448-454
[7] Muhsin M H and AbdElzahra I H 2016 Studying the combination effect of additives and micro steel fibers on cracks of self-healing concrete, *Journal of Engineering*, **22**(1), pp. 49-67.
[8] Fadhil B, Al-Rumaithi A and Al-Sherrawi M H 2018 Properties of reactive powder concrete with different types of cement, *International Journal of Civil Engineering and Technology*, **9**(10), pp. 1313-1321.
[9] Li V C and Herbert E 2013 Robust self-healing concrete for sustainable infrastructure, *Journal of Advanced Concrete Technology*, pp. 207-218.
[10] Palin D, Wiktor V and Jonkers H M 2013 Bacteria-based self-healing concrete for application in the marine environment, *Proc. 5th Int. Conf. on Self-Healing Materials* (Durham, USA) pp. 2–6.
[11] Tsangouri E 2019 A decade of research on self-healing concrete. *Sustain. Constr. Build. Mater.* Sayed Hemeda, IntechOpen, DOI: 10.5772/intechopen.82525.
[12] Iraqi Standard Specification No 5, "Portland Cement (I.O.S. 5/1984)", Baghdad, 1984.
[13] Iraqi Standard Specification No 45, "Aggregates from Natural Sources for Concrete and Building Construction (I.O.S. 45/1984)", Baghdad, 1984.
[14] Iraqi Standard Specification No 1703, "Water Used in Concrete (I.O.S. 1703/2000)", Baghdad, 2000.
[15] ACI Committee 211 1991 Standard Practice for Selecting Proportions for Normal, Heavyweight, and Mass Concrete, ACI Manual of Concrete Practice.
[16] Janotka I 2001 Hydration of the cement paste with Na_2CO_3 addition, ceramics − *Silikáty* **45**(1), pp. 16-23.
[17] Huang H and Ye G 2013 Self-healing of cracks by using saturated calcium hydroxide solution to activate slag in slag cementitious materials, *3rd International Conference on Sustainable Construction Materials and Technologies*. In Kyoto, Japan, pp. 18-21.
[18] Reddy C S T and Ravitheja A 2019 Macro mechanical properties of self healing concrete with crystalline admixture under different environments, *Ain Shams Engineering Journal*, **10**, pp. 23–32.

[19] BS EN 12390-3:2019 Testing hardened concrete. Compressive strength of test specimens, British Standards Institution.
[20] ASTM C293 / C293M-16, Standard Test Method for Flexural Strength of Concrete (Using Simple Beam with Center-Point Loading), ASTM International, West Conshohocken, PA, 2016.
[21] ASTM C496 / C496M-17, Standard Test Method for Splitting Tensile Strength of Cylindrical Concrete Specimens, ASTM International, West Conshohocken, PA, 2017.
[22] ACI Committee 224 2009. Control of Cracking in Concrete Structures (ACI 224R-09), ACI Manual of Concrete Practice.
[23] Zhong L Z and Chen H 2013 Self-healing efficiency of unhydrated cement nuclei for dome-like crack mode in cementitious materials, *Materials and Structures*, **46**, pp.1881–1892.
[24] Zhong W and Yao W 2008 Influence of damage degree on self-healing of concrete, *Construction and Building Materials*, **22**, pp. 1137–1142.
[25] Ferrara L, Krelani V and Brac E G 2013 Assessing the self-healing capacity of cementitious composites, In K. Sakata and e. T.R. Naij (Eds.), *Proceedings of SCMT3* - Professor R. Gettu sessions, August 19-22, Kyoto, Japan, pp. 103–112.
[26] Gupta R 2014 Crystalline technology for concrete waterproofing, *New Building Materials and Construction World [crystalline]*. https://const96.rssing.com/chan-62964530/all_p1.html.
[27] Jimenez A F and Puertas F 2003 Effect of Activator Mix on The Hydration and Strength Behavior of Alkali-Activated Slag Cements, Adv. Cem. Res., 15, pp. 129-136.
[28] Temuujin J, van Riessen A, and Williams R 2009 Influence of calcium compounds on the mechanical properties of fy ash geopolymer pastes, *Journal of Hazardous Materials*, pp. 82–88.

IOP Conf. Series: Materials Science and Engineering 1117 (2021) 012027 doi:10.1088/1757-899X/1117/1/012027

Synthesis and characterization of linear multi-functional phosphazene structures for polymer cross-linking

Anna I Chernysheva[1], Andrey S Esin[1], Mikhail A Soldatov[1], Nikolay S Bredov[1], Vyacheslav V Kireev[1], Volodymyr V Oberemok[1, 2], Igor S Sirotin[1] and Mikhail V Gorlov[1, 3]

[1] Department of Polymer Chemistry, Mendeleev University of Chemical Technology of Russia, 9 Miusskaya square, Moscow 125047, Russia
[2] V.I. Vernadsky Crimean Federal University, Simferopol, Russia

mikgorlov@gmail.com

Abstract. Phosphazenes are a well-studied class of organometallic compounds with perspective characteristics, already tested in various applications. However, until now, three-dimensionally crosslinked structures based on them are primarily obtained by irradiation (that is UV and Cobalt-60). It is generally accepted that such processes proceed via the mechanism of the cleavage of C-H bonds present in the organic substituents, which clearly indicates the lack of selectivity and the impossibility to control the crosslinking degree and distribution. Within this article, multifunctional organosubstituted structures based on the short-chain penta-functional trichlorophosphazodichlorophosphonyl with eugenol and methacrylic fragments were obtained. All products were characterized by ^{1}H and ^{31}P NMR spectroscopy and MALDI-TOF mass spectrometry. The tendency of compounds with linear methacrylic substituents to undergo the phosphazene-phosphazane rearrangement, so that the dominant reaction product turns to the tetrasubstituted derivative, has been shown. All the obtained compounds can be used as the independent monomers to obtain rigid hybrid organo-inorganic matrices, as well as polyfunctional crosslinking agents for various polymers.

1. Introduction

Phosphazene materials have found a wide range of applications in various fields of science [1, 2]. They possess such unique and diverse properties as high thermal and chemical stability, good biocompatibility and the ease of derivatization, which makes them promising starting materials for elastomers [3, 4], polyelectrolytes [5-7], dentistry [8, 9], polymer modification [10], adsorption and extraction processes [11-13], catalysis [14], and, thanks to the discovery of chain length precise control via $Cl_3P{=}NSiMe_3$ synthesis and polymerization [15-17], drug delivery [18, 19]. However, most of these applications demand the formation of relatively hard three-dimensional polymeric matrixes with the possibility for tuning and control of cross-linking degree [20, 21]. For example, in case of polyelectrolytes the cross-linking not only improves the dimensional stability of the final material [22] but also prevents the formation of undesired dendrites [23]. As well, considering catalysts, the cross-linked ones can be recycled in comparison with non-cross-linked [24, 25]. The traditionally exploited methods for cross-linking of phosphazene chains, such as UV-irradiation, in some cases are not quite appropriate, since they can lead to chemical bond cleavages [26] and usually do not provide sufficient control of final product hardness. On the other hand, the chemical routes are more effective because the amount and the distribution of linkable functional groups can be regulated and so the cross-linking degree can be well tuned, excluding the existence of the side processes. Hence, the investigation of novel phoshazene structures capable for further cross-linking and methods to obtain them is still of utmost importance. In this work we synthesized low-molecular-weight model multi-functional compounds on the basis of trichlorophosphazodichlorophosphonyl, which can be both monomers for three dimensional phosphonitrilic matrixes and cross-linking agents for other functional polymeric systems. These compounds contain methacrylic and eugenol groups and are promising for cross-linking by such reactions as hydrosilylation, thiol-ene reaction and radical polymerization.

2. Experimental part

2.1. Materials

Unless otherwise noted, all the reagents and solvents were purchased from commercial suppliers and used without further purification.

Trichlorophosphazodichlorophosphonyl $Cl_3P{=}N{-}P(O)Cl_2$ (TCDP) was synthesized according to reported method [27]. After vacuum distillation white crystals ($T_m = 32–35°C$) were obtained with the 74% yield. ^{31}P NMR, $CDCl_3$: –3 ÷ –4 ppm (d, $Cl_3P{=}N–$), –11 ÷ –12 ppm (d, $–P({=}O)Cl_2$).

All the solvents were purified by standard methods and were used freshly distilled.

2.2. Characterization methods

^{31}P and ^{1}H NMR spectra were recorded by Bruker CXP-200 spectrometer at 81 and 200 MHz respectively. MALDI-TOF mass-spectra were recorded by Bruker Auto Flex II spectrometer.

2.3. Synthesis of eugenol derivative of TCDP (TCDP-Eug)

7.22 g (0.044 mol) of eugenol and 20 mL of dioxane were charged in the three-necked flask, equipped with a magnetic stirrer and reflux condenser. Then 1.36 g (0.0444 mol) of sodium were added gradually in the form of thin plates. After the full dissolution of sodium, the solvent was totally removed. Then 30 mL of toluene was introduced, followed by the dropwise addition of 2 g (0.0074 mol) of TCDP in 5 mL of toluene solution. Reaction was carried out at room temperature for 14 h, washed with water and dried with $MgSO_4$. Afterwards solvent was evaporated and the product was dried under vacuum. ^{31}P NMR, $CDCl_3$: –15 ÷ –17 ppm (d, $(EugO)_3P{=}N–$), –22 ÷ –23 ppm (d, $(–P({=}O)(OEug)_2)$, –10 ÷ –12 ppm (d, $Cl(EugO)_2P{=}N–$), –22 ÷ –23 ppm (d, $(–P({=}O)(OEug)_2)$. The product was obtained with a 80% yield.

2.4. Synthesis of methacrylic derivatives of TCDP (TCDP-2-HEM)

TCDP-2-HEM-1: 86.83 g (0.668 mol) of 2-hydroxyethylmethacrylate (2-HEM), 377,66 g (424,72 mL) of THF and 30 g (0.111 mol) of TCDP were charged in the three-necked flask. Then 52,76 g (0.667 mol) of pyridine were added to the reaction mixture. Reaction was stirred for 6 h at room temperature. Then the mixture was poured into the solution of HCl and the precipitated product was dissolved in chloroform and washed with water. The organic layer was dried with $MgSO_4$ and then the solvent was rotary evaporated, yielded the desired product with a 73% yield.

TCDP-2-HEM-2: The synthesis was carried out according to the same procedure with the use of toluene as a solvent. The product was obtained with a 78% yield.

Both methacrylic products represent transparent yellowish viscous liquids.

3. Results and discussion

Synthesis of eugenol derivative TCDP-Eug was performed via two step method (**Figure 1**) with preliminary sodium eugenolate preparation and its further reaction with TCDP in toluene medium at room temperature. For the preparation of aryloxy derivative with the maximum substitution degree the excess of eugenolate was used and molar ratio TCDP:EugONa was 1:6. However, the final product along with the desired pentaeugenoxyphosphazophosphonyl (characteristic doublets for $(EugO)_3P{=}N–$ group at –15 ÷ –17 ppm and doublets for $–P({=}O)(OEug)_2$ group at –22 ÷ –23 ppm) also include some amount of tetrasubstituted compound (doublets at –10 ÷ –12 ppm for $Cl(EugO)_2P{=}N$– group and –22 ÷ –23 ppm for $–P({=}O)(OEug)_2$), which cannot be purified by recrystallization.

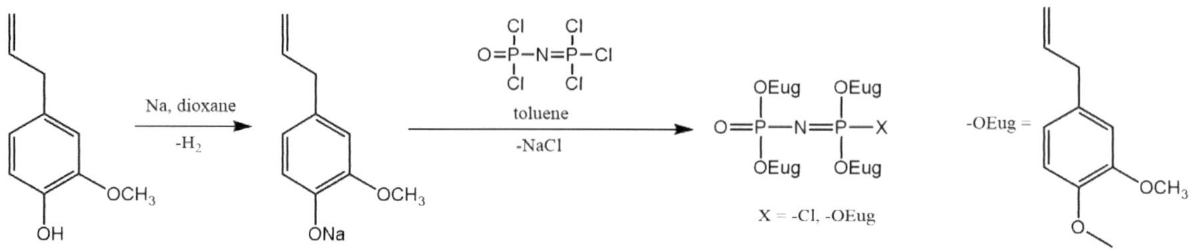

Figure. 1 Synthesis of TCDP-Eug.

^{1}H NMR spectra of the obtained product includes the signals of the aromatic protons at 6.5 – 7.4 ppm, methoxy groups signals at 3.7 ppm and peaks at 6.0 (m, $-CH_2-CH{=}CH_2$), 5.0 (d, $-CH_2-CH{=}CH_2$) and 3.2 ppm (d, $-CH_2-CH{=}CH_2$), corresponding to allyl group protons. From the splitted character of aromatic and methoxy group signals we can conclude that these protons are not equal for eugenol substituents at phosphazenyl and phosphonyl phosphorus atoms in pentaeugenoxy- and tetraeugenoxychlorophosphazophosphonyl.

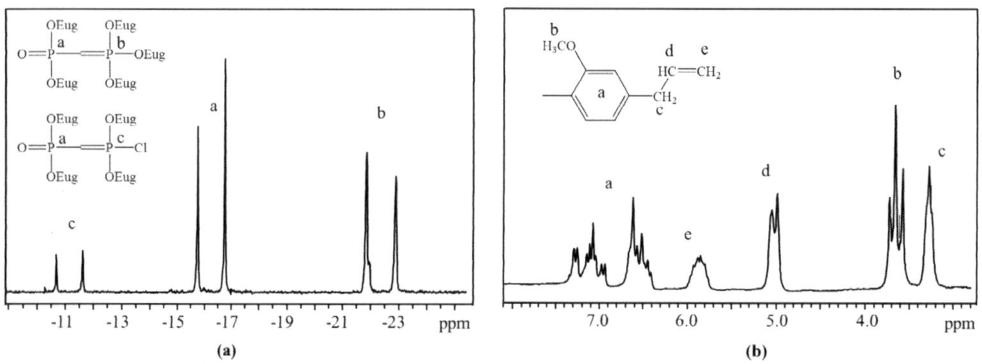

Figure. 2 ^{31}P (a) and ^{1}H (b) NMR spectra of TCDP-Eug.

The presence of both penta- and tetrasubstituted TCDP was additionally confirmed by MALDI-TOF analysis (**Figure 3**).

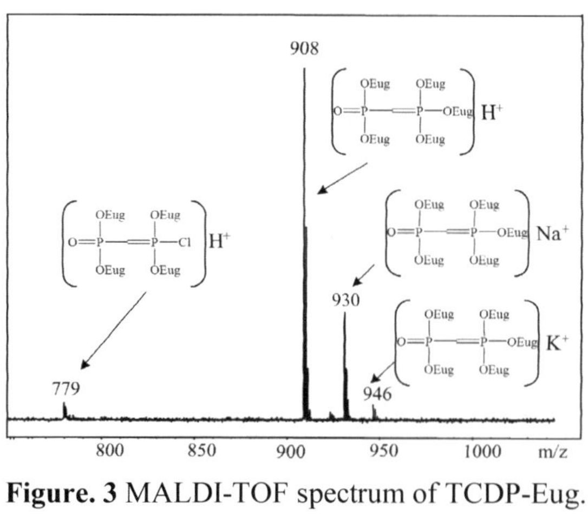

Figure. 3 MALDI-TOF spectrum of TCDP-Eug.

A low-intensity signal of tetrasubstituded compound (m/z = 779) is observed as well as the signals of the targeted pentasubstituted derivative (m/z = 908, 930, 946). Tetrasubstituted product formation even in the presence of sodium eugenolate excess most probably might be explained by existing sterical hindrances due to the eugenol structure.
It should be noted that if the reaction of sodium eugenolate with TCDP is carried out at reflux, the side product bis(dieugenoxyphosphonyl)amine can be formed by the reaction of phosphazene-phosphazane rearrangement (**Figure 4**) [28].

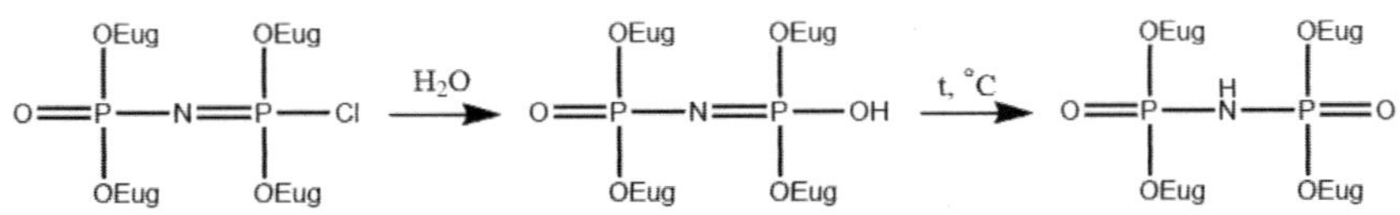

Figure. 4 Formation of side product bis(dieugenoxyphosphonyl)amine.

This product exhibits one singlet at –6 ppm in the ^{31}P NMR spectrum and two signals at m/z = 762 and 785 in the MALDI-TOF spectrum (**Figure 5**).

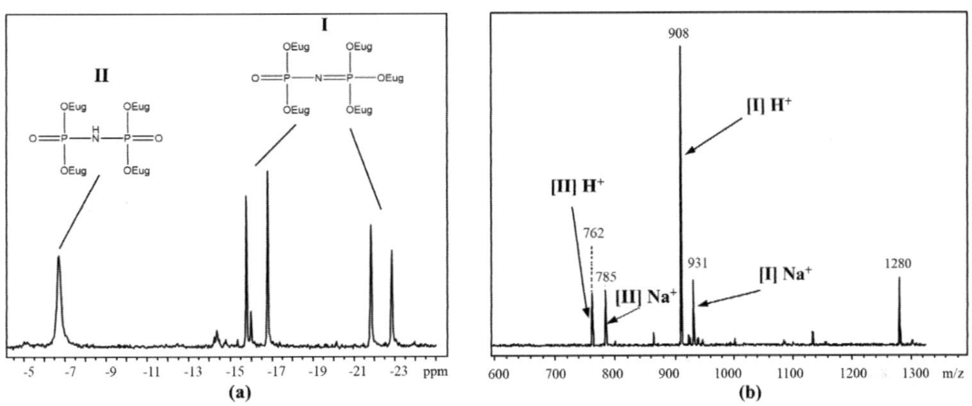

Figure. 5 ^{31}P (a) and MALDI-TOF (b) spectra of eugenol derivative product, obtained after the reaction of TCDP with eugenol in toluene medium at 55 °C for 14 h.

Synthesis of methacrylate derivatives was conducted by the following scheme (**Figure 6**) with the use of THF and toluene as solvents:

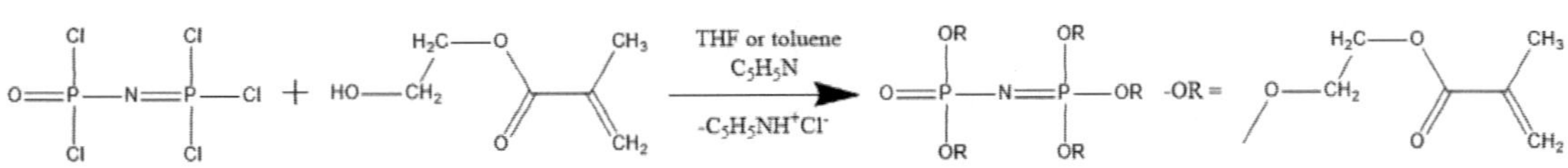

Figure. 6 Synthesis of TCDP-2-HEM.

Analysis of the obtained compounds has shown that the final products, regardless of the solvent used, also consist of tetra- and pentasubstituted derivatives, despite the fact that nucleophilic 2-HEM was taken in excess. There are two singlets at +0.3 and –2.5 ppm in the ^{31}P NMR spectra (**Figure 7**) of the product, corresponding for tetra- and pentasubstituted derivatives, respectively.

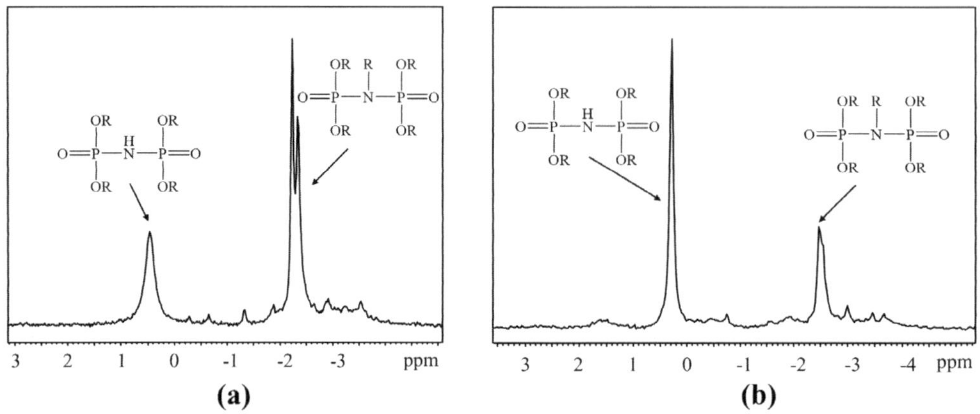

Figure. 7 ^{31}P NMR spectra of TCDP-2-HEM-1 (a) and TCDP-2-HEM-2 (b).

Unlike the eugenol derivatives, the singlet character of these peaks indicates that phosphazene-phosphazane rearrangement proceeds quantitatively with the formation of the more stable phosphazane structure. The yields of the products, obtained in THF are 62 and 73%, whereas in toluene medium – 70 and 83% when calculated over penta and tetrasubstituted derivatives respectively.

4. Conclusions

In conclusion, the functional eugenol and methacrylate derivatives of trichlorophosphazophospholyl were synthesized and characterized. It was shown that, regardless of the structure of substituent, both tetra- and pentasubstituted products are formed. Moreover, for methacrylic derivative, due to its aliphatic structure, the phosphazene-phosphazane rearrangement process takes place. The obtained products, due to the presence of the functional groups, in further, are promising agents for the preparation of phosphazene-based cross-linked polymeric materials.

5. References

[1] Gleria M, De Jaeger R 2004 *Phosphazenes: a Worldwide Insight* (New York: Nova Science Publishers, Inc.)
[2] Allcock H R, Chen C 2020 *J. Org. Chem.* **85** 14286
[3] Li Z, Chen C, McCaffrey M, Yang H and Allcock H R 2020 *ACS Appl. Polym. Mater.* **2** 475
[4] Zhang S, Wang J, Zhu H, Ali S, Ma H, Zhang L, Wu D, Wu Z 2017 *High Perform. Polym.* **29(4)** 450
[5] Gutru R, Peera S G, Bhat S D, Sahu A K 2016 *Solid State Ionics* **296** 127
[6] Zhou M, Qin C, Liu Z, Feng L, Su X, Chen Y, Xi L, Xi Y, Liu Z 2017 *Appl. Surf. Sci.* **403** 260
[7] Chawla N, Bharti N, Singh S 2019 *Batteries* **5(1)** 19
[8] Chistyakov E M, Kolpinskaya N, Posokhova V, Chuev V 2020 *Polymers* **12(5)** 1176
[9] Khan R U, Wang L, Yu H, Muhammad Akram Z, Wu J, Haroon M, Ullah R S, Deng Z and Xia X 2018 *Russ. Chem. Rev.* **87** 109
[10] Kireev V V, Bilichenko Y V, Borisov R S, Mu J, Kuznetsov D A, Eroshenko A V, Filatov S N, Sirotin I S 2019 *Polymers* **11** 1914
[11] Zhang M, Li Y, Bai C, Guo X, Han J, Hu S, Jiang H, Tan W, Li S, Ma L 2018 *ACS Appl. Mater. Interfaces* **10** 28936
[12] Soldatov M, Liu H 2019 *Chem. Asian J.* **14** 4345
[13] Bredov N S, Gorlov M V, Esin A S, Bykovskaya A A, Kireev V V, Sinegribova O A and Ryabochenko M D 2020 *Appl. Sci.* **10(4)** 1229
[14] Boileau S, Illy N 2011 *Prog. Polym. Sci.* **36** 1132
[15] Allcock H R, Crane C A, Morrissey C T, Nelson J M, Reeves S D 1996 *Macromolecules* **29** 7740
[16] Zhang Y, Huynh K, Manners I, Reed C A 2008 *Chem. Comm.* **4** 494

IOP Conf. Series: Materials Science and Engineering 1117 (2021) 012027 doi:10.1088/1757-899X/1117/1/012027

[17] Gorlov M V, Bredov N S, Esin A S, Kireev V V 2016 *J. Organomet. Chem.* **818** 82
[18] Teasdale I, Bruggeman O 2013 *Polym.* **5** 161
[19] Lakshmi S, Katti D S, Laurencin C T 2003 *Adv. Drug Deliv. Rev.* **55** 467
[20] Liu J, He Z, Wu G, Zhang X, Zhao C, Lei C 2020 *Chem. Eng. J.* **390** 124620
[21] Poscher V, Teasdale I, Salinas Y 2019 *ACS Appl. Nano Mater.* **2** 655
[22] Tsao C-H, Ueda M, Kuo P-L 2016 *J. Polym. Sci. Part A* **54** 352
[23] Stalin S, Johnson H E N, Biswal P, Vu D, Zhao Q, Yin J, Abel B A, Deng Y, Coates G W and Archer L A 2020 *Macromolecules* **53** 5445
[24] Liras M, Verde-Sesto E, Iglesias M, Sanchez F 2017 *Eur. Polym. J.* **95** 775
[25] Ren C, Zhu X, Zhao N, Shen Y, Chen L, Liu S, Li Z 2019 *Eur. Polym. J.* **119** 130
[26] Allcock H R, Kwon S, Riding G H, Fitzpatrick R J and Bennett J L 1988 *Biomaterials* **9** 509
[27] Emsley J, Moore J, Udy P B 1971 *J. Chem. Soc. A* 2863
[28] Allcock H 1976 *Phosphorus Nitrogen Compounds: Cyclic, Linear, and High Polymeric Systems* (New York: Academic Press)

Acknowledgments
The work was supported by Mendeleev University of Chemical Technology of Russia. Project Number 2020-035.

IOP Conf. Series: Materials Science and Engineering 1117 (2021) 012028 doi:10.1088/1757-899X/1117/1/012028

Genetic Algorithm and Particle Swarm Optimization for Parameter Optimization of Least-Square Support Vector Regression Model in Electricity Load Demand Forecasting

Irhamah[1*], K H Gusti[2], H Kuswanto[3], N. A. Firdausanti[4]

[1,2,3,4]Department of Statistics, Institut Teknologi Sepuluh Nopember Surabaya
Jl. Raya ITS, Kampus ITS Sukolilo, Surabaya, Indonesia 60111

Email: irhamah@statistika.its.ac.id[1*]

Abstract. Accuracy is very important in time series forecasting where the model obtained depends on historical, linear, or nonlinear data patterns. This study aims to forecast short term electricity load in East Java, Indonesia, whereas electricity load is one of major and vital need in our daily life. The linear approach is carried out using the ARIMA method, while the nonlinear approach used in this study consist of the SVR and LSSVR models. The parameter selection greatly affects the results of accuracy, so an optimization method is needed. The usual grid search optimization method does not guarantee an optimum solution and more importantly, it is not efficient in some practical applications. Therefore metaheuristic optimization methods are needed, including GA and PSO which can find the entire possible space for the search for solutions. PSO is easier to apply but is prone to have premature convergence due to trapped in the minimum locale. To overcome this, modified PSO is developed to seek the optimal position according to the specified criteria. The result of this research is that the accuracy of the nonlinear approach is much better than the linear approach. The addition of an optimization method to the SVR provides a significant change in accuracy compared to LSSVR. Meanwhile, MPSO is the best optimization method because it produces a lowest RMSE value.

1. Introduction

Accuracy is the most important thing in forecasting which is a fondation of decision making process. There are two kinds of time series forecasting methods, which are forecasting model based on statistics model and forecasting model based on artificial intelligence, such as SVR. The forecasting result of ARIMA method tends to be less good because it tends to be flat in the range of its average value [1]. Meanwhile, the SVR method produces more accurate forecasting results than other artificial intelligence-based methods because it can overcome overfitting [2]. The solution of the SVR method is computationally complex, so the LSSVR method is proposed which only requires solving one set of linear equations without eliminating the principle of SVR [3]. The results of the accuracy of the SVR and LSSVR methods depend on the selection of parameter values that can be done by trial and error but are not effective because there are too many combinations of numbers and experiments [4].

The traditional optimization method is grid search which works by traversing all points in the parameter range and is guided by a performance matrix [5]. However, grid search optimization does not guarantee a global optimal solution and is impractical in some applications [6]. Alternative optimization methods that can be used are metaheuristic methods that mimic social behavior or strategies that exist in nature to find effective and comprehensive solutions. Metaheuristic methods have a superior ability to avoid local optima due to the ability to search the entire search space extensively thereby avoiding stagnation of local solutions [7]. Genetic Algorithm is one of the metaheuristic methods which can be implemented in a wider variety of problems and solution space [8]. Another metaheuristic method is the PSO method which is easier to apply because it does not require many parameters to be specified [9]. However, PSO can be trapped at a local minimum or premature convergence. Therefore, MPSO adds a parameter of PSO that will seek to find the optimal particle position according to predetermined criteria and show better accuracy than PSO [10].

The case study in this study is a short-term electricity load that has the characteristics of a complicated series pattern due to an overlapping repetitive pattern generated by linear and nonlinear processes [11]. In this study, a comparison of the SVR and LSSVR methods along with the grid search, GA, PSO, and MPSO optimization methods were carried out so that short-term power load forecasting values were obtained. The best model was determined based on the minimum RMSE value.

2. Literature Review

2.1. *Genetic Algorithm.* The GA method is an optimization technique based on the principles of genetics and natural selection. The population is formed from many individuals who develop according to specific selection rules by maximizing fitness. This algorithm is done by looping Darwin's concept. The main stages of the GA procedure are (1) Defining variables as chromosomes, (2) Forming an initial population consisting of several chromosomes, (3) Evaluating the ability of chromosomes based on fitness, (4) Selecting chromosomes with maximum fitness as parents, (5) Combining genetic information in parents by crossing over, (6) Carrying out a process of mutation to introduce new genes randomly, (7) The process is repeated from stage 2 until it reaches convergence [12].

2.2. *Particle Swarm Optimization.* The PSO method was developed based on the behaviour of birds or fish in search of food sources. The direction of movement towards food is influenced by instinct, which in the algorithm is identical to that of exploration to obtain optimum results in different regions [13]. In the PSO algorithm, a set of m moving particles is influenced by (1) Inertia (2) Best local position (3) Best global position. In general, the motion of particles is formulated as follows:

$$v_{id}^{k+1} = v_{id}^{k} + c_1 r_1^{j}\left(pbest_{id}^{k} - x_{id}^{k}\right) + c_2 r_2^{k}\left(gbest_{d}^{k} - x_{id}^{k}\right) \tag{1}$$

$$x_{id}^{k+1} = x_{id}^{k} + v_{id}^{k+1} \tag{2}$$

The greater the value of ω, the greater the ability of PSO to find possible solutions in the search space, and vice versa. In general, if ω is one, then in the next iteration there is a partial lack of search capability. The position and velocity of the particles will be updated until convergence is reached [14].

2.3. *Modified Particle Swarm Optimization.* In complex problems, the PSO algorithm risks premature convergence which results in particles trapped in the local optimum solution. The MPSO algorithm adds a parameter of mean particle distance and operator mutation to guide population selection and evaluate convergence. The formulation of the average particle distance is written in equation (3) and the operator mutation capable of controlling the convergence speed of the PSO algorithm is formulated in equation (15).

$$D(t) = \frac{1}{mL}\sum_{i=1}^{m}\sqrt{\sum_{d=1}^{n}\left(x_{id} - \bar{x}_{d}\right)} \tag{3}$$

$$\sigma^2 = \sum_{i=1}^{m} \left(\frac{f_i - \bar{f}}{f} \right)^2$$

$$f = \begin{cases} \max\left|f_i - \bar{f}\right|, \max\left|f_i - \bar{f}\right| > 1 \\ 1 \quad , \min\left|f_i - \bar{f}\right| \leq 1 \end{cases} \tag{4}$$

With the increasing number of iterations, the values $D(t)$ and σ^2will get smaller. When $D(t) < \alpha$ and $\sigma^2 < \beta$ (α, β are the specified threshold values), the algorithm is considered to be in the final search stage and the population will experience premature convergence. The MPSO algorithm maintains the optimal location of the current group of particles and recovers the particle solution space so that it can get out of the optimal locale [10].

3. Methodology

3.1. Data Sources

The data used in this study is secondary data obtained from PT PLN Distribution East Java regarding the demand for electricity loads in the districts and cities of East Java. Electricity load is recorded per 30 minutes from January 2016 to December 2016. The data is divided into two groups: training, and testing. The training data used to build the model is from January 2016 to November 2016 and testing data in 2016.

3.2. Data Structure

The data used in this study are short-term electricity loads recorded per 30 minutes at a 20kV substation (distribution section) in East Java. The research data is arranged in a structure which is presented in Table 1.

Table 1. Electricity Loads at 20kV Substation

Time	Day	Electricity Loads	Description
1	1	$Z_{i,1}$	Training
...	...	...	...
16031	334	$Z_{i,16031}$	Training
16032	334	$Z_{i,16032}$	Training
...	...	...	...
17520	365	$Z_{i,17520}$	Testing

3.3. Steps of Analysis

The analysis steps in this research are as follows:

- Obtain an electric load forecasting model based on the ARIMA, SVR, and LSSVR models
- Get the best optimization results on the SVR and LSSVR models with the GA, PSO, and MPSO optimization methods. The selection of the best model is done by evaluation using RMSE and SMAPE

4. Result and Discussion

4.1. Characteristics of Electricity Load

The characteristics of the demand for electricity in the East Java region are obtained through descriptive statistical analysis by exploring the information contained in the data. Figure 1(a) shows that the data is non-stationary in mean. The consumption pattern of public electricity loads is presented in Figure 1(b). It can be seen that the lowest average consumption of electricity load occurs at 07:00 in line with the commencement of community activities to work outside the home. The average consumption of electric loads reaches the highest peak at 18:30 as the end of industrial activities and work outside the home. After that, the average electricity load gradually drops along with the entry of public rest hours and agencies that do not operate at night. Based on Figure 1 (b) it is known that during the day, the consumption of electric loads has high variance. This is related to the consumption pattern which is dominated by the industrial sector.

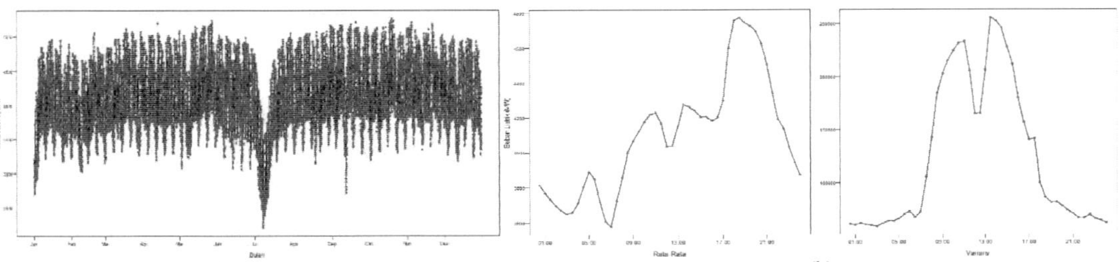

(a) (b)

Figure 1.(a) Time Series Plot of Electricity Loads (b)Mean and Variance Plot of Electricity Loads

The characteristics of the seasonal pattern from the data are shown in Figure 2 (a) which shows the amount of electricity consumption from January 1 to March 31, 2016. This illustrates the weekly seasonal pattern where on Saturdays and Sundays the consumption of electricity loads is lower than during the active day because it was contributed by industrial consumption which was operating on that day. Meanwhile, Figure 2 (b) shows the amount of electricity consumption from 1 January to 7 January 2016, which illustrates a pattern of low electricity consumption at night to early morning and then an increase in the morning to evening. This phenomenon shows the presumption of a seasonal daily pattern in the data. Based on the information above, it is estimated that there are daily variations (per 48 hours) and weekly variations (per 336 hours).

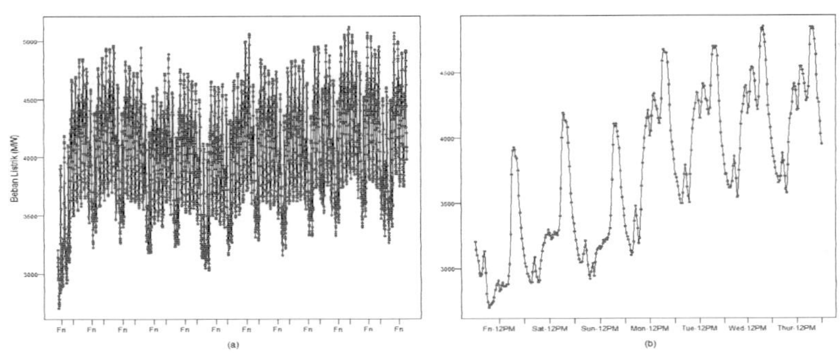

Figure 2. Seasonal Plot of Electricity Loadsin East Java

4.2. *ARIMA Estimation*

The stationary in mean can be identified by ACF and PACF. The data is not stationary in mean because the ACF plot decays very slowly, so it is necessary to do differencing on 1st lag. However, ACF plot after differencing still shows repeating patterns at multiples of 48, so it is necessary to do differencing at 48thlag. After the second differencing, the data is still not stationary. The high significance of the PACF lag at the multiple of 336 indicates that the data needs to be differencing at 336thlag. Therefore, the possible models are as follows:

$$ARIMA([28,31,47],1,[11,12,31])(0,1,1)^{48}(0,1,2)^{336}$$

$$ARIMA([28,47],1,[11,12])(0,1,1)^{48}(0,1,2)^{336}$$

The parameter estimation is shown in Table 2.Both models have significant parameters so that forecasting was done and compared with testing data. This procedure resulted in RMSE values of 618.1697 and 646.465 with sMAPE values of 13.76 and 14.48 for model 1 and model 2, respectively. The best model is the 1st model that can be written as follows.

$$\left(1-\phi_{28}B^{28}-\phi_{31}B^{31}-\phi_{47}B^{47}\right)\left(1-B\right)\left(1-B^{48}\right)\left(1-B^{336}\right)\left(1-B^{672}\right)Z_t$$
$$=\left(1-\theta_{11}B^{11}-\theta_{12}B^{12}-\theta_{31}B^{31}\right)\left(1-\Theta B^{48}\right)\left(1-\Theta B^{336}\right)\left(1-\Theta B^{672}\right)a_t$$

Table 2. Parameter Estimation of ARIMA Models

	Model 1		Model 2	
	Est	p-value	Est	p-value
θ_{11}	-0,0519	<.0001	-0,0525	<.0001
θ_{12}	-0,0292	0.0002	-0,0274	0.0006
θ_{31}	-0,2297	0.0012	-	-
$\Theta_{1(48)}$	0,7072	<.0001	0.7057	<.0001
$\Theta_{2(336)}$	0,7862	<.0001	0.78644	<.0001
$\Theta_{3(672)}$	-0,0311	0.0002	-0.03102	0.0002
ϕ_{28}	-0,0199	0.0103	-0.0192	0.0167
ϕ_{31}	-0,2403	0.0006	-	-
ϕ_{47}	0,0783	<.0001	0.08006	<.0001

4.3. Parameter Optimization using Grid-search

Grid-Search optimization is done by describing the ARIMA model and then a significant lag is obtained used as an exogenous variable. The significant lag focuses on the AR model so that there are 111stdata lag. Two input parameters, which are the cost parameter (C) and the RBF kernel function parameter (σ^2) for the LSSVR method and three parameters for the SVR method, which are cost (C), epsilon (ε) and RBF kernel function parameters (σ^2) were optimized using the Grid-Search. The parameter ranges for LSSVR used are {(50,200), (1, 25.75, 50.5, 75.25, 100)} for σ^2 and C, while the SVR is {($1e$ - 2,1), ($1e$ −4,10), ($1e$ −2,50)} for ε, C, and σ^2. Furthermore, the model obtained was used to model the next 1488 and a half hours. The accuracy criteria used are the minimum RMSE and the optimal parameters obtained for σ^2 and c are 50 and 25.75, and the SVR method results in optimal parameters for ε, C, and $\sigma 2$ are 0.01, 10, and 0.01, respectively. The RMSE values obtained for LSSVR and SVR were 51.114845812 and 58.11766134.

4.4. Parameter Optimization using Genetic Algorithm

Optimization with GA is done by making the parameters from the Grid-Search as a good initial value. The procedure was carried out by generating 15 chromosomes for 5 generations with a cross over the chance of 0.8, the chance for mutation of 0.1, the number of chromosomes that survive in the next generation, namely 1, the solution space for c = [49,250] and σ^2 = [25,100]. The selection of parents uses the RWS method, the combination of genes uses a single point crossover process, the mutation process uses a uniform mutation that replaces one of the genes with a uniformly distributed random number (0.1), the fitness value used is RMSE. Based on the optimum fitness, the optimal parameters are 197.2273 and 43.88546 for σ^2 and C with an RMSE value of 43.25871.

In the SVR method, the procedure performed for GA optimization is the same as for the LSSVR method. The size of genes in chromosomes in the SVR method is under many optimized parameters, namely cost (C), epsilon (ε), and RBF kernel function parameters (σ^2) so that there are 3 genes. The fitness value used in the SVR is -RMSE. Based on the optimum fitness, the optimal parameter is 0.00839; 11,29062; 0.001119 for ε, C, and σ^2 with an RMSE value of 42.901.

4.5. Parameter Optimization using Particle Swarm Optimization

Optimization with PSO is done by making the parameters from the Grid-Search as the good initial value. The procedure is carried out by generating 15 particles with an inertia value of 0.9, the coefficient of each particle and the overall particle of 2, the maximum speed limit of 2% of the upper limit of each particle position, the fitness used is RMSE and the convergence criterion by limitedby 5 iterations. Based on the optimum fitness, the optimal parameters are 196,1141 and 40,597 for σ^2 and C with an RMSE value of 41.87447.

In the SVR method, the procedure performed for PSO optimization is the same as for the LSSVR method. The particle size in the SVR method corresponds to many optimized parameters, which are cost (C), epsilon (ε), and the RBF kernel function parameter (σ^2). The fitness value used in the SVR is RMSE. Based on the optimum fitness, the optimal parameter is 0.005567992; 9,481282; 0.002875503 for ε, C, and $\sigma 2$, with an RMSE value of 43.6688

4.6. Parameter Optimization using Modified Particle Swarm Opimization

Optimization using MPSO is done by making the parameters of the Grid-Search as the good initial value. The procedure is carried out by determining the same parameter values as PSO, plus the MPSO parameters, which are alpha and beta as the algorithm guides to retrieve the optimal position and diagonal length of the search space. The values assigned to each alpha, beta, and diagonal length are 0.5, 0.05, and 100. Based on the optimum fitness, the optimal parameters are 188.0168 and 47.67838 for σ^2 and C with RMSE values of 45.403822.

In the SVR method, the procedure carried out for MPSO optimization is the same as for the LSSVR method, including the values of alfa, beta, and diagonal length. The particle size in the SVR method corresponds to many optimized parameters, which are cost (C), epsilon (ε), and the RBF kernel function parameter (σ^2) so that there are 3 sizes. The fitness value used in the SVR is RMSE. Based on the optimum fitness, the optimal parameter is 0.0001051; 8,3459835649; 0.0044634277 for ε, C, and $\sigma 2$, with an RMSE value of 45.403822.

4.7. The Comparison of the Optimization Algorithms

The performance comparison of each optimization algorithm were measured based on the RMSE values shown as in Table 3.

Table 3. The Performance of Each Model

Metod	RMSE	Metod	RMSE	rank	Metode	RMSE	rank
DSARIMA	618,1697	LSSVR	42,02479	4	SVR	48,7939	4
		LSSVR-GA	41,86809	2	SVR-GA	42,4174	3
		LSSVR-PSO	41,87447	3	SVR-PSO	41,4365	2
		LSSVR-MPSO	41,85998	1	SVR-MPSO	40,1367	1

It can be seen that the LSSVR method the best optimization method is the GA algorithm. In contrast to [16] in this study, MPSO did not improve the accuracy of PSO, this was possible because of an error in the selection of the guiding parameters. Overall, the accuracy of the SVR method is better than the LSSVR when combined with the optimization method, this can be seen from the high difference in RMSE values, while LSSVR has a not too significant difference in accuracy between optimization methods. When compared to DSARIMA, artificial intelligence-based methods have better accuracy because of the complex patterns in the short-term time series cannot be accommodated by the linear model.

5. Conclusions and Suggestions

5.1. Conclusions

Forecasting using the LSSVR method is carried out based on the results of the ARIMA model first. Short-term electricity load data for the East Java area is known to have a Double Seasonal model

ARIMA ([28,31,47], 1, [11,12,32]) (0,1,1) 48 $(0,1,2)^{336}$ that has 111 significant lags. Forecasting using the LSSVR method produces a much better accuracy of forecast results than ARIMA.

Overall, the optimization method that has the best accuracy is MPSO, GA, then PSO, respectively. This is because the MPSO optimization results are ranked first in each model. As for the results of the accuracy in the comparison of the SVR and LSSVR methods, it is concluded that the SVR method has the best accuracy when combined with the optimization method, while in the LSSVR method the addition of the optimization method does not give a significant change in the accuracy results.

Acknowledgement
The authors greatly appreciate Ministry of Research, Technology, and Higher Education, Republic of Indonesia and Institut Teknologi Sepuluh Nopember Surabaya Indonesia for financial support under "Penelitian Tesis Magister".

References

[1] D. Wiyanti and R. Pulungan, "Peramalan Deret Waktu Menggunakan Model Fungsi Basis Radial (RBF) dan Auto Regressive Integrated Moving Average (ARIMA)," Jurnal MIPA, vol. 35, no. 2, pp. 175-182, 2012.
[2] M. Amin and M. Hoque, "Comparison of ARIMA and SVM for Short-Term Load Forecasting," pp. 205-210, 2019.
[3] J. Suykens and J. Vandewalle, "Least Square Support Vector Machine Classifiers," Neural Processing Letters, vol. 9, pp. 293-300, 1999.
[4] S. Prangga, "Optimasi Parameter pada Support Vector Machine Menggunakan Metode Taguchi untuk Data High-Dimensional," Institut Teknologi Sepuluh Nopember, Surabaya, 2017.
[5] L. Septiningrum, H. Yasin and Sugito, "Prediksi Indeks Harga Saham Gabungan Menggunakan Support Vector Regression (SVR) dengan Algoritma Grid Search," vol. 4, no. 2, pp. 315-321, 2015.
[6] P. Cortez, "Modern Optimization with R," Gurmaraes.
[7] A. Pangestu, B. Arie and E. Jovi, "Optimizing Neural Network Classifier for Diabetes Data Using Metaheuristic Algoritms," ITSMART : Jurnal Ilmiah Teknologi dan Informasi, vol. 6, no. 2, pp. 85-91, 2017.
[8] S. Sivandam and S. Deepa, Introduction to Genetic Algorithm, Berlin, Heidelberg: Springer, 2008.
[9] W. Hu, L. Yan, K. Liu and H. Wang, "PSO-SVR : A Hybrid Short-Term Traffic Flow Forecasting Method.," International Conference on Parallel and Distributed Systems, pp. 553-561, 2015.
[10] D. Niu and S. Dai, "A Short-Term Load Forecasting Model with A Modified Particle Swam Optimization Algorithm and Least Square Support Vector Machine Based on the Denoising Method of Empirical Model Decomposition and Grey Relational Analysis," Energies, vol. 10, no. 408, pp. 1-20, 2017.
[11] F. Fahmi and H. Sofyan, "2017," International Conference on Electrical Engineering and Informatics (ICELTICs) , pp. 97-102, 2017.
[12] Z. Ismail and Irhamah, "Solving the Vehicle Routing Problem with Stochastic Demands via Hybrid Genetic Algorithm-Tabu Search," Journal of Mathematics and Statistics, vol. 4, no. 3, pp. 161-167, 2008.
[13] Ghalia and B. Mounir, "Particle Swarm Optimization with An Improved Exploration-Exploitation Balance," Midwest Symposiu on Circuit and Systems, pp. 759-762, 2008.
[14] Q. Bai, "Analysis of Particle Swarm Optimization Algorithm," Computer and Information Science, vol. 3, no. 1, pp. 180-184, 2010.

IOP Conf. Series: Materials Science and Engineering 1117 (2021) 012029 doi:10.1088/1757-899X/1117/1/012029

Release Analysis of Nano-Titanium Dioxide (TiO_2) from Paint: An Accelerated Weathering Experiment

M J Llana[1,5,a], M S Tolentino[2,3,6,b], N C C Valeza[4,6], J P Reyes[6] and B A Basilia[1,2,6]

[1]School of Graduate Studies, Mapúa University, 658 Muralla St., Intramuros, Manila City 1002, Philippines
[2]School of Chemical, Biological, and Materials Engineering and Sciences, Mapúa University, 658 Muralla St., Intramuros, Manila City 1002, Philippines
[3]General Education Department, Colegio de Muntinlupa, Mayor J. Posadas Ave., Sucat, Muntinlupa City 1770, Philippines
[4]Department of Mining, Metallurgical, and Materials Engineering, College of Engineering, University of the Philippines, Diliman, Quezon City, 1101, Philippines
[5]Nanotech Analytical Services and Training Corporation (NASAT Labs), Katarungan Village, Brgy. Poblacion, Muntinlupa City 1781, Philippines
[6]Materials Science Division, Industrial Technology Development Institute, Department of Science and Technology, DOST Cpd., Gen. Santos Ave., Bicutan Taguig City 1631, Philippines

Email: marlon_llana@nasatcorp.com[a], marvstolentino@yahoo.com[b]

Abstract. Engineered nanomaterials (ENM's), particularly TiO_2 nanoparticles are being incorporated on paints due to their capabilities to enhance optical properties and to scatter UV radiation, which eventually protects the paint from discoloration. In this study, the release of these nano-TiO_2 from paint was investigated. Two (2) types of paint were used, one containing pigment- and nano-TiO_2 (P1), while the other only contains pigment-TiO_2 (P2). The paints were subjected to accelerated weathering wherein the paints were exposed to cycles of UV irradiation and water condensation. The SEM-EDX results showed the distribution of the pigment- and nano-TiO_2 on the surface of P1 and P2, and the changes in the morphologies before and after weathering. The photomicrograph revealed the formation of cracks on the surface of P1 brought about by the photocatalytic degradation due to the presence of nano-TiO_2. Leachate samples were also collected weekly in the duration of the weathering test and were analysed using ICP-OES. The presence of Ti, which can be correlated to the release TiO_2, on the samples was detected. The release of Ti from P1 and P2 showed a decreasing trend throughout the weathering experiment, having the following initial amounts of release: 1.38 mg/L and 2.10 mg/L for P1 and P2, respectively. Moreover, a graphical release mechanism for the nano-TiO_2 release was conceptualized based on the results of the study. In general, potential release of nano-TiO_2 from paints could happen since these TiO_2 nanoparticles serve as a photocatalyst in the paint degradation when exposed to prolonged weathering conditions.

1. Introduction

Surface coatings serve as a protective layer to prevent degradation and/or hinder further deterioration of their corresponding substrates, especially when exposed to extreme environmental conditions [1]. Due to this significant role of surface coatings, several studies are being conducted to enhance their properties and to promote desirable attributes. These include the incorporation of additives, such as engineered nanomaterials (ENM's), on the coatings' polymer matrices. Using ENM's, with dimensions less than 100 nm, as additive on polymeric coatings is an area of interest in this field, since ENM's have better properties in comparison to their bulk counterparts [2]. One important area of application of ENM's, specifically on surface coatings, is on paints. ENM's are usually incorporated on paints to improve properties such as chemical resistance, protection against physical deterioration like abrasion and erosion, promote surface-cleaning, anti-fouling, or anti-microbial property, and most importantly, ultraviolet light protection [2–4].

Nanoscale titanium dioxide (nano-TiO_2) is one of the ENM's that has a wide range of industrial applications. It is commonly used on surface coatings, particularly on paints, as whitening agent and integrates the above-mentioned desirable properties. Furthermore, nano-TiO_2 on paint enhances its photodynamic properties [5]. In this regard, increased volumes of nano-TiO_2 are being manufactured, together with the rising demand for paints and other surface coatings. But aside from the advantages brought about by the incorporation of nano-TiO_2 on paints, its potential environmental risk must also be considered. Numerous researches have been conducted and predicted that relevant amounts of these nano-TiO_2 could end-up in the environment and the release of nano-TiO_2 can possibly affect living organisms to some extent [2,4]. Paints and other polymeric coatings containing nano-TiO_2 are often exposed to natural degradation or weathering that are initiated by different mechanical stresses and/or ultraviolet radiation from the sun. Consequently, exposure to the said conditions can result to the release of nano-TiO_2 from paints and other coatings to the environment, which can eventually cause environmental hazards. In the study conducted by Al-Kattan et al. (2013), it was found that after 113 cycles of weathering, only 0.007% of the total Ti (1.0 to 1.5 μg/L) released from paint with pigment- and nano-TiO_2 [4]. This signifies that the nano-TiO_2 particles were strongly embedded on the paint matrix. In addition, Zhang and his co-researchers (2017) also observed in their study the release of nano-TiO_2 from paints under various weathering and rainfall durations [2]. The total nano-TiO_2 released ranged from 0.6 to 2.3 μg/L. The release of nano-TiO_2 was found to be evident at low pH. Similarly, Azimzada and his co-workers (2019) studied the release of nano-TiO_2 from paints in cold climates [6]. Interestingly, they found that the release was greater during winter, specifically under freeze-thaw conditions. Though, only <0.001% of the total nano-TiO_2 were released. In relation to these release studies, the research conducted by Sun et al. (2016) showed that almost half of the yearly production of nano-TiO_2 (39,000 tons/year) was just being released directly into the environment [7].

Several studies have been done to investigate the quantities of nano-TiO_2 that are being released from various sources such as paints. Surface characterizations and morphological studies have also been employed in relation the release of nano-TiO_2. But majority of the studies were done by means of natural weathering which are more challenging to control and cannot be generalized due to climate differences across countries. Moreover, regardless of how the weathering experiment was conducted, the discussions on the release of nano-TiO_2 from the polymer matrix, such as paint, in relation to surface morphologies are still insufficient. The release mechanism of nano-TiO_2 is needed in detail to understand the concept of release in correlation to the factors that were considered in the study.

The main objective of this undertaking was to determine the amount of nano-TiO_2 released from paints through accelerated weathering conditions. The specific objectives of this study were as follows: a.) to utilize two (2) types of paint (P1-paint with pigment- and nano-TiO_2, and P2-paint with pigment-TiO_2 only); b.) to subject the painted panels with P1 and P2 under accelerated weathering conditions; c.) to characterize and compare the surface morphologies of the paints (P1 and P2) before and after weathering using scanning electron microscopy (SEM) with energy dispersive x-ray analysis (EDX) and digital microscopy; d.) to quantify the amount of nano-TiO_2 released from the painted panels due to weathering by means of inductively coupled plasma – optical emission spectrometry (ICP-OES); e.) to determine the trend for the release of nano-TiO_2; and lastly, f.) propose a graphical release mechanism of nano-TiO_2 from paints.

Understanding the details of the release of nano-TiO_2 from paints is important and can be one of the key aspects to further evaluate its ecological effects. In addition, this study could contribute to the life cycle assessment of paints with nano-TiO_2.

This study only considered two (2) types of local paint. Elements other than titanium (Ti) or compounds other than TiO_2 were not quantified in the experiments performed. The study only focused on correlating the results of surface morphology characterization and the release of TiO_2 from the paints' surfaces, which was further used to generate the graphical release mechanism.

2. Methodology

Two types of paint were used in this study: first was a special type of paint which contains pigment- and nano-TiO_2 (P1), and the other one was a regular acrylic paint which contains pigment-TiO_2 only (P2). Fiber cement panels (155 mm × 70 mm, $l \times w$) were prepared and painted on one side using P1 and P2 (3.80 g of paint/mm^2 of panel), and the other side of the panels were coated with wax. The painted panels were then subjected to accelerated weathering machine (Suga Test Instruments Co. Ltd.) for seven (7) weeks using a UV lamp with 388 nm wavelength and irradiance of 24 W/m^2. Exposure conditions were based on the established weathering standards for paint and related coatings and materials (ASTM D4587). In the duration of the accelerated weathering test, weekly collections of leachates were done. In this process, panels were pulled out of the weathering machine and deionized water was sprayed (with 150 mm distance) on the surface of the painted panels. The resulting leachates were then collected. Afterwards, the leachates were digested using aqua regia. Approximately 10 mL of the digested leachates for each type of paint per week were then submitted to ICP-OES (Prodigy7, Teledyne Leeman Labs) testing to determine the concentration of Ti that can be correlated to the released TiO_2. Furthermore, samples of the painted panels (using P1 and P2) before and after weathering were subjected to SEM with EDX (Helios NanoLab 600i) to characterize the surface morphologies. The photomicrographs of the painted samples were also taken using a digital microscope (Keyence VHX-7000). A graph was also generated to determine the relationship of the release of Ti with the duration of the weathering experiment. Moreover, a proposed conceptual release mechanism for nano-TiO_2 was generated.

3. Results and Discussion

3.1. Surface morphology characterization

3.1.1. Scanning electron microscopy with energy dispersive x-ray (SEM-EDX). This characterization technique was done to investigate observable changes with regards to the surface morphologies of the painted samples. The result of SEM with EDX mapping for P1 before and after weathering were presented in Figure 1. In this study, detected Ti was correlated with the presence of TiO_2 on the painted surface. It can be seen from the SEM image of P1 before weathering (Figure 1.a) that TiO_2 nanoparticles can be observed, which was supported by the result of the EDX mapping – represented by blue dots (Figure 1.d). Initially, the weight % of TiO_2 is 15.7%. After P1 underwent accelerated weathering, the top polymer layer of the paint surface wore off, making the nanoparticles of TiO_2 exposed and more visible (Figure 1.e). Considering the EDX result after weathering, the weight % of TiO_2 was reduced to 13.5% after weathering, which can be observed with the slight reduction of blue dots on Figure 1.h.

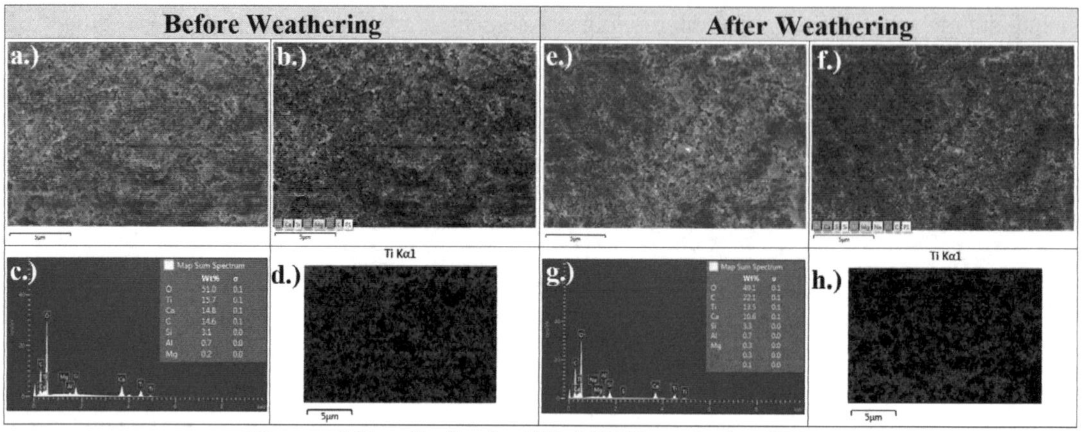

Figure. 1 Results for the SEM-EDX analysis of P1.

Due to accelerated weathering, the exposed TiO_2 nanoparticles tend to be washed off from the surface of the paint which resulted to the reduction of TiO_2 weight %. Since the EDX mapping was done on a point basis, five (5) more points for each panel used were considered for the EDX data to be statistically significant. The average weight % TiO_2 nanoparticles was found to be 16.2% before weathering, whereas after weathering, the weight % if TiO_2 was reduced to 10.2%. As the top layer of P1 was exposed to weathering conditions, degradation of paint happened which contributed to the release of around 5% of TiO_2 nanoparticles.

For P2 (Figure 2), same observation with P1 was observed but in contrary, the initial concentration of TiO_2 on the paint is low, around 1.2 wt % (Figure 2.c). As P2 underwent accelerated weathering, the visible particles (blue dots) tend to decrease (comparison between Figure 2.d & h). This can be related to the reduction of the TiO_2 concentration due to weathering. Based on the EDX result, the concentration of TiO_2 was significantly reduced by half (0.6 wt %) after weathering. But upon considering five (5) more points on the surface of the panels coated with P2, the average weight % of TiO_2 was found to be 11.96% and 16.4%, before and after weathering, respectively. Instead of decreasing weight % of TiO_2, there was the increase in the weight %. This can be accounted to the exposure or surfacing of more pigment-TiO_2 on the surface of the P2 due to weathering, and instead of being released from the surface, pigment-TiO_2 remained intact on the surface, which cause the increase in its weight % [4].

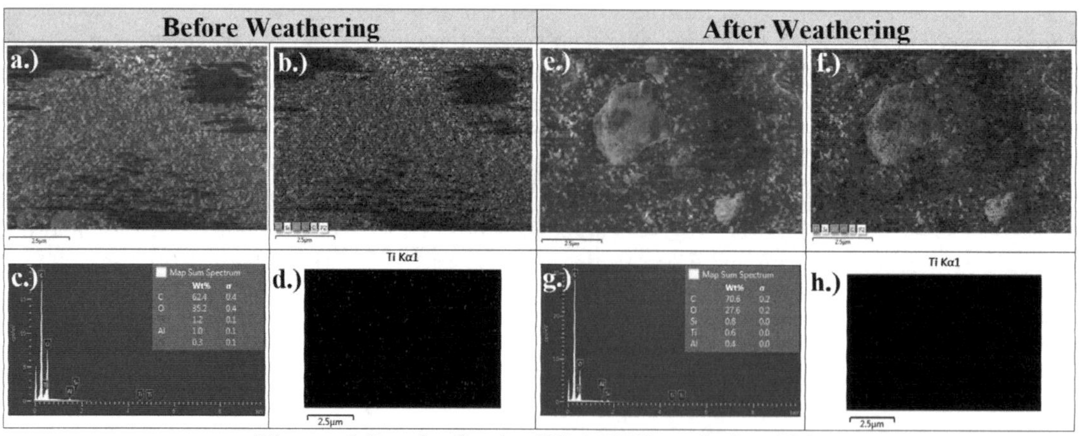

Figure. 2 Results for the SEM-EDX analysis of P2.

Figure 3 presents the SEM images of P1 and P2 before and after weathering at a higher magnification (25 000 x). The SEM image for P1 before weathering clearly shows that the nano-TiO_2 that was observed on the EDX analyses were strongly embedded on the polymer matrix of the paint. After weathering, it can be seen that the surface morphology of P1 changed. More spherical shaped particles were formed. This can be attributed to the degradation of paint, specifically the photodegradation of the organic binder due to the presence of nano-TiO_2. The particles of nano-TiO_2 that were incorporated on the paint matrix acts as a photocatalyst in the process of weathering; hence, resulting to the degradation of the top layer (as shown on Figure 3, after weathering of P1). The spherical shaped particles could be regarded as the chalking phenomenon, which was due to the paint degradation [8].

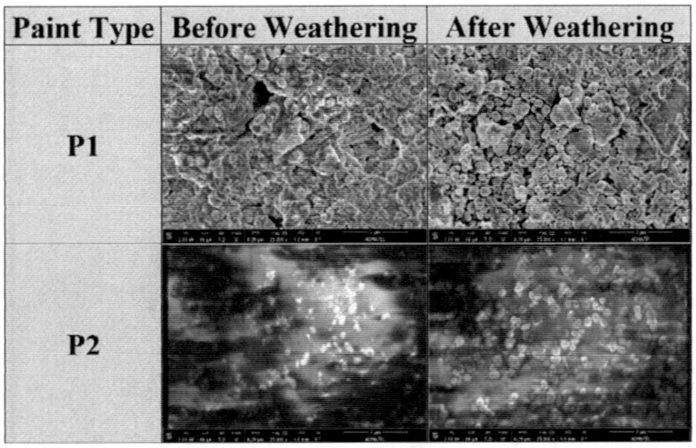

Figure. 3 SEM images of the surfaces of P1 and P2, before and after weathering at high magnification.

Formation of spherical particles were also observed in P2 after weathering, but more intact on the polymer matrix or organic binder. Even though pigment-TiO_2 were present on P2, their corresponding photo-reactivity, specifically their reaction to UV light, was different from that of nano-TiO_2 that were present on P1 due to their differences in sizes.

3.1.2. Photomicrography. The samples painted with P1 and P2 were also observed under a digital microscope before and after weathering to observe structural or surface deterioration of the paints. After weathering for seven (7) weeks, there was an evident formation of cracks on the surface of P1 (Figure 4). Several portions of the P1's surface exhibited crack development and potential branching of cracks. On the contrary, the surface of P2 remained the same and no significant changes were observed. The cracking of the surface of P1 was primarily due to the gradual degradation of the polymeric matrix, specifically the organic binder on the paint [2,9], in addition to the photodegradation brought about by the photocatalytic activity of nano-TiO_2 [8].

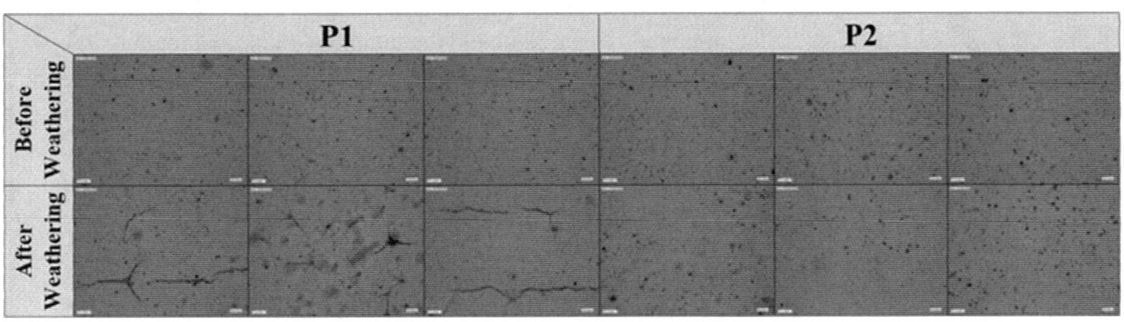

Figure. 4 Photomicrograph of P1 and P2 surfaces.

3.2. Release of nano-TiO_2 from paint

The release TiO_2 particles were quantified in this study, particularly the nano-TiO_2 from P1. Collection of leachates were done weekly to determine the leached amount of Ti from the painted panels using ICP-OES. The obtained concentrations of released Ti in the duration of the weathering cycles were presented on Figure 5. After seven (7) days of weathering, 1.38 mg/L of Ti was detected on the leachate for P1. The amount of the released Ti significantly decreased thereafter, having a decreasing trend in the duration of the weathering experiment. The release stabilized after 28 days of weathering with an average release of 0.031 mg/L. Similar observations were seen with the release of Ti from P2. The initial amount of Ti release was 2.10 mg/L, but unlike P1, the release stabilized after 42 days.

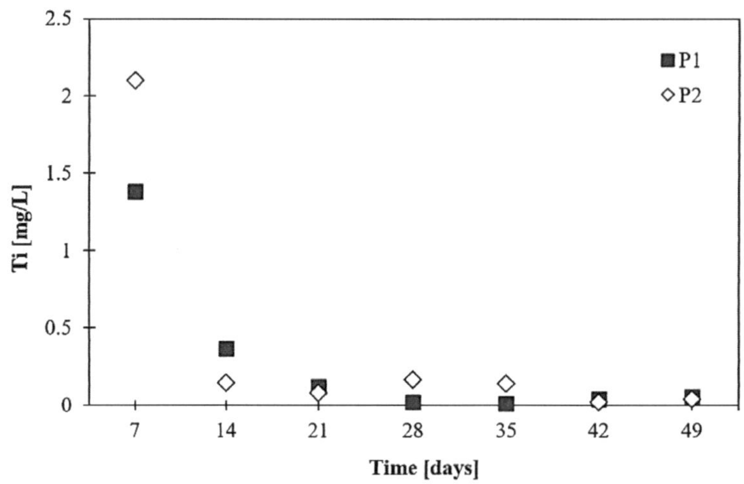

Figure. 5 The graph of the release of Ti from P1 and P2 during the weathering cycles.

3.3. Graphical release mechanism

Based on the results of the surface morphology characterizations that were done, the release mechanism of nano-TiO_2 from P1, which was the focus of the study, was conceptualized. Figure 6 shows that due to the exposure of the paint's surface to UV light or irradiation from the weathering chamber, and moisture or water condensed on the surface during weathering, the polymer matrix or organic binder tend to degrade (chalking phenomenon). This degradation resulted to the surfacing of more nano-TiO_2 that are embedded on the paint matrix, and eventually be released due to continuous weathering.

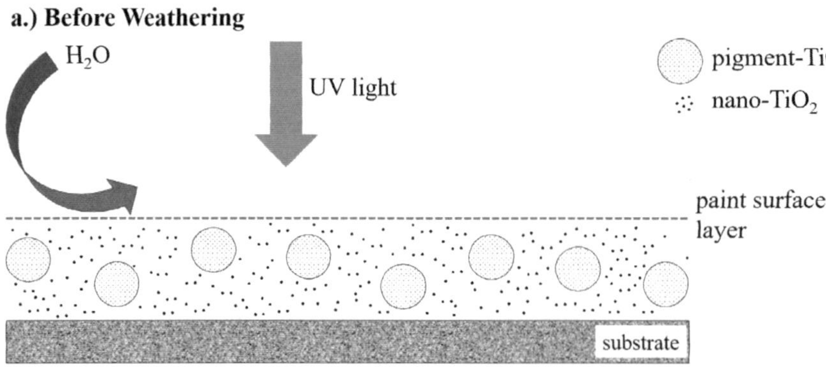

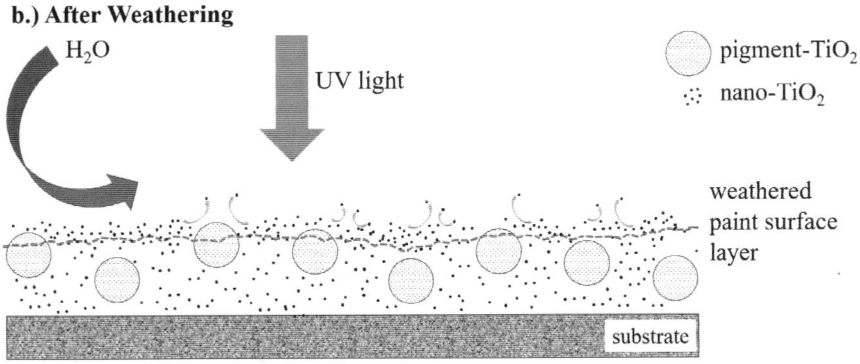

Figure. 6 The conceptualized release mechanism of the nano-TiO_2 from paints due to accelerated weathering.

4. Conclusion

In this undertaking, the objectives of the study were satisfied. Two (2) types of paint were used, and the accelerated weathering experiment was successfully carried out using the appropriate standard. Upon the performance of the experiment, it was observed that there was the evident change in the surface morphologies of the paints used (P1 and P2), particularly for P1. Formation of spherical particles can be seen on the SEM image of P1 after weathering, which can be related to the chalking of P1. For P2, surfacing of spherical particles were also observed but somehow still embedded on the surface. This could be the reason why chalking was not visible (by naked eye) on the surface of P2, unlike P1. SEM-EDX mapping analysis further showed that weight % of Ti on selected points on the surface of P1 decreased by around 5%. On the contrary, weight % of Ti on P2 increased by 4.44%. The result of SEM-EDX analyses for P1 and P2 only showed that upon subjecting to accelerated weathering conditions, the polymer matrix or organic binder on the paints tend to degrade and wear off, exposing or showing more Ti particles bounded in the paint matrix. The increase or decrease in the weight % of Ti on the selected points only show that weathering can cause changes on the surface morphologies, which can potentially result to the release of TiO_2 from the paints.

Furthermore, photomicrographs showed the formation of cracks on the surface of P1, but none was seen on P2. The organic binder of P1 experienced severe degradation when subjected to accelerated weathering. Further degradation happened on P1 due to the presence of nano-TiO_2 which acted as photocatalyst. This resulted to the chalking of P1's surface as seen on the SEM image. The appearance of cracks on the surface of P1 was also due to the photocatalytic degradation of the organic binder brought about by nano-TiO_2 photocatalyst. In this regard, cracks on the surface of P1 offers a larger surface area for the release of nano-TiO_2 particles which can be correlated to the detected Ti concentration on the leachates collected. In addition, although P2 only contains pigment-TiO_2, it still served as a photocatalyst to the degradation that happened during weathering. Although no visible chalking or cracks were observed on the surface of P2, there was still the release of TiO_2 particles.

References

[1] Tator K B 2018 Coating Deterioration *Prot. Org. Coatings* **5** 462–73

[2] Zhang X, Wang M, Guo S, Zhang Z and Li H 2017 Effects of weathering and rainfall conditions on the release of SiO_2, Ag, and TiO_2 engineered nanoparticles from paints *J. Nanoparticle Res.* **19** 5–10

[3] Lankone R S, Ruggiero E, Goodwin D G, Vilsmeier K, Mueller P, Pulbere S, Challis K, Bi Y, Westerhoff P, Ranville J, Fairbrother D H, Sung L P and Wohlleben W 2020 Evaluating performance, degradation, and release behavior of a nanoform pigmented coating after natural and accelerated weathering *NanoImpact* **17** 1–9

[4] Al-Kattan A, Wichser A, Vonbank R, Brunner S, Ulrich A, Zuin S and Nowack B 2013 Release

of TiO_2 from paints containing pigment-TiO_2 or nano-TiO_2 by weathering *Environ. Sci. Process. Impacts* **15** 2186–93

[5] Diamond S A, Kennedy A J, Melby N L, Moser R D, Poda A R, Weiss C A and Brame J A 2017 Assessment of the potential hazard of nano-scale TiO_2 in photocatalytic cement: application of a tiered assessment framework *NanoImpact* **8** 11–9

[6] Azimzada A, Farner J M, Hadioui M, Liu-Kang C, Jreije I, Tufenkji N and Wilkinson K J 2020 Release of TiO_2 nanoparticles from painted surfaces in cold climates: Characterization using a high sensitivity single-particle ICP-MS *Environ. Sci. Nano* **7** 139–48

[7] Sun T Y, Bornhöft N A, Hungerbühler K and Nowack B 2016 Dynamic Probabilistic Modeling of Environmental Emissions of Engineered Nanomaterials *Environ. Sci. Technol.* **50** 4701–11

[8] van Driel B A, Wezendonk T A, van den Berg K J, Kooyman P J, Gascon J and Dik J 2017 Determination of early warning signs for photocatalytic degradation of titanium white oil paints by means of surface analysis *Spectrochim. Acta - Part A Mol. Biomol. Spectrosc.* **172** 100–8

[9] Shandilya N, Le Bihan O, Bressot C and Morgeneyer M 2015 Emission of titanium dioxide nanoparticles from building materials to the environment by wear and weather *Environ. Sci. Technol.* **49** 2163–70

IOP Conf. Series: Materials Science and Engineering 1117 (2021) 012030 doi:10.1088/1757-899X/1117/1/012030

A Study on the Release of Nanoparticles to the Environment from Nano-Enabled Asphalt by Weathering Experiment

M J Llana[1,5,a], M S Tolentino[2,3,6,b], N C C Valeza[4,6], J P Reyes[6] and B A Basilia[1,2,6]

[1]School of Graduate Studies, Mapúa University, 658 Muralla St., Intramuros, Manila City 1002, Philippines
[2]School of Chemical, Biological, and Materials Engineering and Sciences, Mapúa University, 658 Muralla St., Intramuros, Manila City 1002, Philippines
[3]General Education Department, Colegio de Muntinlupa, Mayor J. Posadas Ave., Sucat, Muntinlupa City 1770, Philippines
[4]Department of Mining, Metallurgical, and Materials Engineering, College of Engineering, University of the Philippines, Diliman, Quezon City, 1101, Philippines
[5]Nanotech Analytical Services and Training Corporation (NASAT Labs), Katarungan Village, Brgy. Poblacion, Muntinlupa City 1781, Philippines
[6]Materials Science Division, Industrial Technology Development Institute, Department of Science and Technology, DOST Cpd., Gen. Santos Ave., Bicutan Taguig City 1631, Philippines

marlon_llana@nasatcorp.com[a], marvstolentino@yahoo.com[b]

Abstract. Researches in the field of nanotechnology is continuously gaining a lot of interest due to its wide range of applications. This study focused on the incorporation of halloysite nanotubes (HNT's) on asphalt and on the investigation of the release of the said nanoparticles from the asphalt to the environment through accelerated weathering experiment. Two (2) samples were prepared: asphalt with HNT (A1) and asphalt only (A2). SEM-EDX results showed that A1 initially had nanoparticles on the surface which can be attributed to the added HNT. SEM images further showed the changes on the surface morphologies of A1 and A2 after accelerated weathering experiment. These changes can be further correlated to the intercalation and exfoliation reactions that happened on the surface of asphalt during UV irradiation. Release studies revealed that Al and Si nanoparticles were being released on the surface due to the layer by layer degradation of the asphalt matrix. Hence, continuous exposure of asphalt could result to further degradation of asphalt, leading to the surfacing and release of more nanoparticles.

1. Introduction

In the recent years, nanotechnology has been continuously regarded as a promising field of research due to its numerous industrial applications. Nanotechnology deals with particles having dimensions ranging from 1 nm to 100 nm, which offers a large potential in enhancing the properties of materials [1]. Nanoscale materials incorporated on polymers forming nanocomposites possess excellent physical and chemical properties which include high stress/strain resistance, increased compressive strength, weathering protection, and improved thermal and electrical properties [2,3]. Due to impending applications of nanoscale materials, this gave rise to engineered nanoparticles (ENP's) which are purposely designed and engineered to meet the desired material properties for specific applications [3].

Among the materials being modified by employing ENP's, pavement materials are given interest by researchers. Studies in the field of pavement materials tend to focus on utilizing nanomaterials to modify asphalt [4]. Commonly used as a binder to several applications such as corrosion protection, waterproofing and resistance against moisture, asphalt is a black organic material made up of complex hydrocarbons with diverse molecular weights [2]. Being widely utilized in the pavement industry, asphalts are required to be resistant to different environmental and weathering conditions, creep, fatigue, as well as to heavy thermal and traffic loads. Hence, cost-effective modification of asphalt is needed. Numerous ENP's serve as reinforcements or additives to asphalt in order to enhance its properties and performance [5,6]. ENP's incorporated to asphalt include nanoclay, carbon nanoparticles, carbon nanotubes, carbon nanofibers, subnano-sized hydrated lime, nano-phosphorus and nano-SiO_2 [6,7]. Although incorporation of ENP's on asphalt can enhance its properties, their stability and compatibility must be considered; otherwise, the asphalt will be damaged prematurely due to aging [2]. For this reason, ENP's that are incorporated in the asphalt matrix could be released, which can pose potential health and environmental hazards.

Most studies tend to investigate the effects of adding ENP's on asphalt and analyze the corresponding increase in its properties. In addition, researches on the potential hazards of asphalts were about occupational safety during asphalt processing, and asphalt particles are only in the form of dust or fumes. Very few studies focus on the release of nanoparticles from asphalt when exposed to extreme weathering conditions.

The main goal of this study was to quantify the release of the nanoparticles from asphalt by performing an accelerated weathering experiment. Specifically, the objectives of this study were as follows: a.) to produce a nano-enabled asphalt by incorporating halloysite nanotubes (HNT's); b.) to subject the nano-enabled asphalt (A1) and regular asphalt (A2) to accelerated weathering conditions; c.) to characterize and observe changes in the surface morphologies of A1 and A2 after accelerated weathering using scanning electron microscopy (SEM) with energy dispersive x-ray analysis (EDX); and lastly, d.) to quantify and observe the trend of the release of nanoparticles, specifically HNT's, from A1 and A2 in the form of aluminum (Al) and silicon (Si) using inductively coupled plasma – optical emission spectrometry (ICP-OES).

This undertaking could serve as a fundamental study in developing standards for the potential health and environmental hazards brought about by the release of nanomaterials from asphalt when used in pavement applications.

The release of Al and Si (potentially from HNT added) were the only point of detection for the release of nanoparticles from asphalt. This was correlated to the changes in the surface morphologies of A1 and A2, as seen from the SEM images.

2. Methodology

HNT (Sigma-Aldrich), 2.0 wt. %, was added to asphalt (Shelby, Bayer AG) and mixed mechanically for 1 hour to produce the nano-enabled asphalt (A1). An asphalt without HNT (A2) was also used as reference. Fiber cement panels (155 mm × 70 mm, l × w) were prepared and coated on one side using A1 and A2 (3.80 g of paint/mm^2 of panel), and the other side of the panels were coated with wax. The painted panels were then subjected to accelerated weathering machine (Suga Test Instruments Co. Ltd.) for nine (9) weeks using a UV lamp with 388 nm wavelength and irradiance of 24 W/m^2. Exposure conditions were based on the established weathering standards for bituminous materials (ASTM D4799). In the duration of the accelerated weathering test, weekly collections of leachates were done. In this process, panels were pulled out of the weathering machine and deionized water was sprayed (with 150 mm distance) on the surface of the asphalt-coated panels. The resulting leachates were then collected. Afterwards, the leachates were digested using aqua regia. Approximately 10 mL of the digested leachates for each type of asphalt per week were then submitted to ICP-OES (Prodigy7, Teledyne Leeman Labs) testing to determine the concentration of Al and Si that can be correlated to the released nanoparticles, specifically from HNT's added. Furthermore, the coated panels (using A1 and A2) before *and after weathering were subjected to SEM with EDX (Helios NanoLab 600i) to characterize the surface morphologies.*

3. Results and Discussion

3.1. Surface morphology characterization

Scanning electron microscopy with energy dispersive x-ray (SEM-EDX). The surface morphologies of the A1 and A2 were analyzed based on the SEM images obtained, as well as the elemental composition through EDX (Figure 1). The surface morphology of A1 shows agglomeration of particles on the surface, which can be potentially HNT nanoparticles. For A2, the image obtained could be due to the uneven application of the asphalt onto the substrate. HNT contains Al and Si elements, and since it was incorporated on A1, EDX results showed the detection of the said elements. For A2, Al and Si were also detected; hence, the asphalt itself already contains Al and Si. In this regard, for the release analysis, Al and Si could be from HNT and asphalt.

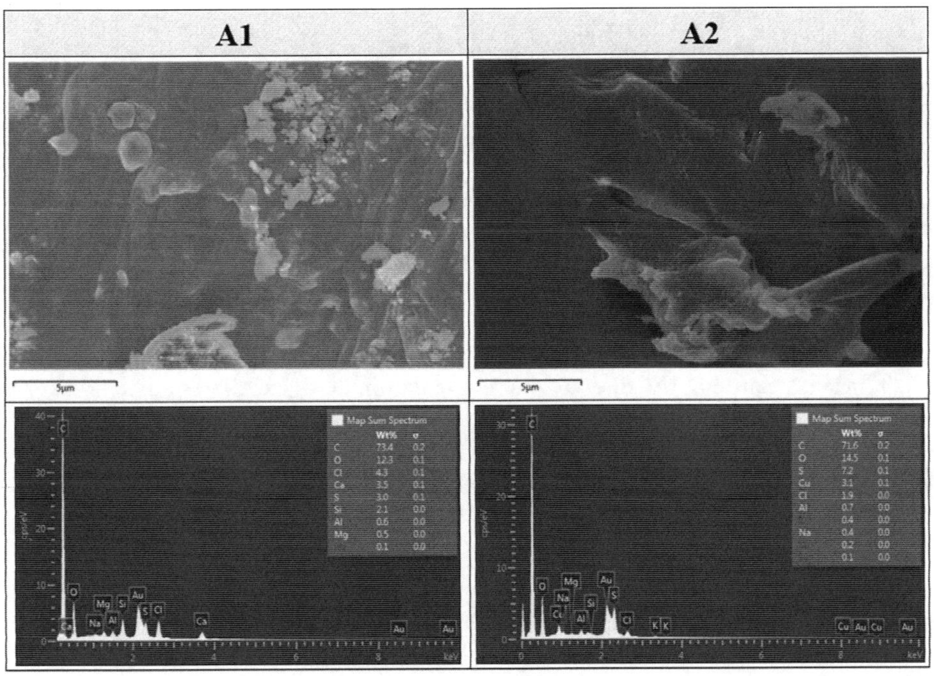

Figure 1. Results for the SEM-EDX analysis of A1 and A2.

For the comparison of the surface of A1 and A2 before and after weathering, their SEM images were presented on Figure 2. It can be observed that for both A1 and A2, changes in the surface morphologies were evident. There was the rearrangement of molecules on the top layer of the surfaces due to UV irradiation and water condensation inside the weathering chamber. Theoretically, considering the weathering process, the microstructures that can be observed for A1 and A2 after weathering were due to the oxidation and polymerization that happened on the surface molecules [8]. Moreover, based on the SEM image of A1 after weathering, it can be said that there was the intercalation and exfoliation reaction between HNT nanoparticles and hydrocarbon chains of the asphalt. In addition, even though A1 underwent weathering and although some HNT nanoparticles in the form Al and Si nanoparticles were released, most particles were still embedded on the surface of A1. This was due to the enhanced property of asphalt with HNT against high-temperature conditions [4]. Unlike in P2 (without HNT), formation of particles due to the oxidation and polymerization reactions brought about by weathering can be observed.

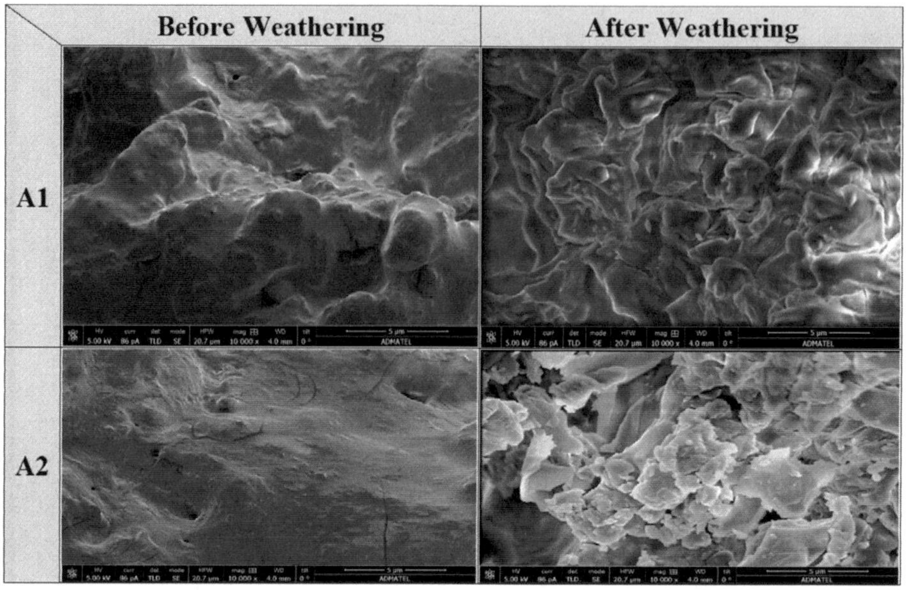

Figure 2. SEM images of A1 and A2 before and after weathering.

3.2. Release of nanoparticles from paint

A1 and A2 were subjected to accelerated weathering, and release of nanoparticles was expected. As can be seen on the plot in Figure 3 and Figure 4 for the release of Al and Si, respectively, there was no significant trend that can be observed. This was probably due to uneven distribution of the nanoparticles in the asphalt matrix, particularly due to the intercalation and exfoliation reactions. An evident observation for both A1 and A2 was the sudden decrease in the release after 5 weeks. This can be correlated to the degradation of asphalt by layer. The nanoparticles on the first exposed layer of the asphalt were released up to week 5. After the degradation of first layer, nanoparticles on the second layer appeared on the surface, then released due to continuous weathering.

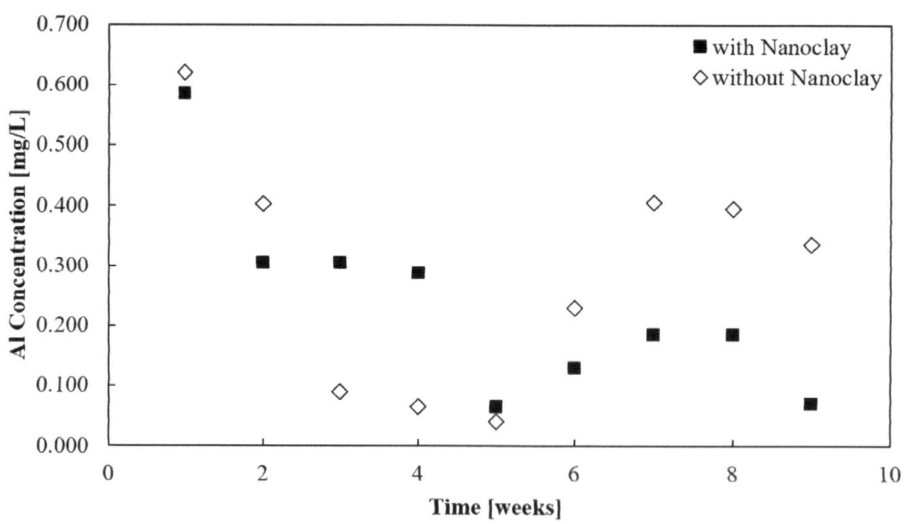

Figure 3. Release of Aluminum (Al) from A1 and A2 during weathering.

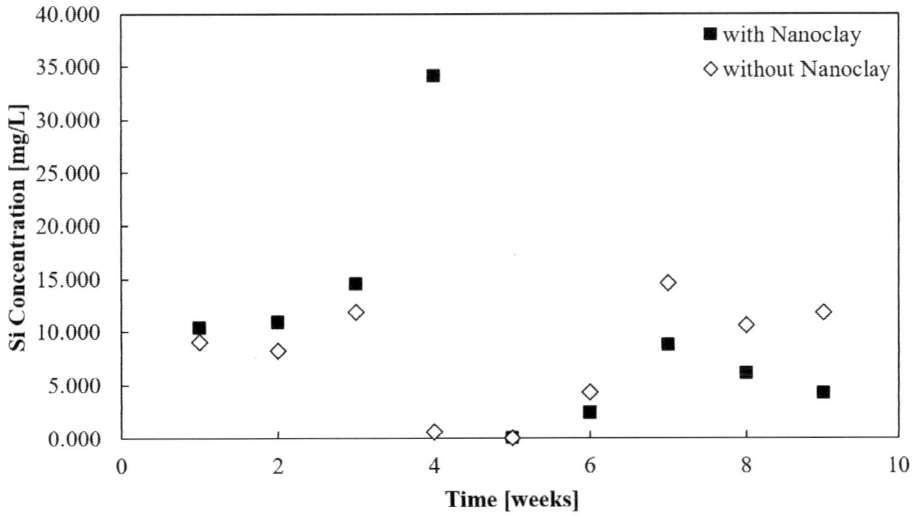

Figure 4. Release of Silicon (Si) from A1 and A2 during weathering.

4. Conclusion

After performing the experiment, the objectives of the study were successfully satisfied. Weathering of the A1 and A2 were performed considering the necessary standard to be used. SEM-EDX showed the surface morphologies of A1 and A2, and their respective elemental compositions. Furthermore, the SEM images evidently show that there were changes in surface morphologies of A1 and A2, which can be correlated to the oxidation and polymerization reactions producing microstructures on the surface. In addition, intercalation and exfoliation reactions between the nanoparticles and asphalt matrix potentially happened. The release of Al and Si nanoparticles (potentially from HNT nanoparticles) were also investigated. The degradation of asphalt was observed to happen on a by layer basis, so as the release of nanoparticles.

References

[1] Yang J and Tighe S 2013 A Review of Advances of Nanotechnology in Asphalt Mixtures *Procedia - Soc. Behav. Sci.* **96** 1269–76
[2] Fang C, Yu R, Liu S and Li Y 2013 Nanomaterials applied in asphalt modification: A review *J. Mater. Sci. Technol.* **29** 589–94
[3] Mohajerani A, Burnett L, Smith J V., Kurmus H, Milas J, Arulrajah A, Horpibulsuk S and Kadir A A 2019 Nanoparticles in construction materials and other applications, and implications of nanoparticle use *Materials (Basel).* **12** 1–25
[4] Ezzat H, El-Badawy S, Gabr A, Zaki E S I and Breakah T 2016 Evaluation of Asphalt Binders Modified with Nanoclay and Nanosilica *Procedia Eng.* **143** 1260–7
[5] Nejad F M, Tanzadeh R, Tanzadeh J and Hamedi G H 2016 Investigating the effect of nanoparticles on the rutting behaviour of hot-mix asphalt *Int. J. Pavement Eng.* **17** 353–62
[6] Yao H and You Z 2016 Effectiveness of Micro-and Nanomaterials in Asphalt Mixtures through Dynamic Modulus and Rutting Tests *J. Nanomater.* **2016**
[7] Faruk A N M, Chen D H, Mushota C, Muya M and Walubita L F 2014 Application of Nano-Technology in Pavement Engineering: A Literature Review 9–16
[8] Menapace I, Yiming W and Masad E 2017 Chemical analysis of surface and bulk of asphalt binders aged with accelerated weathering tester and standard aging methods *Fuel* **202** 366–79

IOP Conf. Series: Materials Science and Engineering 1117 (2021) 012031 doi:10.1088/1757-899X/1117/1/012031

Experimental analysis of the performance characteristics of the regenerative fuel cell system based on serpentine flow-field design

Nang Xuan Ho, Duy Vinh Nguyen

Faculty of Vehicle and Energy Engineering, Phenikaa University, Yen Nghia, Ha-Dong District, Hanoi, Viet Nam

Vinh.nguyenduy@phenikaa-uni.edu.vn

Abstract. The energy demand has recently been increasing dramatically; thus, it is evident that scientists worldwide need to find out the solution to solve the above problem. Among these technologies, solar and wind power energy are the most potential source as alternatives to fossil fuels. However, these energy sources are not continuous and are therefore considered as additional sources. A regenerative fuel cell (RFC) is a type of fuel cell that can function as an electrolyzer and fuel cell; therefore, this system is an independent power source and overcome the drawbacks of the original fuel cell system. This research has built a system to control the entire regenerative fuel cell system's flow, temperature, and energy. The system uses electricity to produce hydrogen and oxygen for storage; when the electrical power is insufficient, the system can automatically reverse the operation using hydrogen and oxygen to create electrical energy. The result showed that the power density in fuel cell mode ranges from 0.1 to 0.69W/cm^2 in the fuel cell mode, and the flow rate of hydrogen and oxygen reaches a stable value of 1100 and 2200 ml/min after 10 minutes, respectively.

1. Introduction

Exhaust emissions from internal combustion engines are the primary polluted source and responsible for over 65 percent of global GHG emissions worldwide [1]. As global energy demand is expected to continue to increase in the future, there are many comments that fundamental measures are urgently needed to meet growing energy demand. The UN Framework Convention on Climate Change was adopted by the United Nations Conference on Environment and Development in 1992 to prevent global warming. As the Kyoto Protocol, including concrete action plans, has entered into force, international environmental regulations to curb greenhouse gas emissions have been strengthened [2]. Generally, the energy forms to be transferred to the final users, the always convenient, clean, safe, efficient, and flexible energy carriers are necessary. Electricity is a traditional form of energy, created by various energy sources, easy to transport, and comfortable to transfer to the final users. And thus, automobile electricity has been developed worldwide due to its advantage in energy consumption prices and exhaust emission, as mentioned in [3]. However, the drawback of the automobile electricity related to the short driving range and long recharge time has limited its applications

Hydrogen is also a clean, efficient, and flexible energy carrier and easily compensates the electricity. The two energy carriers, electricity and the hydrogen, can satisfy all energy demands in the future and build individual first energy sources and an independent permanent energy system [4]. Indeed, hydrogen generated from renewable resources can be used for various applications such as transportation, electrical production, and heat for industry, buildings, and primary industry. Related to this point, proton exchange membrane fuel cell (PEMFC) using hydrogen has been used to cope with the energy crisis and air pollution because of its fewer emissions and high efficiency compared [5-6]. It works as a battery transferring the chemical energy into electricity; if hydrogen is fed continually, it will prevail battery in low energy storage. However, hydrogen production is expensive and not widely available, and its riskiness in the storage process. An RFC system can solve this problem due to its capability to function either as an electrolyzer or a fuel cell [7]. As a result, the reversible reaction most commonly used in an RFC is as follows:

$$H_2O + \text{energy} \leftrightarrow H_2 + 1/2O_2 \qquad (1)$$

Int the electrolyzer mode, electricity is supplied to dissociate H_2O into H_2 and O_2. These storage gases are fed back into the RFC in the fuel cell mode to create electricity. Thus, RFC has potential development in areas of transportation [8]. This research conducted an experimental procedure to test the UFC system based on the serpentine flow-field design. Accordingly, the flow-field of the bipolar plate was manufactured and assembled with a membrane exchange assembly (MEA) for the experiment.

2. Experimental procedure for the regenerative fuel cell system based on serpentine flow-field design

In this research, the simulation work has been conducted to optimize the flow-field design based on a 3-D mass transfer model, including heat transfer aspects of PEMFCs, as mentioned in [1]. Consequently, the optimized flow-field is a 5-passes 40turns serpentine channel flow field fabricated by the steel material. The 3D drawing was performed in 3-D drawing software, and all the flow-fields using for the fuel cell assembly were manufactured by a highly accurate CNC machine to ensure high precision, as shown in Fig. 1.

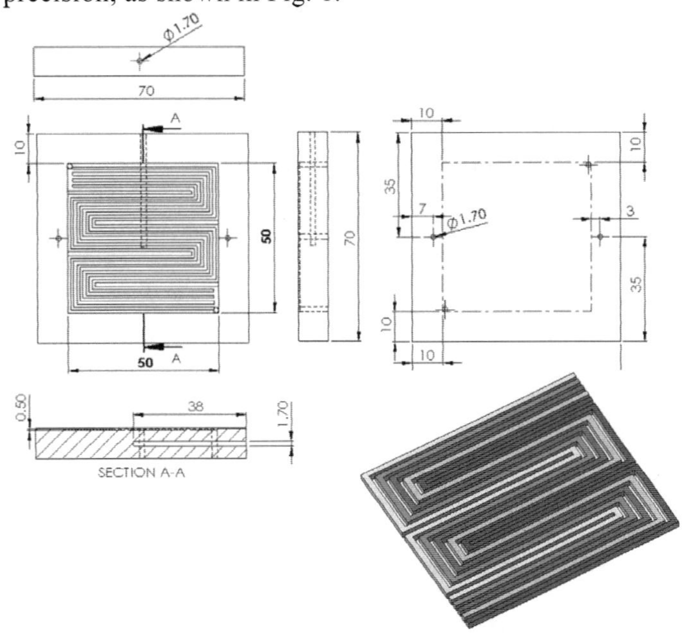

Figure. 1 Flow-field design with 5-pass 4-turn serpentine type

The fabricated flow-fields are assembled with the MEA to form a single fuel cell. MEA composes of the solid polymer membrane playing a role as the electrolyte. The electrodes are in contact with the membrane and must be porous to allow the flow of gases and liquid water to and from reaction sites. In an RFC, fuel cell and electrolyzer device are both integrated into a single system. This system includes a water control system and a gas storage system to function as a standalone energy storage unit. One key advantage of an RFC compared with a separate electrolyzer and fuel cell is integrating all functions in a single compact system; thus, it can conduct the energy storage mission and compete directly with batteries.

Additionally, an RFC allows the closing of the water cycle, whereby product water in the fuel cell mode can be returned to the electrolyzer's water storage. It may become advantageous in stationary power supply applications where the supply of H_2O is challenging. The different design of the electrolyzer and fuel cell allows these two components to be optimized their performance and cost-effectiveness. It permits the cell's capacities in the fuel cell and electrolyzer mode to be conducted separately, and thus it can operate in both ways simultaneously. In electrolyzer mode, an RFC converts water into gaseous hydrogen and oxygen; otherwise, hydrogen and oxygen gases electrochemically combine to form water, with the generation of heat and electricity in the fuel cell mode as mentioned in Fig. 2.

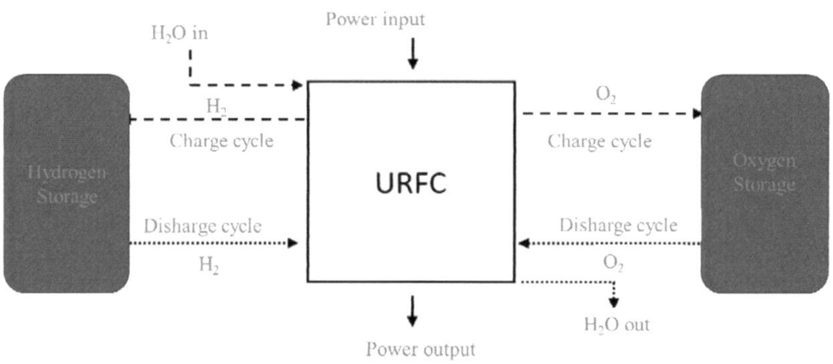

Figure. 2 Fundamental work of the RFC system

In the experiment, the system was designed with the ability to automatically switch between electricity storage mode and electricity generation mode composing of the following parts: main water tank, an automatic water pumping system that is controlled by relays, water supply pipes for supplying water when the procedure operates in electrolytic mode, air pipes for providing air when the system works in power generation mode, 1-way solenoid valves and solenoid control system, flow meters for accurately measuring water and airflow supplied to the system. The full system test is described in Fig. 2.

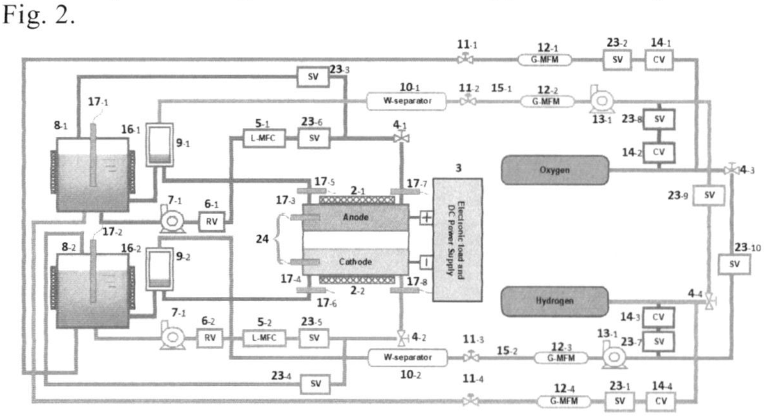

Figure. 3 Schematic diagram of the RFC system

Where (2) is cartridge heater with the thermocouple, (3) is electronic load and DC power, (4) is a one-way valve, (5) is liquid mass flow controller, (6) is a relief valve, (7) is the water pump, (8) is the water tank, (9) is water and gas separation tank, (10) is water separator, (11) is pressure regulator, (12) and (13) are high-pressure gas pump, 14 is a gas check valve, (15) is the gas cooling chamber, (16) is heating tape

3. Results and Discussion

The experimental RFC's performance data with both electrolyzer and fuel cell modes are given in Fig. 4 and Fig. 5. Consequently, the power density in fuel cell mode changes according to the current density and reaches the maximum value of 0.69 W/cm^2 at the current density of 1.4 W/cm^2.

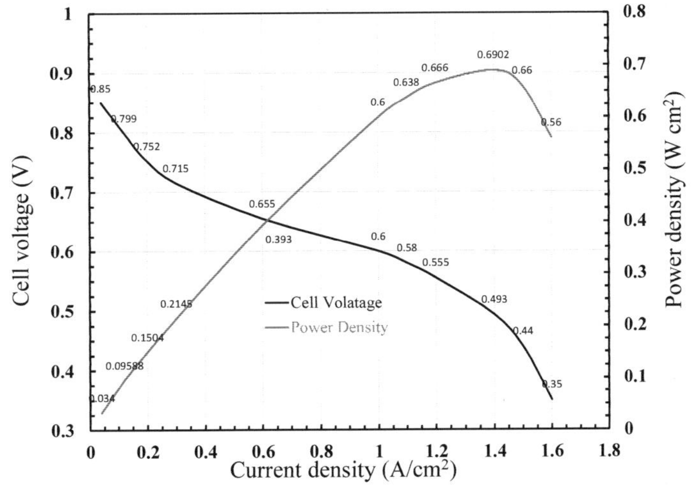

Figure. 4 Polarization and power density curve of URFC at fuel cell mode

The hydrogen and oxygen production rates by the URFCs trialed in electrolyzer mode are presented in Fig. 5. It can be observed that they are small in the first period and rapidly increase after 5 minutes. After 10 minutes, the flow rate of hydrogen and oxygen reach a stable value of 1100 and 2200 ml/min, respectively.

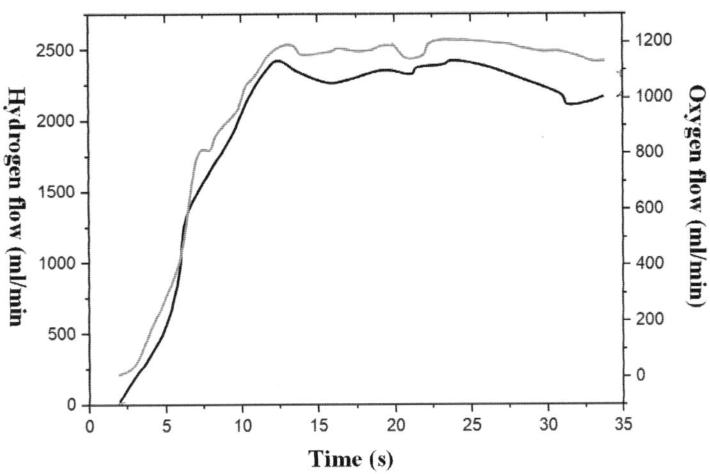

Figure. 5 Hydrogen and oxygen flow rate of URFC at electrolyzer mode

These results are expected a priori since a given electrode's electrochemical performance would be expected to stay the same if its composition remained constant, even though changes are made on the other electrode. The system is well-operated during the longtime of the experiment. However, RFC's power density at the fuel cell mode is a little smaller than that of previous research studies with the same fuel cell design configuration. It can be resulted by the restriction of the membrane protection parts added in the anode side and the cathode side. These parts help protect the membrane under electrochemical corrosion; however, they hider the flow channel and reduce the fuel cell performance. And thus, future research will be conducted to solve this problem to apply worldwide applications.

4. Conclusions

This research shows the experimental result of the RFC system. It can be observed that the system worked smoothly, corresponding to both the fuel cell and electrolyzer mode. The maximum value of the power density is about 0.69 W/cm^2 at the fuel cell mode; meanwhile, the hydrogen flow production can be reached to 2.2 l/min at the electrolyzer mode corresponding to 25 cm^2 active area MEA. This work's findings provide a foundation for optimizing the flow-field design for efficient PEMFCs and applying the RFC system worldwide.

Acknowledgment: This research is funded by PHENIKAA University under grant number 01.2020.07

References

[1] Lee, J., Kim, H.-M., Kim, T., Duy, V. N., Ahn, J., Park, S., & Kim, K. (2015). Dynamic simulations of under-rib convection-driven flow-field configurations and comparison with experiment in polymer electrolyte membrane fuel cells. Journal of Power Sources, 293, 447–457. https://doi.org/10.1016/j.jpowsour.2015.05.107

[2] Park, K., Kim, H. M., & Choi, K. S. (2013). Numerical and experimental verification of the polymer electrolyte fuel cell performances enhanced by under-rib convection. Fuel Cells, 13(5), 927–934. https://doi.org/10.1002/fuce.201200127

[3] Zhang, L., & Shi, Z. (2021). Optimization of serpentine flow field in proton-exchange membrane fuel cell under the effects of external factors. Alexandria Engineering Journal, 60(1), 421–433. https://doi.org/10.1016/j.aej.2020.09.007

[4] Diep Tran, T., Ha Vu, D., Huang, S., & Vinh Nguyen, D. (2020). Various Flow-Field Designs for Enhancing Fuel the Cell Performance of Proton Exchange Membrane Fuel Cells. IOP Conference Series: Materials Science and Engineering, 739(1), 4–9. https://doi.org/10.1088/1757-899X/739/1/012012

[5] Duy, V. N., Lee, J. K., Park, K. W., & Kim, H. M. (2015). Enhancement of PEM fuel cell performance by flow control. https://doi.org/10.4028/www.scientific.net/MSF.804.75

[6] Duy, V. N., Kim, K., Lee, J., Ahn, J., Park, S., Kim, T., & Kim, H. M. (2015). Parametric simulations of optimum flow-field configuration for efficient proton exchange membrane fuel cell. International Journal of Electrochemical Science.

[7] Vinh, N. D., & Kim, H. M. (2016). Comparison of numerical and experimental studies for flow-field optimization based on under-rib convection in polymer electrolyte membrane fuel cells. Energies, 9(10), 1–17. https://doi.org/10.3390/en9100844

[8] Duy, V. N., & Kim, H. M. (2017). Effect of gravity and gas flow direction on the operation of Polymer Electrolyte membrane fuel cells. International Journal of Electrochemical Science. https://doi.org/10.20964/2017.12.78

IOP Conf. Series: Materials Science and Engineering 1117 (2021) 012032 doi:10.1088/1757-899X/1117/1/012032

Material Risk Pooling: Strategies for Precast Fabrication

Chien-Ho Ko[1, a *]

[1]Lee Hall 2-312, Clemson, SC 29634, USA

achienhoko1@gmail.com

Abstract. Materials used in precast fabrication are different from those cast in the construction site. When fabricators lack production materials, they have to await the materials until the supplier transports them to the precast fabricator. The primary objective of the study is to enhance on-time delivery with relatively low inventory by leading risk-pooling concept with lateral transshipment strategy. To achieve the goal, this research develops lateral transshipment strategies by considering the characteristics and requirements of precast production. The computer simulation technique is then used to analyze the material risk pooling strategies. The proposed method can be used to analyze optimum transshipment policy for shipping the required materials from manufacturing plants that have surplus to others starving for materials. Risks of late delivery can thus be reduced.

1. Introduction

For a competitive market in the construction industry, material supply has become an important issue to enhance a company's competitiveness [1]. Traditional precast fabricators make material schedules depending on the rules of the thumb. The unsystematic manner may not fulfill the material requirements for precast fabrication [2].

Precast concrete design is different from that of a regular reinforced concrete structure. Therefore, embedded parts and materials for precast manufacturing are different from those cast in the construction site. Since the sizes and properties of precast materials are specialized for precast elements, these materials are hard to be dispatched when fabricators have a material shortage [3]. In the current practice, precast fabricators must wait for the materials until they have been transported to the precast factory. However, projects in the construction industry cost tremendous. There would be a retard penalty if precast manufacturers fail to ship precast elements to the site on time [4].

Supply chain management in precast manufacturing endeavors on optimizing single factory's profits. Construction projects are known for a high price, complicated interfaces, various subcontractors, and long duration. Although previous studies [5-7] have developed lateral transshipment ideas, their methods focusing on retailers may not be suitable for the situation in the construction industry. The research on principles analysis of lateral transshipment for material risk pooling has not been fully investigated. The objective of the study is to analyze transshipment strategies with relatively low inventory costs to help precast fabricators for delivering products on time.

IOP Conf. Series: Materials Science and Engineering 1117 (2021) 012032 doi:10.1088/1757-899X/1117/1/012032

2. Supply Chain Management

Supply Chain Management (SCM) could be regarded as a series of actions used to manage the product life cycle from raw material to an end customer. The precast fabrication supply chain is shown in Fig. 1. Cooper et al. [9] define SCM as a decision process from production, delivery, and dispatch. Ross [8] regards SCM as a philosophy linking internal and external stakeholders in the company to enhance customer value. The objective of SCM is to reduce costs and increase the margin of profits.

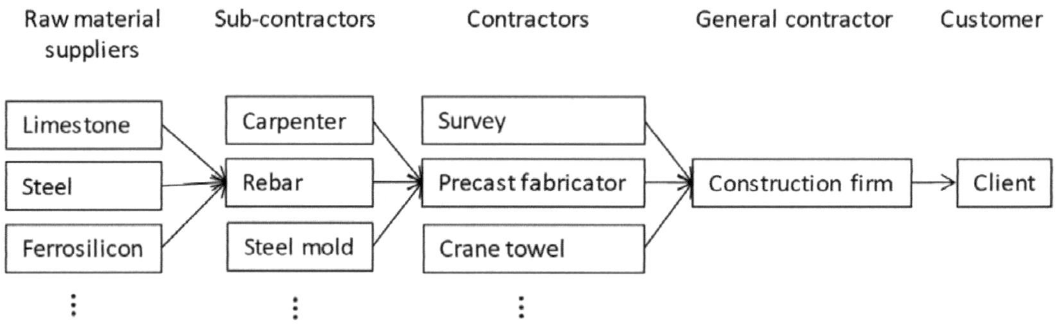

Figure. 1 Precast fabrication supply chain

3. Risk Pooling Strategies

The main purpose of implementing risk pooling strategies is to provide materials to those who short of materials from other manufacturers with surplus materials. A cooperative manufacturer's network needs to be established before launching material risk pooling strategies. Therefore, the impact of the material shortage could be reduced in the network. Axsater [10] and Banerjee et al. [11] proposed strategies to pool material shortage risks.

1) No lateral shipments (NLS)

The function of the NLS strategy is not to start the mutual assistance mode regardless of whether there is a shortage of materials or not until the inventory checking point is reached. This strategy is set as a baseline for a comparison with other risk-pooling strategies.

2) Random Policy (RA)

When a manufacturer faces a material shortage and other manufacturers have surplus inventory, a manufacturer with surplus materials is randomly selected to supply the shortage alliance to eliminate the risk of material shortage. If the material shortage still exists, the next alliance is selected for assistance until a certain inventory level is reached.

3) Priority Policy I (PR-I)

When material shortage occurs among cooperating manufacturers and some manufacturers have surplus materials, the party with the most surplus materials is selected to transship the materials to the manufacturer with the most material shortage. This policy is repetitive until the material shortage is solved, or no surplus materials are left.

4) Priority Policy II (PR-II)

When material shortage occurs among cooperating manufacturers and some manufacturers have surplus materials, the party with the most surplus materials is selected to transship the materials to the manufacturer with the least material shortage. This policy is repetitive until the material shortage is solved, or no surplus materials are left.

5) Lateral transshipments for inventory equalization policy (TIE)

The main idea of this strategy is to take the daily consumption and existing inventory of the cooperative manufacturer as an overall average. Then, the expected inventory of all cooperative manufacturers is calculated and evaluated whether any of a manufacturer will fall into a material shortage.

6) Moving Average (MA)

This strategy uses the moving average method to evaluate the next demand to determine whether there is a shortage of materials or whether there are more materials that can be provided to the manufacturers who are short of materials, so as to eliminate the provider's risks of material shortage.

4. Simulation

This study uses computer simulation to analyze the best risk pooling strategy for the allied precast fabricators. To verify the effectiveness of the system, three precast fabricators are allied. During the simulation process, multi-manufacturer systems, demand variability, order costs, and lead time are considered to explain real-world conditions. The objective of the computer simulation is to reduce the total cost including holding costs, shortage costs, and transportation fees. Simulation results are shown in Fig. 2. In the figure, PR-I outperforms the other strategies, showing that PR-I would be an appropriate strategy for material transshipment in the precast fabrication.

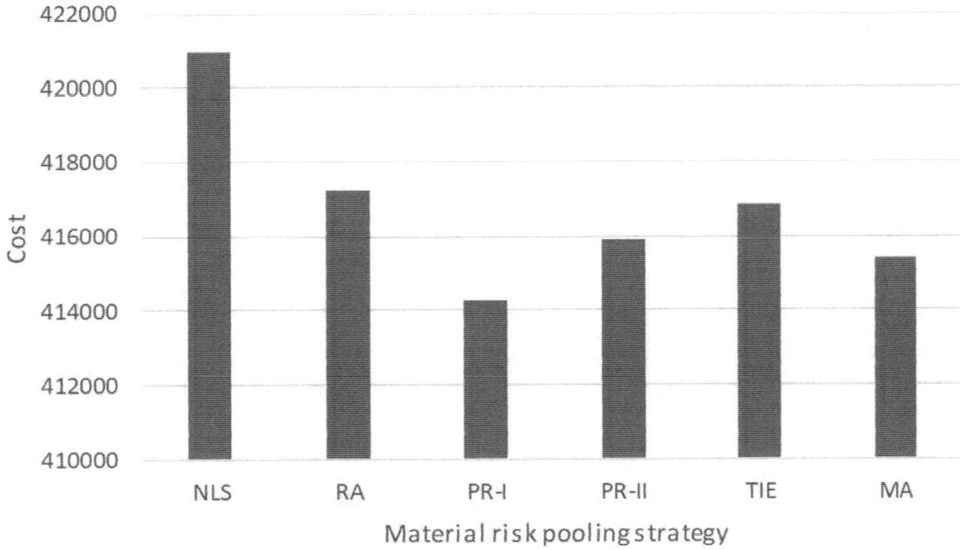

Figure. 2 Simulation results

5. Conclusions

Precast fabrication is inherent manufacturing. The research applies methods of industrial engineering in the construction industry enhancing efficiency in precast production management. In addition, the proposed material risk pooling model with lateral transshipment strategies can reduce risks of late delivery as well as unnecessary inventory wastes in production. Simulation results show that PR-I is an appropriate strategy while pooling risk in material shortage among allied precast fabricators. One reason is that the PR-I strategy can solve the major material shortage issue, and thus reduces holding costs, transportation costs, and shortage costs. The proposed material risk pooling method is established upon an allied precast fabricator network. Future research can focus on how to establish a system for the allied network to function normally, so that network members can cooperate rather than compete with each other.

Acknowledgments

This study is funded by the Ministry of Science Technology (Taiwan) with grant numbers: NSC 94-2218-E-212-011, NSC 96-2221-E-020-030, and NSC 95-2221-E-212-051. The author thanks the investigated precast fabricators for their assistance and the graduate student LI, Yu-Dian for his assistance in programming and analysis.

References

[1] B. Yan, Z. Chen, H. Li. Evaluation of agri-product supply chain competitiveness based on extension theory. Operational Research. 19 (2019) 543-570.

[2] S. Ahmad, R. Soetanto, C. Goodier. Lean approach in precast concrete component production. Built Environment Project and Asset Management (2019).

[3] C.H. Ko. An integrated framework for reducing precast fabrication inventory. Journal of Civil Engineering and Management. 16(3) (2010) 418-427.

[4] C.H. Ko. Material Transshipment for Precast Fabrication. Journal of Civil Engineering and Management. 19(3) (2013) 335-347.

[5] G. Tagaras. Pooling in multi-location periodic inventory distribution systems. Omega the International Journal of Management Science. 27(1) (1999) 39-59.

[6] A. Banerjee, J. Burton, S. Banerjee. A simulation study of lateral shipments in single supplier, multiple buyers supply chain networks. International Journal of Production Economics. 81-82 (2003) 103-114.

[7] Y.T. Herer, M. Tzur. The dynamic transshipment problem. Naval Research Logistics. 48 (2001) 386-408.

[8] D.F. Ross. Competing through supply chain management: creating market-winning strategies through supply chain partnerships. Springer Science & Business Media (1997).

[9] M.C. Cooper, D.M. Lambert, J.D. Pagh. Supply chain management: more than a new name for logistics. The international journal of logistics management 8.1 (1997) 1-14.

[10] S. Axsater. Simple solution procedures for a class of two-echelon inventory problems. Operations Research, 38 (1990) 64-69.

[11] A. Banerjee, J. Burton, S. Banerjee. A simulation study of lateral shipments in single supplier, multiple buyers supply chain networks. International Journal of Production Economics. 81-82 (2003) 103-114.

AUTHOR INDEX

Institute of Physics
Dirac House, Temple Back
Bristol BS1 6BE UK

ISSN: 1757-8981
ISBN 978-1-7138-3403-8

9 781713 834038

CURRAN ASSOCIATES INC.
proceedings.com